Water, Knowledge and the Environment in Asia

The dramatic transformation of our planet by human actions has been heralded as the coming of the new epoch of the Anthropocene. Human relations with water raise some of the most urgent questions in this regard. The starting point of this book is that these changes should not be seen as the result of monolithic actions of an undifferentiated humanity, but as emerging from diverse ways of relating to water in a variety of settings and knowledge systems.

With its large population and rapid demographic and socioeconomic change, Asia provides an ideal context for examining how varied forms of knowledge pertaining to water encounter and intermingle with one another. While it is difficult to carry out comprehensive research on water knowledge in Asia due to its linguistic, political and cultural fragmentation, the topic nevertheless has relevance across boundaries. By using a carefully chosen selection of case studies in a variety of locations and across diverse disciplines, the book demonstrates commonalities and differences in everyday water practices around Asia while challenging both romantic presumptions and Eurocentrism.

Examples presented include class differences in water use in the megacity of Delhi, India; the impact of radiation on water practices in Fukushima, Japan; the role of the King in hydraulic practices in Thailand, and ritual irrigation in Bali, Indonesia.

Ravi Baghel is a postdoctoral researcher at University of Trier and is affiliated to the South Asia Institute, Heidelberg University, Germany.

Lea Stepan is a researcher at the Cluster of Excellence: "Asia and Europe in a Global Context" at Heidelberg University, Germany.

Joseph K. W. Hill is an assistant professor (visiting faculty) at the Department of Rural Management, Xavier Institute of Social Service, Ranchi, Jharkhand, India.

Earthscan Studies in Water Resource Management

Legal Frameworks for Transparency in Water Utilities Regulation
A comparative perspective
Mohamad Mova Al'Afghani

Water Regimes
Beyond the public and private sector debate
Edited by Dominique Lorrain and Franck Poupeau

Negotiating for Water Resources
Bridging Transboundary River Basins
Andrea Haefner

Environmental Water Markets and Regulation
A comparative legal approach
Katherine Owens

The Politics of Fresh Water
Access, conflict and identity
Edited by Catherine M. Ashcraft and Tamar Mayer

Global Water Ethics
Towards a Global Ethics Charter
Edited by Rafael Ziegler and David Groenfeldt

Water, Knowledge and the Environment in Asia
Epistemologies, practices and locales
Edited by Ravi Baghel, Lea Stepan and Joseph K. W. Hill

For more information and to view forthcoming titles in this series, please visit the Routledge website: http://www.routledge.com/books/series/ECWRM/

Water, Knowledge and the Environment in Asia

Epistemologies, Practices and Locales

Edited by Ravi Baghel, Lea Stepan and Joseph K. W. Hill

LONDON AND NEW YORK

First published 2017
by Routledge

2 Park Square, Milton Park, Abingdon, Oxfordshire OX14 4RN
52 Vanderbilt Avenue, New York, NY 10017

Routledge is an imprint of the Taylor & Francis Group, an informa business

First issued in paperback 2019

British Library Cataloguing-in-Publication Data
A catalogue record for this book is available from the British Library

Library of Congress Cataloging in Publication Data
Names: Baghel, Ravi, editor. | Stepan, Lea, editor. |
Hill, Joseph K. W., editor.
Title: Water, knowledge and the environment in Asia:
epistemologies, practices and locales/edited by Ravi Baghel,
Lea Stepan and Joseph K.W. Hill.
Description: London; New York: Routledge is an imprint
of the Taylor & Francis Group, an Informa Business, 2017. |
Includes bibliographical references and index.
Identifiers: LCCN 2016043451| ISBN 9781138685550 (hbk) |
ISBN 9781315543161 (ebk)
Subjects: LCSH: Hydrology–Asia–Case studies. | Hydrology–
Social aspects–Asia–Case studies. | Water use–Asia–Case studies. |
Water use–Social aspects–Asia–Case studies.
Classification: LCC GB773 .W37 2017 | DDC 333.910095—dc23
LC record available at https://lccn.loc.gov/2016043451

ISBN: 978-1-138-68555-0 (hbk)
ISBN: 978-0-367-33514-4 (pbk)

Typeset in Bembo
by Deanta Global Publishing Services, Chennai, India

Contents

List of contributors   vii
Preface   xi

1   Contextualising the Anthropocene: The cultures, practices and
    politics of water knowledge in Asia   1
    RAVI BAGHEL AND LEA STEPAN

2   'If over a hundred Becquerels is no good, then what does fifty
    Becquerels mean?' Governing fisheries and marine radiation in
    Japan after the Fukushima nuclear accident   19
    LESLIE MABON AND MIDORI KAWABE

3   Trans-disciplinary analysis of Australian–Indonesian monsoon
    epistemologies and their implications on climate change
    adaptation strategies   36
    SARAH CASSON

4   An epistemological re-visioning of hybridity: Water/lands   52
    KUNTALA LAHIRI-DUTT

5   Science as friend or foe? Development projects undermining
    farmer managed irrigation systems in Asia's high
    mountain valleys   70
    JOSEPH K. W. HILL

6   Competing epistemologies of community-based groundwater
    recharge in semi-arid north Rajasthan: Progress and lessons
    for groundwater-dependent areas   87
    CHAD STADDON AND MARK EVERARD

7 Traditional knowledge and modernization of water:
   The story of a desert town Jaisalmer   108
   CHANDRIMA MUKHOPADHYAY AND DEVIKA HEMALATHA DEVI

8 The hydro-ecological self and the community of water:
   Anupam Mishra and the epistemological foundation of water
   traditions in Rajasthan   125
   DANIEL MISHORI AND RICKI LEVI

9 Epistemological undercurrents: Delhi's water crisis and
   the role of the urban water poor   142
   HEATHER O'LEARY

10 'Being-in-the-water' or socialisation through interactions with
   water in the thermal baths of Taipei   157
   NATHALIE BOUCHER

11 In the eye of the storm: Water in the cross-currents of
   consumerism, science and tradition in India   173
   NEERAJ VEDWAN

12 Balinese wet rice agriculture in transition: Water knowledge
   between a sentient ecology and the pursuit of development   192
   LEA STEPAN

13 Water flows uphill to power: Hydraulic development
   discourse in Thailand and power relations surrounding
   kingship and state making   210
   DAVID J. H. BLAKE

14 Waterscapes in transition: Past and present reshaping
   of sacred water places in Banaras   230
   VERA LAZZARETTI

15 Resettling a River Goddess: Aspects of local culture,
   development and national environmental movements
   in conflicting discourses on Dhārī Devī Temple and Srinagar
   Dam Project in Uttarakhand, India   246
   FRANCES A. NIEBUHR

   Index   261

List of contributors

Ravi Baghel is a geographer working as a postdoctoral researcher at the Governance & Sustainability Lab at Trier University, Germany. His PhD focused on large dams, flood control and water governance in India. He later conducted research on human relations with glaciers in the Himalayas, with a focus on their geopolitical meaning, such as in the case of the Siachen conflict, and as sites of development intervention, as in Ladakh. As a Research Fellow of the Earth System Governance Project (a core initiative of International Human Dimensions Programme on Global Environmental Change), his emerging research interest is in the role of knowledge in the governance of large dams at the scale of the Global South.

David J. H. Blake is an independent scholar based in the United Kingdom, with a professional background in water resources management and rural development in mainland Southeast Asia. He has a PhD from the School of Development Studies, University of East Anglia with a thesis examining the social and political drivers of irrigation development in Thailand's Northeast region. In autumn 2016, he will be the Luce Visiting Scholar in Environmental and Urban Studies at Trinity College, Hartford, USA.

Nathalie Boucher is an anthropologist who holds a PhD in Urban Studies from the Institute of National Scientific Research in Montreal, Canada. Her research interests include sociability, socialization and public spaces in the neoliberal era, with a particular focus on cities of the Pacific Rim. She is currently exploring the social life of the St Lawrence River in Montreal, at the Planning Faculty of the Université of Montréal as a postdoctoral researcher.

Sarah Casson is a Peter and Patricia Gruber Fellow in Global Justice at the Yale Law School in the United States. She holds a Master in Environmental Science from the Yale School of Forestry and Environmental Studies and a Bachelor of Arts in Anthropology from Grinnell College. Her research focuses on how human society might adapt to changes caused by climate change with a focus on monsoons.

Devika Hemalatha Devi is a former project officer at the UNESCO-created Indian Heritage Cities Network, and has worked on several heritage-planning projects over the past year. She recognises the conflict posed by pitting modernisation against cultural identity, and aims to be a strong advocate against it. She is a graduate of a Masters in Urban and Regional Planning programme from CEPT University, Ahmedabad, India.

Mark Everard has undertaken applied research, policy development and capacity-building relating to the ways in which people connect with eco-systems. The author of 14 other books, including *Common Ground* (Zed Books, 2011) and *The Hydropolitics of Dams* (Zed Books, 2013), over 60 peer-reviewed scientific papers and over 250 technical magazine articles, Mark is also a communicator on sustainability and wider environmental and resource use matters on TV and radio. He has served on numerous government advisory and expert groups in the UK and has advised other governments and multinational corporations on sustainability matters. His speciality is systemic thinking, particularly around connections between the water environment and other environmental media and the human activities that depend on and influence them.

Joseph K. W. Hill is Assistant Professor (Visiting Faculty) at the Department of Rural Management, Xavier Institute of Social Service, Ranchi, Jharkhand, India. His teaching interests include writing and research skills, research ethics and qualitative research methodologies, and his research focuses on rural development, especially small-scale irrigation and, more recently, seed biotechnologies in agriculture.

Midori Kawabe is a Professor in Marine Policy at Tokyo University of Marine Science & Technology, Japan. Her research focuses on coastal and ocean management, with a particular interest in social learning of stakeholders in collaborative management. Since the 2011 nuclear accident, Midori has been closely working with the fisheries sectors of Fukushima by having participatory workshops and *café scientifique* with natural scientists, fishermen and citizens to discuss ways for rehabilitation of the Fukushima coastal area.

Kuntala Lahiri-Dutt is a Senior Fellow in the Resource, Environment & Development Group of the Crawford School of Public Policy at the ANU College of Asia & the Pacific, Australian National University. She has been involved in critical research on the role of gender and community livelihoods in two areas of natural resources: water and extractive industries (mining). Her research is informed by feminist scholar-activist research methodologies.

Vera Lazzaretti has a background in Cultural Anthropology and Indology. She is a post-doctoral Fellow at the University of Milan, and is currently based in Heidelberg as a DAAD visiting scholar at the South Asia Institute. Her work draws on the anthropology of space and place and on critical approaches to cultural heritage. In her research, Vera explores the transformations of sacred spaces, including water places, and religious practices in modern and contemporary South Asia.

Ricki Levi is a PhD candidate at the Porter School of Environmental Studies, Tel-Aviv University. She has a degree in philosophy and South Asian studies, and wrote an MA thesis on 'Time and Temporality in Indian Buddhism (Theravada)'. She was an intern at the Navdanya Bija Vidyapeeth learning center in Dehradun, India, where she studied agro-ecology and related topics. Her research focuses on modern Indian environmental philosophy.

Leslie Mabon is a Lecturer in Sociology at Robert Gordon University in Aberdeen, Scotland. He researches the societal dimensions of environmental risk and uncertainty in coastal regions, and is especially interested in working collaboratively alongside physical science researchers. Since 2014 Leslie has been researching the governance of coastal fisheries in Iwaki City, Fukushima Prefecture, Japan, as they move towards rehabilitation and restarts following the 2011 nuclear accident. Regular research updates on Leslie's work are available at energyvalues.wordpress.com.

Daniel Mishori (PhD) is a faculty member at The Porter School of Environmental Studies and at The Department of Geography and the Human Environment in Tel Aviv University. He is an environmental and social activist, and was a fellow at the 'Environmental Fellows Leadership Program' at the Heschel Center for Sustainability. Dr Mishori co-edited (with Anat Maor) a book on 'Precarious Employment' (2012). Topics of recent publication and research include environmental philosophy, the commons, business ethics, argumentation, epistemology, bioethics, workers' rights, environmental justice and the ecology of physical activity.

Chandrima Mukhopadhyay was working as an Assistant Professor at Faculty of Planning, CEPT University, Ahmedabad, India. Her core research interest is in the area of private sector participation in both infrastructure sector and urban development, and Urban Politics and Governance. For last more than a year, she has been engaged with the literature on water and ecology. She received her PhD from School of Architecture, Planning and Landscape, Newcastle University, Newcastle upon Tyne, UK. She is one of the founding editors of the 'Conversations in Planning Theory and Practice' booklet series project of Association of European School of Planning–Young Academics network.

Frances A. Niebuhr has been conducting research in the Indian Himalayan region since 2009. At the time, she focused on impacts and perceptions of climate change in the state of Himachal Pradesh. In 2013 she joined the Cluster of Excellence: "Asia and Europe in a Global Context" at Heidelberg University, Germany as a PhD candidate. In her current research, she is analysing environmental conflicts and disasters, as well as local meanings and practices connected to water in the hill state of Uttarakhand, India.

Heather O'Leary is a lecturer of Environmental Anthropology at Washington University in Saint Louis (WashU), US. Dr O'Leary is an interdisciplinary specialist in justice, intersectionality, and environmental development. Her past research has focused on environmental politics and gender in the waterscapes of Delhi, India, which informed her approaches to comparative research on development pathways to global urban water security. She is the Chair of the Commission for Anthropology and Environment of the International Union for Anthropological and Ethnological Sciences. Her work has appeared in *Society and Natural Resources* (2016) and *Cambridge Scholar Series* (2015).

Chad Staddon is Professor of Resource Economics and Policy in the Department for Geography & Environmental Management at the University of the West of England, Bristol. Chad's research focuses on urban water services and water–energy relations understood as dynamic coupled social-ecological systems He received his PhD in Geography from the University of Kentucky in 1996 for research on the political economy of water (mis) management in post-communist Bulgaria and has since pursued projects in Europe, the Middle East and Southern and Eastern Africa.

Lea Stepan is a member of the Cluster of Excellence: "Asia and Europe in a Global Context" at Heidelberg University, Germany and the project 'Waterscapes in a transcultural perspective'. She has a background in Anthropology and Religious Studies and works on transformations of local perceptions of the environment under changing conditions through the expansion of the neoliberal economy in Southeast Asia and Oceania. Currently, she is completing her PhD, which examines emerging tensions around water use practices and concepts of ecology, economy and ritual practice in Bali, Indonesia.

Neeraj Vedwan is an Associate Professor in Anthropology at Montclair State University in New Jersey, USA. The focus of his research is on human–environment relations and their relationship to broader socio-political developments in different areas of India and the US. More specifically, he has examined impacts and perceptions of climate change, environmental vulnerability, and policy and cultural politics of water resources. Over the last several years he has worked on the role of consumerism and dominant ideologies of nationalism in shaping perception and practices pertaining to public environmental goods in India.

Preface

A vibrant international community of researchers is engaged in water research, yet an understanding of the role of water knowledge in everyday practices has so far received inadequate attention. This edited volume was conceived during an international workshop entitled *Epistemologies of Water in Asia*, held at Heidelberg University, Germany in December 2014. We highly appreciate the efforts of the book's contributors to bring together their latest work and the essence of the workshop, the result of which is a book with the quality and form of an interdisciplinary conversation about water knowledge in Asia. We would also like to thank the reviewers who invested considerable time and energy in critiquing and improving the chapters.

We are grateful for the generous funding provided by the Cluster of Excellence: "Asia and Europe in a Global Context" at Heidelberg University, which made the initial workshop possible. Beyond financial support, this edited volume benefited from the intellectual stimulation and exchange between experienced and early career researchers working together in the interdisciplinary sub-project 'Waterscapes in a transcultural perspective' under the umbrella of the Cluster's research area on 'knowledge systems'. Among the numerous people who offered advice and encouragement, we would particularly like to thank Marcus Nüsser, Jörg Gengnagel and Annette Hornbacher. We also thank the entire publishing team of Routledge Earthscan, especially Tim Hardwick and Ashley Wright, for their support and patience throughout the publication process.

Ravi Baghel, Lea Stepan and Joe Hill
Heidelberg and Ranchi

1 Contextualising the Anthropocene

The cultures, practices and politics of water knowledge in Asia

Ravi Baghel and Lea Stepan

Introduction

As the world population is projected to rise to 8.5 billion by 2030 and 9.7 billion by 2050, Asia is at the centre of major demographic change with a predicted population of 5.2 billion by 2050 (UNDESA, 2015). Countries in the region are seeing dramatic socioeconomic change, growth in consumption per capita driven by high gross domestic product (GDP) growth, runaway urbanisation and high mobility. The combination of population growth and economic growth with an increased use of water for agriculture and industry are likely to exacerbate water stress in Asia (Shiklomanov 2000; Rosegrant *et al.* 2009; Hijioka *et al.* 2014, p. 1338). These patterns emerge against the broader picture of global change, and the strong domination of Earth systems by humans, that is increasingly subsumed under the metaphor of the 'Anthropocene', the epoch where humans have overtaken other processes to become the most important geological agents (Vitousek *et al.* 1997; Crutzen and Stoermer 2000). The neologism 'Anthropocene' derives from the name of the present geological epoch, the Holocene, with *anthropo-* indicating human and *-cene*, derived from Greek *kainos*, indicating new or recent.

The metaphor of the Anthropocene is useful in that it helps us examine the planetary scale of human impact, and in that it subsumes not only climate change but also changes to other Earth systems. This is especially important because it has been suggested that the impact of changes on the global water system are likely to outweigh the effects of climate change, at least over the decadal time scale (Vörösmarty *et al.* 2000). These changes are driven by human alteration of the water system through the construction of infrastructure such as dams and embankments (Baghel 2014a, Nüsser and Baghel 2017), urbanisation, agriculture, alteration of drainage and unsustainable extraction of groundwater. However, even as the Anthropocene draws attention to the transformation of the water system at the planetary scale, it tends to present humanity as a monolithic agent acting upon an undifferentiated planet. This leads Biermann *et al.* (2016, p. 342) to warn that 'using the Anthropocene lens must not mask the diversity of local and regional contexts and situations, nor the diversity and disparities in the conditions, contexts, and distribution of wealth, consumption and environmental impact across human societies'. This is certainly relevant to Asia with its large human population characterised by significant disparities and highly differentiated contributions to global environmental change.

It is difficult to carry out comprehensive research on water knowledge, a topic that has relevance across linguistic and national boundaries in Asia with its diversity of practices and languages. By using a carefully chosen selection of case studies in a variety of locations and across diverse disciplines, this book attempts to introduce this research to a global readership, making it easier to see commonalities and differences. While challenging existing paradigms and avoiding orientalism, it not only offers a critique in the tradition of science and technology studies, but also addresses everyday practices and in terms of environmental philosophy raises questions about epistemology. In this respect, Hulme (2011, p. 245) argues that

> the new climate reductionism is driven by the hegemony exercised by the predictive natural sciences over contingent, imaginative, and humanistic accounts of social life and visions of the future. It is a hegemony that lends disproportionate power in political and social discourse to model-based descriptions of putative future climates.

In 2015, Steffen *et al.* (2015) refined the large-scale framework of 'planetary boundaries', a concept for sustainable development introduced in 2009 by a group of Earth systems and environmental scientists. The planetary boundaries concept distinguishes nine boundaries and defines a 'safe operating space for humanity' where these limits are not exceeded. However, as this framework dismisses the regional heterogeneity and scales of operation of some of the boundaries, Steffen *et al.* (2015) introduce a two-fold approach that acknowledges the sub-global level within the overarching model. Accordingly, there are regional biochemical, land-system and freshwater use boundaries, which impact the Global Earth System, if transgressed. This refinement is useful to avoid generalizations of the global and to lay emphasis on the complexity of the local. It enables us to recognise Asia as a sub-global level and as a useful scale for contextualising the Anthropocene.

The point of departure of this edited volume is the premise that there is something distinct to water, to water experiences and water knowledges in Asia. Some of these appear to be linked to particular spaces – when associated with specific local cultures or religions – whereas others are structured by functional and symbolic differentiations, such as expert, political or sacred knowledge. In Asia, well tested practices surrounding water, snow and ice are often inseparable from ritual or cosmological symbolism and performance. Therefore, it cannot be assumed that the latter necessarily conflict with 'objective' understandings of water, which also brings into question the epistemological status of water as a mere 'resource'. We would like to examine how varied forms of knowledge pertaining to water flow, encounter and intermingle with one another. This volume is focused on attempts to trace the circulation and transformation of environmental knowledge fragments and practices across the boundaries of diverse knowledge systems.

Through the use of contemporary case studies of the role of knowledge in water practices across Asia, we address criticisms of Eurocentrism, but also avoid romantic notions of ancient traditions and indigenous knowledge. Further, this is a way of challenging the idea of water as a mere resource, by showing how people use and relate to water based on their knowledge obtained in a number of ways, from a multiplicity of sources. We take a bottom-up approach and avoid judging the relative worth of different kinds of knowledge, such as local, global, expert, scientific or religious. Our focus is less on theorizing abstract relations of water knowledge, and more on real world examples that show water knowledge as it is used in everyday life. Some examples of the empirical case studies of contemporary topics included in this volume are: water use in the megacity of Delhi, the impact of radiation on water practices in Fukushima, and ritual irrigation in Bali. However, the case studies highlight that there is no single generalizable epistemology, but a diversity of ways in which water and water-use is understood across Asia. The contributing authors come from a variety of disciplines such as anthropology, human geography, cultural studies, philosophy, sociology, area studies and development studies.

Along with acknowledging the severity of the impact, and appreciating the disparity of its distribution, it is especially relevant to appreciate one of the most important attributes of humans that distinguishes them from other geological processes. This is the ability of humans to understand, explain and alter their behaviour in response to a recognition of ongoing changes. Examining our ability to understand and respond to changes in the global water system (Kerkhoff and Lebel 2006; Kiparsky *et al.* 2012) is where the importance of understanding water knowledge comes to the fore.

Epistemic frames and communities

Haas (1992) proposed the influential concept of 'epistemic community' to describe communities or 'networks of knowledge-based experts' defined by their shared set of normative and principled beliefs; shared causal beliefs; shared notions of validity; and a common policy enterprise. This concept of epistemic communities primarily focused on transnational policy making, international regimes and expert groups. Considering that water is often an object of policy making and transboundary environmental governance, this concept can easily be extended to water. The different domains discussed here, even when not explicitly aimed at policy making, do have an indirect effect by changing the available knowledge base, introducing new considerations and drawing attention to overlooked problems.

Miller and Fox (2001, p. 669) define epistemic communities as 'a group of inquirers who have knowledge problems to solve' with a focus purely on epistemology and the process of knowledge production. According to them, 'members of the epistemic community share norms (albeit contested and revisable) about how good research should be conducted' (ibid.) with these norms varying from community to community. They also point out that every such community identifies some aspects of the problem as pertinent, while ignoring others. Bird (1987, p. 255) argues that scientific knowledge should not be regarded as a representation of nature, but rather as a socially constructed interpretation with an already socially constructed natural-technical object of inquiry. Barnes and Alatout (2012, p. 484) suggest that science and technology studies scholars can contribute to the study of water in two important ways, 'first, by looking at water, as … a singular object with multiple ontologies … and second, by seeing social realms not as being separate from water, but rather, as being built, at least partially, in and through engagements with water'.

This implies that not only is water socially constructed as an object of inquiry, but further that the knowledge thus produced goes on to reconstruct it in representing it. In other words, the epistemic frames that are used to study water produce both the 'problems' and the 'solutions' through a shared understanding of cause and effect and by reducing water to a certain set of attributes while ignoring others. This reductionism should not necessarily be decried as it is also necessary to produce useful (though contingent and revisable) knowledge; it only becomes harmful when reductionism becomes combined with a claim to an exclusive and superior form of knowledge. We discuss several such communities identifying the way they frame water as an object of knowledge; and the kinds of problems and solutions they come up with based on a shared understanding of cause and effect. The aim of this exercise is not to produce an exhaustive list of epistemic approaches to producing water knowledge, rather the intention is to identify their multiplicities and overlaps. By placing these communities in context and contrast to one another we identify emerging trends in the way knowledge of water is created, the aspects of human–water relations that become salient, and attempts at boundary crossing, collaboration and disciplinary exchange.

Integrated Water Resource Management (IWRM) is, arguably the most influential policy driven managerial approach to water (Rahaman and Varis 2005). Its history can be traced to the first United Nations Educational, Scientific and Cultural Organization (UNESCO) International Conference on Water in 1977. This management framework can be considered one that produces 'prescriptions regarding how knowledge should be produced and used (modes of knowledge production and use) to achieve specified desirable (natural resource management) outcomes' (Medema *et al.* 2008, p. 2).

As discussed previously, just like any other epistemic community, IWRM has its own set of premises, a specific understanding of cause and effect, and a goal towards which the knowledge created should be applied. One of the most fundamental elaborations of IWRM, developed in 1992 through international consensus under the auspices of the United Nations (UN), is known as the four Dublin principles. Principle one identified fresh water as a finite and vulnerable resource, essential to sustain life, development and the environment. Principle two proposed that water development and management should be based on a participatory approach, involving users, planners, and policy-makers at all levels. Principle three saw women as playing a central part in the provision, management and safeguarding of water. The fourth, and perhaps the most contentious, principle suggested that water has an economic value in all its competing uses and should be recognised as an economic good (Allan 2006). Building upon these principles, in 2000, the Global Water Partnership (GWP) defined IWRM as:

> a process which promotes the co-ordinated development and management of water, land and related resources, in order to maximize the resultant economic and social welfare in an equitable manner without compromising the sustainability of vital ecosystems.
>
> (GWP-TAC 2000, p. 22)

In terms of water knowledge, IWRM identified not just availability, but also variability as a key source of uncertainty in water management. Therefore one of the aims of IWRM is to reduce knowledge gaps and integrate the knowledge of multiple actors in order to improve coordination and ensure sustainability. Here an epistemic goal is to reduce complexity and measure water in terms of its quantifiable spatio-temporal availability, to identify a baseline or equilibrium that can be sustained over the long term. In addition, the solution to the problems identifed by IWRM are to be achieved through equally sustainable institutional and political arrangements; creation of laws and governance structures based on insights into the ecosystem as well as the system of human actors that interact with it. However, the reduction of uncertainty, and long-term predictability themselves are impossible goals as neither ecosystems nor human institutions are unchanging. Based on these insights, a new paradigm emerged to support the management of natural resources under uncertainty. 'Adaptive management' has been described as an integrated, multidisciplinary and systematic approach to improving management and accommodating change by learning from the outcomes of management policies and practices (Holling 1978).

Thus, it can be seen how IWRM failed in its agenda of holistic integration of various kinds of knowledge, which led to the recognition of the impossibility of certainty in the realm of water and the emergence of the adaptive management paradigm focused on learning (Medema *et al.* 2008). IWRM also faced an epistemological crisis in its inability to incorporate non-expert knowledge due to its *prescriptive* nature. This led to suggestions for its reinvention by including additional epistemic forms: *discursive*, to examine issues of power and control in water management, and *practical* or experiential to incorporate a context-based understanding of water management to 'create a shared space for multiple conflicting epistemologies and allow ways of knowing of non-expert stakeholders' (Mukhtarov and Gerlak 2013, p. 101). This epistemological crisis also led to the emergence of alternative paradigms, such as 'water security' (Cook and Bakker 2012).

Other epistemic frames eschew a managerial and resource-based approach and look at the political aspect of water. One example of this is the epistemic community of political ecology which 'combines the concerns of ecology and a broadly defined political economy' (Blaikie and Brookfield 1987, p. 17). Robbins (2004) describes the role of political ecology as being that of a 'hatchet' in its use as critique, and also that of a 'seed' in its commitment to equitable and sustainable solutions to environmental problems. Both these roles are not only appropriate but also complementary in the case of water knowledge. Another epistemic frame derives from the long tradition of hazards research, especially work related to the framing of floods and drought (Blaikie *et al.* 1994). An example of such work in South Asia is that on 'hazardscape', which builds upon insights from hazards research (especially pragmatism), political ecology and 'socionature' (Mustafa 2005).

The 'socionature', or social nature tradition, has given rise to another epistemic frame that is relevant to water research. The notion of a 'socially constructed nature' builds upon the idea that there is no nature separate and external to man, and that all nature is known only through social and discursive practices (Swyngedouw 2004, 2009; Kaika 2006). The importance given to the social aspect of water is based upon the idea that the water cycle is not an inert natural resource contested by humans but rather something with which humans have an inevitably social relationship (Robbins 2004; Swyngedouw 2007, 2009). One of the most important elements of this tradition, as opposed to the view that natural resources are sites of human contestation, is the notion of a dialectically produced socionature. This implies that natural entities such as rivers, lakes and seas are not to be seen as inert objects but as having agency of their own. Moreover, the very centrality of water to human existence arguably means that changes in its use and management always co-evolve with social changes; changes in the one are systematically reflected in the other (see Strang 2004). Even ideas of scarcity are anything but objectively quantifiable, for they can also be interpreted as social and political constructs (Mehta 2001). There has been recent interest among natural scientists and engineers in including social aspects in hydrological research (Sivapalan *et al.* 2014). However this appears to be partly a case of reinventing the wheel, as this work has tended to neglect the large body of preexisting research (e.g. Nüsser *et al.* 2012) on the social aspects of water – mentioned above – and has at the same time failed to address criticism of the use of co-evolution approaches to the study of natural resources (Jeffrey and McIntosh 2006).

Another set of epistemic approaches draw upon Michel Foucault's work on governmentality and knowledge/power (Foucault 1980, 1991, 2007) as applied to the environment. This research examines the role of institutions, which through the use of 'expert knowledge' construct an 'environment' that can then receive various forms of management and intervention – all in the interest of governing its constituents (Baghel 2012, 2014b). Typical of this is the work of Michael Goldman (2005, 2007), which examines the role of the World Bank in producing environmental knowledge and relating this to its interventions in the hydropolitics of the Mekong region.

Apart from epistemic frames focusing on these varied aspects of water, there is a large body of research primarily emerging from anthropology on the cultural attributes of water; including explicit critiques of concepts such as IWRM. The cultural meaning of water is often overlooked, but is critical at the scale of the everyday lived experience of water (Baviskar 1995, 2007). Dividing water into three analytical sites, Orlove and Caton (2010) define 'watershed' as an area of land through which water drains downhill to a lowest point usable as a possible management unit; 'water regimes' or the aggregate of institutional rules and practices for managing water resources in a specific setting or watershed; and 'waterscape', 'the culturally meaningful, sensorially active places in which humans interact with water and with each other'. Of these, waterscapes is an especially productive frame to explore the cultural aspect of water. At the same time, culture is just one element in what Hastrup and Rubow (2014) call the 'complex entanglement' of water and society. To address matters such as water inequity or mismanagement, an understanding of the environment and water in particular, within a local actor-network, as 'a product of the social and cultural experiences and values' (Cole 2012, p. 1238), provides essential insights in diverse levels of understanding, attitudes and required actions beyond the identification of interactions between environmental and socio-political factors.

Cultural experiences point to other ways of knowing and valuing the natural environment, which get marginalised by a homogenizing Eurocentric perspective (Gibbs 2010). Such water knowledge might not necessarily be articulated as explicit knowledge but finds expression in cultural practices and aesthetics. The role of water in rituals, myths, classifications or metaphors provides a window to study cultural value systems and human–water relations. Such cultural expressions have tended to be treated as separate and excluded from other forms of water knowledge especially in development projects and state politics. But ritual practices, sacred space and domains of divine power reveal intersections and conjunctions between the local ecology and economic resource/ water use practices (e.g. Lansing 1991). In a similar manner, Alley (2002, p. 55) argues that the sacred in Hindu views 'is not detached from the material realm of ecology and the built environment, but provides a context for understanding the truth of the physical world'. Palmer (2015) shows how meanings, spiritual beings, human identities and material reality attributed to and connected with the flow of water form a complex socio-ecological realm connecting visible and invisible worlds.

Cultural values have long been accepted as socially constructed, mediating between the materiallity and the meanings humans ascribe to that environment. This process takes place within 'a cultural landscape which is the product of specific social, spatial, economic and political arrangements, cosmological and religious beliefs, knowledges and material culture, as well as ecological constraints and opportunities' (Strang 2004, p. 5). Ingold (2000) argues that meanings, values and ethics do not precede but rather emerge through skilled and sensory engagement with the environment, and are thus neatly tied into the web of ecological relations and interactions. Thus, the cultural landscape is not only the product but simultaneously a producer of meanings that emerge from the synaesthetic experiences and engagement in it. Strang (2005, p. 115) therefore, calls for 'a greater appreciation of sensory experience and of the part played by "natural" resources and their characteristics in the generation of meanings'. Meanings are thus not only created by humans but generated from the material characteristics of environments. This offers new ways of thinking about cross-cultural meanings and materiality. At the same time, Helmreich (2011, p. 133) warns that a turn to the 'form of water' might lead to 'new reifications' at the intersection of nature, culture, water and theory.

Hornborg (1996, p. 53) concludes that the recognition of the subjectivity of all species as 'the very constitution of ecosystems ... [implies, that] the destruction of meaning and the destruction of ecosystems are two aspects of the same process'. Greater appreciation of cosmology, metaphysical or spiritual agencies and living beings as co-agents in ecological relations encourages thinking about human-environment relationships which are not solely centered on human epistemology and agency (e.g. Kohn 2013).

The epistemic communities discussed here are all varieties of 'expert' knowledge whose credibility depends upon formal and institutionalised practices. Scott (1998, p. 311) argues that expert knowledge 'dismisses practical know–how as insignificant at best and as dangerous superstitions at worst. The relation between scientific knowledge and practical knowledge is ... part of a political struggle for institutional hegemony by experts and their institutions'. Furthermore, in his argument in favour of *Metis*, which he describes as a practical and experiential form of knowledge, Scott also points out the intangible nature of such knowledge because it 'is often so implicit and automatic that its bearer is at a loss to explain it' (ibid., p. 329). A focus on the organised production of knowledge also tends to overlook certain aspects such as the role of the locale in the production of knowledge, with an implicit acceptance of science as the 'view from nowhere' (Naylor 2005). Where is water knowledge produced? How is it produced? Who produces it? Why counts as knowledge? These are all important questions addressed by contributors to this volume with the goal of contextualising water knowledge.

Water knowledge in context

According to the classic concept of *umwelt* advanced by Jakob von Uexküll (1957) in 1934, the interactions of animals (including humans) with the environment are mediated through knowledge, mental constructions of that environment and the material reality that exists beyond those. Within human societies, the process through which this knowledge of the environment is produced is further influenced by asymmetrical power relations. This is reflected in, for instance, the greater relative credence given to expert knowledge. The epistemological stances towards environmental knowledge range from a positivist belief in ever more accurate, objective scientific knowledge to a radical disbelief in the possibility of any incontrovertible knowledge and an epistemological relativism. Neither of these stances is useful when it comes to water knowledge; the positivist stance fails through its disregard for non-expert knowledge, whereas the relativist stance fails in its disregard for systematic scientific studies at the planetary scale.

Ingold (2000, pp. 215–216) asserts that the difference between 'local' and 'global' perspectives

> 'is not one of hierarchical degree, in scale or comprehensiveness, but one of kind. [T]he local is not a more limited or narrowly focused apprehension than the global, it is one that rests on an altogether different mode of apprehension – one based on practical, perceptual engagement with components of a world that is inhabited or dwelt-in, rather than on the detached, disinterested observation of a world that is merely occupied'.

Similarly Hornborg (1996, p. 54) calls for a 'recontextualisation' of environmental knowledge, because 'the sheer complexity and specificity of ecosystemic interrelationships and fluctuations, [mean] that optimal strategies for sustainable resource management are generally best defined by local practitioners with close and long-term experience of these specificities, and *with special stakes in the outcome*' (emphasis added). These arguments bring to the fore some of the most important attributes of local knowledge, its situatedness, immediacy and practicality.

Local knowledge is by definition site specific, and emerges through people's practical engagement with their environmental setting, including its material resource base. When we speak of the site-specific, place-based character of local knowledge, the most relevant features of place are nature as experienced in daily human life, that is, a particular set of environmental conditions and human interactions, including the process of ascribing meaning to their location. These features contribute to the specificity of local knowledge. However, the dynamism of place further implies that local knowledge cannot be timeless or static. Any such site-specific knowledge is itself part of the ever-changing constellation of human and environmental factors across boundaries of time and space (Baghel 2014a; Nüsser and Baghel 2016).

Nygren (1999) sees a need to describe local knowledge in more nuanced terms to avoid fixing it in dichotomies such as rational/magical, universal/ particular, theoretical/practical, and modern/traditional. Though well intentioned, these binaries are ultimately damaging because they do nothing to change the terms of the debate; they merely cast local knowledge as the threatened underdog in a script of good versus evil (see also Agrawal 1995).

Case studies

This volume's contributions are arranged into four parts: scales of knowledge; community and heritage; experience and everyday practices; and sacredness and progress. The difficulty in structuring the contributions stems from the edited volume's intention of highlighting overlaps in the plurality and multiplicity of water knowledge practices. We have therefore attempted to group the contributions according to the most dominant theme shared amongst them, rather than other possible categorizations such as region or discipline. This demonstrates the diverse epistemic frames that can be applied to understandings of water knowledge, while also highlighting their conjunctions and disparities. We encouraged the authors to make cross-cutting links to different contributions to allow this volume to emerge in the quality and form of an interdisciplinary conversation. This set of contemporary case studies related to water knowledge deliberately crossesdisciplinary and national boundaries. The locations of the various case studies can be seen in Figure 1.1. The original research work was carried out by authors with expertise in different regions and languages; something which tends to fragment the publication of work along disciplinary or regional/area studies lines. Below we elaborate upon the overarching themes of the book and its case-studies.

Scales of knowledge

This part focuses on the structural differentiations of water knowledges and the (lack of) mediation among specific knowledge systems across different scales. These studies are especially relevant for policy and decision makers to deal appropriately with the links between knowledge practices and policies.

Leslie Mabon and Midori Kawabe evaluate how risk and uncertainty are negotiated in the governance of fisheries in Iwaki City, Fukushima Prefecture,

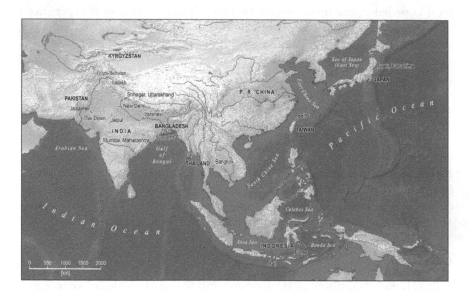

Figure 1.1 Locations of the case studies.

Japan, after the 2011 nuclear disaster. The authors draw attention to the limitations the nature of water places on what can be known with certainty, epitomised by the title that questions the meaning of quantified risk of radiation for local people. Coalescing the water knowledge of experts and local practitioners can bring significant explanatory purchase to the management of situations of high uncertainty and unstable knowledge relations.

Monsoons are highly variable, complex climate patterns that directly and indirectly provide for the water requirements of much of the world's agriculture. Sarah Casson critically explores the different kinds of knowledges and epistemologies that are used to understand and predict a changing Australian–Indonesian monsoon. Her careful analysis draws together a wide range of data, from macro level modelling to local perceptions of individual farmers in East Flores, Indonesia, and has important implications for formulating and communicating climate change adaptation strategies.

In her chapter on the floodplains of Riverine Bengal, Kuntala Lahiri-Dutt proposes an epistemological reframing of the concept of hybridity in considering the environment. In particular, she argues for a flexible or 'wet' theory in thinking about hybrid water-lands characterised by their uncertain existence, their indeterminacy and their fluid and liminal presence as ambiguous temporal, cultural and political terrains. This is especially useful for reframing and drawing attention of water research not just to the socio-historical but equally to the spatial and biophysical setting.

Joe Hill discusses how farmer-managed irrigation systems (FMIS) in the valleys of Ladakh, India, are being undermined by the very development interventions enacted to support them. Though Ladakhi farmers have engineering skills that can surpass modern engineering surveying techniques, have an environmental awareness that surpasses that of modern agricultural scientists, and have organisational skills that surpass the formal committees and standardised 'irrigation service fees' recommended by international donors, their knowledge is sidelined as their objectives and norms are not included in the design of interventions overseen by engineers and contractors. Though development practice is framed in a language of science-based policy, it takes place through political manoeuvring.

Community and heritage

Many communities in arid and semi-arid regions are faced with water scarcity and have developed communal water infrastructure and use strategies over time. These water harvesting systems have become challenged by technical and political changes pertaining to use and ownership. The case studies draw attention to the potential to revive traditional water harvesting and community water infrastructure, and aim to acknowledge the multiplicity of knowledge systems in their spatial and societal context.

While most studies on water scarcity are concerned with the depletion of surface water, Chad Staddon and Mark Everard direct our attention to a community groundwater recharge project in the Arvari River catchment in semi-arid north Rajasthan, led by the non-governmental organization (NGO) Tarun Bharat Sangh (TBS). Applying Ostrom's framework for common pool resource (CPR) systems, they examine the potential of alternative governance systems for stopping ecological, social and economic degradation. TBS initiatives are rebuilding traditional community-designed and maintained water harvesting structures in order to increase groundwater recharge and revitalise aquatic, farmed and natural ecosystems with the goal of social and economic regeneration. The authors argue that, thereby, the TBS are refashioning water epistemologies away from hydrosocial complexes based on nature-culture and commodity-noncommodity axes constitutive of the neoliberal water epistemology.

Traditional knowledge around water and site specific water infrastructures are a crucial element for desert towns such as Jaisalmer, located in the Indian Thar desert region. Chandrima Mukhopadhyay and Devika Devi present ethnographically rich material on traditional water knowledge and practices that reflect the cultural creativity to cope with restricted access to water. The authors critically examine the discrepancy between the sensitive perceptions of (non-) availability of water in a desert setting and shifting meanings of water as abundant resource related to the development of domestic water supply through the Indira Gandhi Canal project.

Daniel Mishori and Rivka Levi offer a critical overview of the epistemological foundations of the Indian author and Gandhian activist Anupam Mishra's philosophy of water. His rejection of new Western knowledge and expertise was based on the claim that their adoption often causes a 'forgetting' of traditional ecological knowledge, exacerbated by the bias of the modern Indian state and the Market for modern methods, materials and conceptions. To counter such 'forgetting', Mishra worked on re-cultivating traditional water practices and expertise, with an emphasis on praxis and craftsmanship, rather than on discursive knowledge. This chapter draws attention to the importance of contextualising water knowledge and the social and spiritual dimensions inherent in water practices.

Experiences and everyday practices

In the third part we shift the focus to sensed experiences and daily practices with water. The contributions focus on examples of everyday water practices and the way these shape actions and generate new knowledge. These contributions address water practices not only in the context of their material and productive dimension but also as a materiality with multiple meanings and forms of agency through which interaction takes place.

As India's capital, New Delhi, faces unprecedented rates of urban growth, questions of water access and justice for the marginalised urban poor have increasingly come to the fore. Drawing on fieldwork among domestic workers employed by middle class households, Heather O'Leary discusses how they navigate through dramatic class differences in access to water. They produce rituals of resistance and alternate water philosophies that include the reincorporation and adaptation of traditional water practices and beliefs. Domestic workers try to reconcile middle-class practices that depend on the availability of excess water with their own daily practices of frugality and reuse resulting from their status as the urban water poor.

Nathalie Boucher offers a specific contribution to forms of cultural and social knowledge of water in the hot springs of Taipei, Taiwan. Her chapter explores ways in which water informs distinct epistemologies and exemplifies how the selves, the bodies, the settings and the image are interwoven. Human–water interactions allow for perception to happen, for the body to sense and for meaning to be discovered. Boucher suggests that water appears not only as the bearer of meaning but as a coded material with and through which bathers interact. Being-in-the-water, she concludes, is a sensorial and interactional experience in which bathers learn, create, participate in, maintain, or challenge tradition, modernity, and the contemporary Taiwanese culture.

The intensifying conflicts over water in urban and rural India, Neeraj Vedwan argues, are as much a tussle over a natural resource as they are about water's meaning and significance. Vedwan provides parallel case-studies from urban and rural areas to show how a certain deterritorialised discourse has gained dominance at the expense of local and regional voices and traditions. He concludes that the figure of the citizen-consumer – ahistorical and animated by a crude economic logic – is the pivot on which this discourse has come to rest.

Lea Stepan's chapter examines competing meanings given to Bali's water resources within political trajectories of agricultural intensification and growing environmental pressure on the local waterscape. These processes create new contexts for an active and creative negotiation according to the framework of the villagers' perceptions of economy and ecology in a sentient waterscape. The site specific water knowledge is reworked against the hegemony of techno-scientific categories of water as an object and a natural resource. Her analysis sheds light on modes of knowing, the adaption of diverse knowledge fragments and the way their interaction shapes meanings, values and practices.

Sacredness and progress

The fourth and final part brings together a variety of case studies that highlight the interlocking aspects of societal progress and sacredness of water. Although these might at times appear as in opposition, they are in a sense that continuation of the same theme of more-than-human water.

Focusing on large-scale hydraulic development in Thailand, David Blake critically investigates the compatibility of Wittfogel's controversial thesis of hydraulic society with state formation in Thailand. He examines some of the cultural, discursive, material and symbolic elements that link the Thai King's reign with water resources control, and argues that this has formed an essential component of his remarkable longevity, power and legitimacy. The analysis exemplifies elite constructions of dominant narratives into a Foucauldian 'regime of truth', providing an example of the interplay between discourse, knowledge and power. These prove not only to rely on the governance of water as a universal form of economic and political power, but are centred on the monarch as the key figure in the discourse.

Vera Lazzaretti deals with the overlapping of distinct epistemologies during crucial phases of transition of the urban waterscape of Varanasi (Uttar Pradesh, India). Her study focuses on the evolution of and changes to water places in the city by documenting the case of the sacred well Jnanavapi. The chapter shows how the element water creates confluences of diverse epistemologies. Lazaretti highlights the interaction of various knowledge systems related to water during colonial times, and the not always inclusive results. On the other hand, through a critical analysis of current transitions of the urban landscape, grounded in heritage discourses, she suggests that multiple perspectives on and of water, and their potentials, are currently being discarded.

Major infrastructural developments like the Indira Gandhi Canal or the construction of large hydropower projects prove to be recurrent contexts for social and political tensions. Frances Niebuhr's comparative analysis examines the epistemological backgrounds of iconic environmental protest movements in India with regard to the present conflict at Srinagar dam (Uttarakhand, India). The opposition between water as commodity and water as sacralised element in the landscape appears to map the underlying framework for conflicting interests. Yet, her close analysis of prevailing discourses reveals a more nuanced view of culturally modified arguments, which are reproduced but also deviate from the previous protest discourses.

Conclusion

The case studies presented here remind us of the potential and possibilities that emerge from human engagements with water. These interactions are not only characterised by environmental constraints but also open up prospects to sense the connection with the lived-in world. The prevalent crisis narratives should not obscure positive developments, overlook agency or potential transfers of knowledge. Further, these case studies suggest that knowledge systems must not be seen as exclusive monoliths, instead they all fit within a composite system of interacting meanings. As Pyne (2007, p. 653) points out, we must refute the idea that 'the failures of science, technology, and humanistic scholarship can be solved only by more science, more technology, more humanism, each by itself. Rather, we need them all, in contrast, in conflict, in context, and given expression'.

Much of contemporary Asia is faced with uncertain availability of water, social conflicts, inequality of access and power differentials. An attempt to replace uncertainty with predictability does nothing to reduce risk but only forecloses possible futures. The future is open and full of possibilities, and therefore modelling and prediction tend to obscure water knowledge rather than improve it because they fail to pay attention to human efforts and the ingenuity, multiplicity, creativity and adaptability of water knowledge. At the same time the real need is not for an increase in water knowledge but for more contextualisation, juxtaposition and the intermingling of diverse water knowledges.

We use the metaphor of the Anthropocene to underscore the role of human agency in bringing about the current state of the global water system. However, it is important to see this agency as opening up possibilities rather than giving in to a determinist account in which humanity has already sealed its fate through its past actions. The multiplicity and richness of the various forms of water knowledge suggest that there are numerous possibilities of adapting to and mitigating foreseeable environmental change. Readers of this book are invited to form their own judgements.

References

Agrawal, A. (1995) 'Dismantling the divide between indigenous and scientific knowledge', *Development and Change*, vol. 26, no. 3, pp. 413–439.

Allan, J. A. (2006) 'IWRM: The new sanctioned discourse?' in P. P. Mollinga, A. Dixit and K. Athukorala (eds) *Integrated Water Resources Management. Global Theory, Emerging Practices and Local Needs*, Sage Publications, New Delhi, pp. 38–63.

Alley, K. D. (2002) *On the Banks of the Gângā: When Wastewater Meets a Sacred River*, University of Michigan Press, Ann Arbor, MI.

Baghel, R. (2012) 'Knowledge, power and the environment: Epistemologies of the Anthropocene', *Transcience: a Journal of Global Studies*, vol. 3, no. 1, pp. 1–6.

Baghel, R. (2014a) 'Misplaced knowledge: Large dams as an Anatopism in South Asia', in M. Nüsser (ed.) *Large Dams in Asia: Contested Environments Between Technological Hydroscapes and Social Resistance*, Springer, Dordrecht, Heidelberg, New York and London, pp. 15–32.

Baghel, R. (2014b) *River Control in India: Spatial, Governmental and Subjective Dimensions*, Springer, Dordrecht, Heidelberg, New York and London.

Barnes, J. and Alatout, S. (2012) 'Water worlds: Introduction to the special issue of Social Studies of Science', *Social Studies of Science*, vol. 42, no. 4, pp. 483–488.

Baviskar, A. (1995) *In the Belly of the River: Tribal Conflicts Over Development in the Narmada Valley*, Oxford University Press, New Delhi.

Baviskar, A. (ed.) (2007) *Waterscapes: The Cultural Politics of a Natural Resource*, Permanent Black, New Delhi

Biermann, F., Bai, X., Bondre, N., Broadgate, W., Chen, C.-T. Arthur, Dube, O. P., Erisman, J. W., Glaser, M., van der Hel, S., Lemos, M. C., Seitzinger, S. and Seto, K. C. (2016) 'Down to Earth: Contextualizing the Anthropocene', *Global Environmental Change*, vol. 39, pp. 341–350.

Bird, E. A. R. (1987) 'The social construction of nature: Theoretical approaches to the history of environmental problems', *Environmental Review*, vol. 11, no. 4, pp. 255–264.

Blaikie, P. and Brookfield, H. (1987) *Land Degradation and Society*, Methuen, London.

Blaikie, P., Cannon, T., Davis, I. and Wisner, B. (1994) *At Risk: Natural Hazards, People's Vulnerability, and Disasters*, Routledge, London and New York.

Cook, C. and Bakker, K. (2012) 'Water security: Debating an emerging paradigm', *Global Environmental Change*, vol. 22, no. 1, pp. 94–102.

Crutzen, P. J. and Stoermer, E. F. (2000) 'The "Anthropocene"', *IGBP Newsletter*, vol. 41, pp. 17–18.

Foucault, M. (1980) *Power/Knowledge: Selected Interviews and Other Writings 1972–1977*, Pantheon Books, New York.

Foucault, M. (1991) 'Governmentality', in G. Burchell, C. Gordon, and P. Miller (eds), *The Foucault Effect: Studies in Governmentality*, University of Chicago Press, Chicago, IL, pp. 87–104.

Foucault, M. (2007) *The Politics of Truth*, Semiotext(e), Los Angeles, CA.

Global Water Partnership – Technical Advisory Committee (2000) 'Integrated water resources management', Background paper, Global Water Partnership, Stockholm, Sweden.

Goldman, M. (2005) *Imperial Nature: The World Bank and Struggles for Social Justice in the Age of Globalization*, Yale University Press, New Haven and London.

Goldman, M. (2007) 'How "Water for All!" policy became hegemonic: the power of the World Bank and its transnational policy networks', *Geoforum*, vol. 38, no. 5, pp. 786–800.

Haas, P. M. (1992) 'Introduction: Epistemic Communities and International Policy Coordination', *International Organization*, vol. 46, no. 1, pp. 1–35.

Hastrup, K. and Rubow, C. (eds) (2014) *Living with Environmental Change: Waterworlds*, Earthscan Routledge, London.

Helmreich, S. (2011) 'Nature/culture/seawater', *American Anthropologist*, vol. 113, no. 1, pp. 132–144.

Hijioka, Y., Lin, E., Pereira, J., Corlett, R., Cui, X., Insarov, G., Lasco, R., Lindgren, E. and Surjan, A. (2014) 'Asia', in Intergovernmental Panel on Climate Change (ed.), *Climate Change 2014 – Impacts, Adaptation and Vulnerability. Part B: Regional Aspects*, Cambridge University Press, Cambridge, UK, pp. 1327–1370.

Holling, C. S. (ed.) (1978) *Adaptive Environmental Assessment and Management, International Series on Applied Systems Analysis, International Institute for Applied Systems Analysis*, Wiley, Laxenburg, Austria; Chichester, NY.

Hornborg, A. (1996) 'Ecology as semiotics: Outlines of a contextualist paradigm for human ecology', in P. Descola and G. Palsson (ed.) *Nature and Society: Anthropological Perspectives*, Routledge, London, pp. 45–62.

Hulme, M. (2011) 'Reducing the future to climate: A story of climate determinism and reductionism', *Osiris*, vol. 26, no. 1, pp. 245–266.

Ingold, T. (2000) *The Perception of the Environment: Essays on Livelihood, Dwelling and Skill*, Routledge, London; New York.

Jeffrey, P. and McIntosh, B. S. (2006) 'Description, diagnosis, prescription: A critique of the application of co-evolutionary models to natural resource management', *Environmental Conservation*, vol. 33, no. 4, pp. 281–293.

Kaika, M. (2006) 'Dams as symbols of modernization: The urbanization of nature between geographical imagination and materiality', *Annals of the Association of American Geographers*, vol. 96, no. 2, pp. 276–301.

Kiparsky, M., Milman, A. and Vicuña, S. (2012) 'Climate and water: Knowledge of impacts to action on adaptation, *Annual Review of Environment and Resources*, vol. 37, no. 1, pp. 163–194.

Kohn, E. (2013) *How Forests Think: Toward an Anthropology Beyond the Human*, University of California Press, Berkeley, CA.

Medema, W., McIntosh, B. S. and Jeffrey, P. J. (2008) 'From premise to practice: A critical assessment of integrated water resources management and adaptive management approaches in the water sector', *Ecology and Society*, vol. 13, no. 2, article 29.

Mehta, L. (2001) 'The manufacture of popular perceptions of scarcity: Dams and water-related narratives in Gujarat, India', *World Development*, vol. 29, no. 12, pp. 2025–2041.

Miller, H. T. and Fox, C. J. (2001) 'The epistemic community', *Administration & Society*, vol. 32, no. 6, pp. 668–685.

Mukhtarov, F. and Gerlak, A. K. (2013) 'Epistemic forms of integrated water resources management: Towards knowledge versatility', *Policy Sciences*, vol. 47, no. 2, pp. 101–120.

Mustafa, D. (2005) 'The production of an urban hazardscape in Pakistan: Modernity, vulnerability, and the range of choice', *Annals of the Association of American Geographers*, vol. 95, no. 3, pp. 566–586.

Naylor, S. (2005) 'Introduction: historical geographies of science – places, contexts, cartographies', *The British Journal for the History of Science*, vol. 38, no. 1, pp. 1–12.

Nüsser, M. and Baghel, R. (2016) 'Local knowledge and global concerns: Artificial glaciers as a focus of environmental knowledge and development interventions', in P. Meusburger, T. Freytag, and L. Suarsana (eds) *Ethnic and Cultural Dimensions of Knowledge*, Springer, Heidelberg, pp. 191–209.

Nüsser, M. and Baghel, R. (2017) 'The emergence of technological hydroscapes in the Anthropocene: Socio-hydrology and development paradigms of large dams, in B.Warf (ed.) *Handbook on Geographies of Technology*, Edward Elgar, Cheltenham: Forthcoming.

Nüsser, M., Schmidt, S. and Dame, J. (2012) 'Irrigation and Development in the Upper Indus Basin: Characteristics and Recent Changes of a Socio-Hydrological System in Central Ladakh, India', *Mountain Research and Development*, vol. 32, no. 1, pp. 51–61.

Nygren, A. (1999) 'Local knowledge in the environment–development discourse', *Critique of Anthropology*, vol. 19, no. 3, pp. 267–288.

Orlove, B. and Caton, S. C. (2010) 'Water sustainability: Anthropological approaches and prospects', *Annual Review of Anthropology*, vol. 39, no. 1, pp. 401–415.

Palmer, L. (2015) *Water Politics and Spiritual Ecology: Custom, Environmental Governance and Development*, Routledge, London.

Pyne, S. J. (2007) 'The end of the world, *Environmental History*, vol. 12, no. 3, pp. 649–653.

Rahaman, M. M. and Varis, O. (2005) 'Integrated water resources management: Evolution, prospects and future challenges', *Sustainability: Science, Practice, & Policy*, vol. 1, no. 1.

Robbins, P. (2004) *Political Ecology: A Critical Introduction*, Blackwell, Malden, Oxford and Victoria.

Rosegrant, M. W., Ringler, C. and Zhu, T. (2009) 'Water for agriculture: Maintaining food security under growing scarcity', *Annual Review of Environment and Resources*, vol. 34, no. 1, pp. 205–222.

Scott, J. C. (1998) *Seeing Like a State: How Certain Schemes to Improve the Human Condition Have Failed*, Yale University Press, New Haven, CT and London.

Shiklomanov, I. A. (2000) 'Appraisal and assessment of world water resources', *Water International*, vol. 25, no. 1, pp. 11–32.

Sivapalan, M., Konar, M., Srinivasan, V., Chhatre, A. Wutich, A., Scott, C. A., Wescoat, J. L. and Rodríguez-Iturbe, I. (2014) 'Socio-hydrology: Use-inspired water sustainability science for the Anthropocene', *Earth's Future*, vol. 2, no. 4, pp. 225–230.

Strang, V. (2004) *The Meaning of Water*, Berg, Oxford and New York.

Strang, V. (2005) 'Common senses: Water, sensory experience and the generation of meaning', *Journal of Material Culture*, vol. 10, no. 1, pp. 92–120.

Swyngedouw, E. (2004) *Social Power and the Urbanization of Water: Flows of Power*, Oxford University Press, New York.

Swyngedouw, E. (2007) 'Technonatural revolutions: The scalar politics of Franco's hydro-social dream for Spain, 1939-1975', *Transactions of the Institute of British Geographers*, vol. 32, no. 1, pp. 9–28.

Swyngedouw, E. (2009) 'The political economy and political ecology of the hydro-social cycle', *Journal of Contemporary Water Research & Education*, vol. 142, no. 1, pp. 56–60.

United Nations Department of Economic and Social Affairs, Population Division (2015) *World Population Prospects: The 2015 Revision, Key Findings and Advance Tables*, Working Paper ESA/P/WP.241, United Nations, New York.

Van Kerkhoff, L. and Lebel, L. (2006) 'Linking knowledge and action for sustainable development', *Annual Review of Environment and Resources*, vol. 31, no. 1, pp. 445–477.

Vitousek, P. M., Mooney, H. A, Lubchenco, J. and Melillo, J. M. (1997) 'Human domination of Earth's ecosystems', *Science*, vol. 277, no. 5325, pp. 494–499.

Von Uexküll, J. (1957) 'A stroll through the worlds of animals and men: A picture book of invisible worlds', in C. H. Schiller (ed.) *Instinctive Behavior*, International Universities Press, Madison, pp. 5–80.

Vörösmarty, C. J., Green, P., Salisbury, J. and Lammers, R. B. (2000) 'Global water resources: Vulnerability from climate change and population growth', *Science*, vol. 289, no. 5477, pp. 284–288.

2 'If over a hundred Becquerels is no good, then what does fifty Becquerels mean?'

Governing fisheries and marine radiation in Japan after the Fukushima nuclear accident

Leslie Mabon and Midori Kawabe

1 Introduction

> *We're told that over a hundred Becquerels is no good, we can understand that fine. But often when the experts do the measurements and give us the data, we see fifty Becquerels coming up a lot. We see sevens and nines as well, and we don't know what that means. If over a hundred Becquerels is no good, then what does fifty Becquerels mean?*
>
> Fisher, Onahama Fish Market, Iwaki,
> Fukushima Prefecture, July 2014.

In this chapter, we discuss governance of fisheries in Iwaki, Fukushima Prefecture after the 11 March 2011 earthquake, tsunami and nuclear accident. The earthquake and subsequent tsunami left over 17,000 people either dead or missing, and disabled cooling systems at the Fukushima Dai'ichi nuclear power plant (FDNPP). This triggered hydrogen explosions which resulted in large-scale radioactive releases. The contamination led to the stoppage of all commercial coastal fisheries in Fukushima apart from restricted trial operations, and for the first time brought terms like Becquerel[1] into broader societal awareness within Japan as concerns over potential for radiation in food emerged. The quote, from a fisher participating in the trial coastal fishing operations, goes to the heart of what we explore in this chapter. Namely, the role of 'expertise' and relationships between actors in a situation of significant complexity and uncertainty. What limits are there on what can be known with certainty when it comes to water, and what might this tell us about the role of knowledge in governing a body of water specifically?

Our approach is distinct in that the 'water' under study is the Pacific Ocean, and our interest is not so much water itself as the things in it – marine produce and radioactive material. The data and thoughts on which this chapter is based come from empirical fieldwork in Iwaki, which has been ongoing since spring

2011, involving interviews, discussion groups and field observation. Fishers, fisheries cooperative staff, prefectural fisheries officers and researchers, local academics, politicians, municipal government staff and science communicators have participated in the research.

We focus on our findings and the questions they raise for future research, both in Fukushima and for epistemologies of water more widely. This chapter is thus largely empirical in nature with reference to aspects that may warrant further theoretical attention, where appropriate. After giving a brief overview of the current situation in Iwaki, we explore two aspects of the Fukushima fisheries situation in which epistemology of water comes to the fore – the role of 'experts', and the limitations to what can be known. We conclude by drawing broader implications from the Iwaki fisheries case for the role of knowledges in the governance of water.

2 Context

Iwaki City[2] is an administrative district on the south-east coast of Fukushima Prefecture in north-east Japan (see Figure 2.1). It is home to over 300,000 people, and outside of its urban core has an economy based on industry, tourism, agriculture and fisheries. Fishing, the focus of this chapter, plays a central role in Iwaki's coastal towns and villages, with deep-sea trawling for tuna and Pacific saury running out of the main port in Onahama, and coastal fisheries for species such as whitebait, abalone, surf clam and sea urchin.

In Iwaki, the earthquake and tsunami directly led to 293 deaths (Iwaki City 2016) and caused significant infrastructural damage in coastal areas. Port facilities, fishers' homes and boats were destroyed. Whilst reconstruction of fishing facilities has taken place, the FDNPP accident has had more severe and lasting implications for fisheries across all of Fukushima Prefecture, including Iwaki. Over 80 per cent of radioactive material emitted from the FDNPP was deposited over the north-west Pacific Ocean (Yoshida and Kanda 2012). Between April 2011 and April 2012 more than 40 per cent of Fukushima fish sampled exceeded the Japanese regulatory threshold for radioactive caesium of 100 becquerels per kilogram (Bq/kg) (Buesseler 2012), leading to the suspension of all Fukushima fisheries. Concerns around radioactive leaks of contaminated water into the ocean continue as plant operator Tokyo Electric Power Company (TEPCO) attempts to store and manage contaminated water used to cool FDNPP's reactors. For example, in early 2015 two separate incidents where observed radioactivity levels rose at drains in the plant led Fukushima fisheries representatives to claim trust in TEPCO had 'collapsed' (Kyodo News 2015).

Although deep-sea fishing resumed in 2012, full-scale coastal fishing remains suspended in Fukushima Prefecture. However, since the disaster monitoring of marine radioactivity has been undertaken to understand the dispersal of radioactive material over time, with a view to moving towards the resumption of coastal fisheries in Fukushima. At present, trial fishing operations are being conducted,

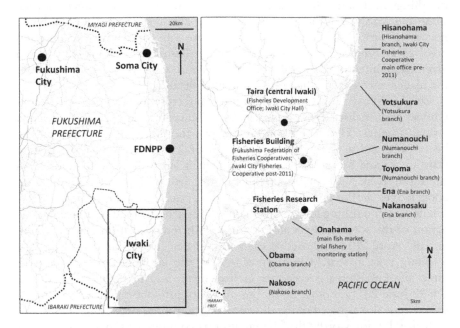

Figure 2.1 Location of Iwaki in Fukushima Prefecture (left) and location of ports/key infrastructure in Iwaki (right).

whereby small amounts of fish species in which radioactive caesium has not been detected for several months are caught, screened for radiation and sold at market. The aim of these trial fisheries is to encourage fishers to resume fishing activities post-disaster, and to monitor the sales of Fukushima fish on their return to markets (see Section 3 for a fuller description of the trial fisheries process).

It is worth mentioning that Iwaki is one of two fishing districts in Fukushima Prefecture, the other being Soma-Futaba to the north. Given our central argument about the complexity of producing knowledge in sea water and its implications for governance, to allow deeper engagement this chapter focuses on Iwaki fisheries. Many of the issues we raise in Sections 3 and 4 are applicable to Soma-Futaba as well, but to be clear, in this chapter we use 'Iwaki' fisheries when talking about our empirical observations and 'Fukushima' fisheries to discuss more generic points. Indeed, in Section 5 we return to reflect on the heterogeneity of experience that water engenders and the difficulties this raises for governance across scales.

3 Relationships – who are the experts, and where are they located?

We first assess the process of deciding the safety of Fukushima marine produce, in particular the question of who makes claims to knowledge about water and the things in it, and whose claims to safety or otherwise are trusted. Several actors are involved in the governance of post-disaster fisheries in Fukushima, hence the process of deciding which fish are safe for human consumption involves a complex web of knowledge. To illustrate this complexity – and given the limited number of English-language sources to which we can refer the reader – we explain the process in a little detail.[3]

On 15 March 2011, the Fukushima Prefecture Federation of Fisheries Cooperative Associations (*ken-gyoren*) 'voluntarily' suspended Fukushima coastal fisheries. The first reason for this was the destruction of most fishing boats, ports and market facilities by the tsunami. The second was the chaotic situation caused by the nuclear disaster, which led to large-scale evacuations and threw Iwaki into confusion. Having consulted with presidents of fisheries cooperatives on the phone, the *ken-gyoren* president, Nozaki, made the decision to suspend fishing for the time being, foreseeing that radioactive contamination of the sea would get all Fukushima fish removed from market. Then, in early April 2011, radioactive caesium was detected in sand eels landed at a fishing port in Ibaraki Prefecture, leading the Japanese prime minister to direct the Fukushima governor to control intake and marketing of sand eels. On 22 June 2011, another 36 species of fishery product caught in Fukushima water were also assigned to the direction of controlling marketing.

It may be noteworthy that the Fukushima commercial fisheries have never actually been 'banned' by government. However, underpinning the governance of Fukushima fisheries are central government maxima – the 'temporary regulation level' of 500 Bq/kg used until March 2012, and the 'criterion value' of 100 Bq/kg used thereafter. Since the disaster, fish stocks in Fukushima coastal waters have been monitored regularly by national government (Fisheries Agency of Japan), local government (Fukushima Prefecture Fisheries Section) and industrial (TEPCO) institutions. Other institutions such as Iwaki-based aquarium Aquamarine Fukushima, working in conjunction with the UmiLabo citizen monitoring group, also monitor. Among participants interviewed for this research, there was consensus that the results obtained by different monitoring institutions were consistent, or at least that any differences could be explained by the monitoring techniques used.

In terms of translating monitoring results into decisions on which fish may be put back on sale, when and from where, it is locally-collected and analysed data which takes the lead. Fukushima Prefecture Fisheries Section continually monitors radioactivity across fish stocks, sea water and bottom sediment, fishers themselves getting involved in this process by catching fish in their boats and passing them across for monitoring. If radioactive caesium is not detected for several months during this monitoring process, fish species are considered for trial fisheries (Wada *et al.* 2013). This process has four steps. First, initial face-to-face discussion between fishers and fish brokers. Second, consensus-building at the District Trial Fisheries Exploratory Committee. Third, consideration at the Fukushima Prefecture Regional Fisheries Rehabilitation Convention, involving scholars from across Japan, local government representatives, brokers and local fisheries representatives. Fourth, discussion at the regional Representatives of Fisheries Cooperatives Conference (Mabon and Kawabe 2015).

National-level 'expertise' sets the absolute standards for radioactive concentration and informs decision-making through briefing visits to Fukushima by national government scientists and the involvement of scholars from across Japan on trial fisheries governance committees. Crucially, however, it is data collected, analysed and disseminated within Fukushima itself that moves trial fisheries forwards in practice. The standards set by the Fukushima Prefecture Federation of Fisheries Cooperative Associations are more stringent than the national government standards. Fish exceeding 25 Bq/kg in the screening test are referred to the laboratories for further precision analysis, and samples exceeding 50 Bq/kg result in the particular fish species being withdrawn from trial fisheries and marketing and returned to the monitoring phase (Fukushima Prefecture Federation of Fisheries Cooperative Associations 2014).

Throughout this process, actors at the local level who could be considered 'experts' play a pivotal role not just in physically collecting data, but in establishing the conditions of trust necessary for post-disaster fisheries to progress. This is key to first, getting fishers involved with the trial fisheries, and second, providing material to allow citizens to make an informed decision as to whether or not to consume Iwaki and/or Fukushima fish.

As far as engagement with fishers goes, fishers' first point of contact for scientific information on marine radioactivity are the officers from Fukushima Prefecture's Fisheries Section; be it the Fisheries Development Office or the Fisheries Research Station. Although fisheries cooperative managers agreed that they valued information given to them by the national and local governments, the description of national government scientists 'coming up from Kasumigaseki once a month'[4] hints at a certain degree of social distance through the mentioning of the far-off place from which the scientists come. This stands in stark contrast to the light-hearted horseplay punctuating meetings with prefectural scientists and officers. Indeed, fishers' reluctance to participate in formalised engagement processes involving 'outsiders' is documented in both a UK (Gray *et al.* 2005; Roberts and Jones 2013) and Japanese (Mabon and Kawabe 2015) context. After the 2011 disaster, vital in overcoming concerns about resuming fisheries (exposure to radiation, effects on the community if contaminated fish caught and sold on) was contact with the prefectural fisheries officers. Prefectural officers' mode of engagement – meeting fishers informally and face-to-face in their own ports, and supplementing formalised information meetings with informal discussion – allows knowledge about marine radioactivity to be transmitted in a way consistent with the informal and closed modes of interaction preferred within Iwaki's fishing communities. Further, many of fisheries cooperative administrators actually managing the trial fisheries and doing the radioactivity screening are local to Iwaki itself. Even fisheries researchers born outside the area become familiar to fishers over time thanks to the lifetime employment system operated by the Fukushima prefectural government.

In short, the conditions of trust between fishers, fisheries cooperative administrators and local governors necessary to move Iwaki fisheries forward may come about because 'expert' knowledge on the techniques and processes necessary to assess the safety of fish is communicated to fishers in a manner consistent with 'local' values, by people perceived simultaneously as both citizens of the local community and scientists or managers. When it comes to engaging with the local community, this notion of locally-situated expertise likewise appears important. Catches of fish landed as part of trial fishing operations are screened in newly-established laboratories in Onahama – the biggest fish market in Iwaki – by fisheries cooperative staff trained in radiation measurement techniques by the prefecture's own scientists. Wider monitoring is also undertaken 'in situ' at the Fisheries Research Station, and the results of both monitoring and screening are uploaded to publicly-viewable websites. In addition to this 'official' data collection, the Iwaki-based non-governmental organization (NGO) UmiLabo carries out its own monitoring. Formed after the disaster by local citizens who trained themselves in monitoring techniques to verify government data, UmiLabo invites members of the public to join cruises or fishing expeditions, makes monitoring results available online and carries out public monitoring/tasting sessions in collaboration with a local aquarium.

Two of the groups carrying out the most direct and visible radiation screening of fish for members of the public – the fisheries cooperative and UmiLabo – possessed virtually no expertise in environmental radioactivity prior to 2011. Potentially, the most 'trusted' sources of information for citizens on marine radioactivity are the sources with the shortest time span of experience. This again may make sense, however, if we consider that these actors engaging mainly at the local level – whether ordinary fisheries cooperative workers and administrators, locally-based fisheries scientists, or independent groups formed by concerned citizens – may be viewed as distinct and separate from the people (TEPCO and central government) perceived as causing the FDNPP accident to happen. Further, as these groups carry out monitoring in the community, making not only the results of their data but also the monitoring *processes* and remaining uncertainties visible, their monitoring procedures and the claims to safety of fish arising from it may appear more credible because they are open to scrutiny. In McKechnie's (1996) study into perceptions of ionising radiation from the Sellafield nuclear power plant on the Isle of Man, 'local experts' seen as distant from nuclear power as an entity and willing to be honest about the limitations of their own knowledge, were likewise seen as being afforded the most trust and credibility by the community. There are here also parallels to Matthews (2011) through the emergence of an alternative narrative from below to counter or supplement 'official' activities, with the community in Iwaki succeeding in producing their own knowledge on which consumers can base consumption decisions.

The situation in Iwaki fisheries, then, reflects the tradition of literature challenging a simple divide between 'experts' and 'non-experts' (e.g. Irwin 1995; Michael 1992). The most trustworthy science appears to be produced by local citizens, or at least locally-based scientists. Central to this, however, may be locals' willingness to open up to scrutiny the processes through which they create knowledge, and to acknowledge the limitations of their findings. It is to these limitations water places on what can be known that water places this chapter now turns.

4 Water – limits to what can be known?

The second aspect emerging from the Iwaki fisheries case is the challenge engendered by monitoring in a marine environment, and the effect that decisions based on knowledge gathered in the sea may have on people. At present, coastal fisheries in Fukushima Prefecture are at the trial stage for some species (catching less than 10 per cent of pre-disaster levels) and remain completely suspended for others. However, the adjoining prefectures of Miyagi to the north and Ibaraki to the south can continue most full-scale fishing, albeit conducting regular monitoring of their own. Interviewed fishers expressed frustration at the perceived difference in regulation and monitoring regimes across prefectures, noting that currents carry radioactive matter – and fish – across prefectural borders. Indeed, the southernmost fishing port in Fukushima, Nakoso, is less than one kilometre from the Ibaraki border (see Figure 2.1).

The flows of water – or, perhaps more importantly, the things water currents carry – across regional and national boundaries, therefore, raise questions about how easily the marine produce of one place can be marked as 'safe' where another can be seen as 'dangerous'. This does not mean fish from Ibaraki or Miyagi are harmful. But in terms of water and epistemology, it illustrates how the movement of water and the things in it makes it difficult to tie knowledge of radioactive contamination to one point in space. Within Fukushima Prefecture, sampling of fish for monitoring and trial fisheries takes place at set points. Fishers travel to a set of coordinates in their boats and fish at that location. However, as was raised during discussion with fisheries research officers, whilst there is relatively good knowledge of how particular species of fish are likely to travel and feed, this knowledge cannot extend down to the level of individual fish (see also Casson, this volume, on the limitations of available data and their implications for decision-making). Each individual fish moves through the sea and feeds in slightly different locations over the course of its life. Hence, all of a batch of fish caught in the same location will not necessarily have been exposed to the same amount of radiation over their life span. Further, adding to this complexity is the issue of contaminated sediment – whilst sampled water may appear to be free from radioactivity, underlying sediments from which some fish species feed may contain radioactive matter. The designation of an area as 'safe' or not may thus also depend on what exactly – fish, water, sediment – is sampled (e.g. both Fukushima Prefecture and UmiLabo sample sediment to build a fuller picture of marine radiation).

The key response by both Fukushima Prefecture and Fukushima's fisheries cooperatives to this indeterminacy has been precautionary. This has entailed starting trials from as far away from FDNPP as possible, imposing lower radioactive caesium limits than the national standard, and mandating a period of monitoring before fish can be released for trial operations. Complementary to this, both UmiLabo and TEPCO monitor as close to FDNPP as possible, the rationale being that the highest levels of radioactivity will be found in the waters and fish closest to the plant. These approaches reveal a salient point about monitoring of invisible substances in water – namely, the question not just of who has access to knowledge, but also who has the technology and permissions to access the water bodies required to carry out monitoring and produce knowledge. Fukushima Prefecture's main research vessel was foundered in Onahama Port in the 2011 tsunami. Although a substitute boat was acquired, fishers have been vital in filling the gaps by helping the prefecture collect data (as an aside, fishers are now involved in collecting survey data for upcoming offshore wind energy installations). Despite not being 'experts', the need to have access to offshore locations for sampling means fishers are therefore indispensable in getting access to data. In contrast, concerns over risk from radioactivity and pragmatic security issues mean that only TEPCO can sample in the area immediately around FDNPP. Other institutions like UmiLabo sample only as close as TEPCO permit them to, and citizens without boats supplement this 'citizen' data by catching fish off Onahama pier with rods. Although the data TEPCO releases is generally considered congruent with data from other sources, the closed nature of TEPCO's land-based operations means the back story to this data is sometimes viewed with caution or concern. The lack of access to FDNPP itself, albeit for pragmatic safety and security reasons, means TEPCO's assertions cannot always be verified.

When it comes to governance based on knowledge, water is therefore challenging. This is not just because currents and flows make spatial demarcation difficult, but also because certain equipment and rights are necessary to access closely controlled spaces of water like that close to FDNPP and collect the data required to produce knowledge. Moving away from water itself, though, it is also important to pay attention to the spaces in which water comes into contact with land. There is significant difference and complexity in the way in which knowledge and understanding of the marine radiation situation plays out at the edge of the sea.

For instance, there is huge heterogeneity within Iwaki fisheries. During the fieldwork on which this chapter is based, it became clear fishers identified strongly with both the port in which they were based and also the fish they caught. The length of Iwaki itself, 50 km from north to south, means there is significant difference in how the effects of both the nuclear accident (FNDPP being a further 30 km north of the uppermost point of Iwaki) and tsunami were felt. Depending on the level of infrastructural damage suffered and how the type of fish caught in each port have been affected by radioactivity, there is difference in the extent to which fishing operations have been able to resume. Outside of Iwaki, there is also Fukushima's other fishing district of Soma-Futaba to the north, where some post-disaster fisheries are buoyant (even more so than those in Iwaki) whilst other ports remain completely uninhabitable and off-limits due to being in the exclusion zone around the FDNPP.

It is thus extremely difficult, if not impossible, to make any claims as to the status of 'Fukushima' fisheries as a whole. Paralleling Dove and Kammen's (2015) challenge to the assumption that 'locals' do not understand ecosystems or environmental protection, Iwaki fishers appear all too aware of the complexity of the marine environment and the challenges this may raise for governing radiation in fish with certainty. One participating fisher in particular pointed out different spatial scales over which fisheries could be imagined, ranging from 'Japanese' fisheries and 'Fukushima' fisheries down to 'Iwaki' fisheries and questioning the spatial scales over which fish were perceived as 'contaminated' because of the accident. The heterogeneity of experience that can exist within a spatial unit is of course an issue on land as well, but the movement of water across boundaries and also the differing ways in which humans may interact with water adds an additional layer of complexity. In Section 5, we return to this question of governing water across scales under conditions of high uncertainty.

The socio-cultural effects that may arise from making decisions about coastal places (and the people living in them) under the conditions of indeterminacy necessarily imposed by water should not be overlooked. The significance of seafood to Japanese diet and culture is well known, however, in Iwaki's ports, before the disaster, fishing was also a source of pride and a crucial focal point for community interaction (Mabon and Kawabe 2015). As well as protecting citizens from contaminated marine produce, at stake when making decisions about progressing Fukushima fisheries (or not) under uncertain conditions are the livelihoods and lifestyles of coastal villages. Hill (this volume) summarises that the way 'scientific knowledge' is incorporated into environmental management can have effects on the normative or organizational aspects of social processes. Indeed, the implications of knowledge about radiation in water have the potential to affect coastal communities in profound ways by determining when – if at all – to what extent and in what way culturally significant fishing practices can be resumed. Oughton (2013) questions how a drive for absolute safety in environmental remediation ought to be balanced with the socio-cultural dimensions of re-settling people or completely outlawing certain foods. In a similar way, the cultural significance of fisheries to Iwaki raises questions about who has the right to make decisions on the governance of a body of water, which knowledge they use to make such decisions, and who is affected as a result of the decisions made. Given the heterogeneity, it may thus be the case that rather than making clear-cut 'yes' or 'no' decisions about the resumption of fisheries from on high, the actors at the local level outlined in Section 3 are better placed to decide what the socio-cultural implications are of incrementally restarting fisheries under uncertain conditions, and how the various risks at play can best be balanced. We draw out these broader questions of how epistemologies of water inform governance and decision-making in the following discussion.

5 Discussion – epistemology in water governance

From the previous sections, for Iwaki fisheries at least it would be fair to say we understand the 'epistemology of water' as encompassing not only *how* different people come to know water, but also *why*. Like Choy (2011), we suggest that environmental politics are inseparable from environmental science – with this 'science' being deployed to inform subsequent decision making. The effects of the nuclear disaster on the fishing communities on Iwaki's coast illustrate that 'knowledge' of water has instrumental purpose in understanding which fish can be safely released for sale for economic gain, and also more intrinsic purpose in shaping the identity and culture of those involved in fisheries. So, an 'epistemology of water' in Iwaki fisheries entails two things. One is knowing – or at least knowing the limits of one's knowledge about – how water and the things it carries interact with one another. The other is extrapolating this knowledge to include how these technical and scientific processes give rise to benefits or risks for humans, and thus understanding what constitutes a socially acceptable way to manage risk. Nowhere is this better illustrated than Iwaki City's *Joban-Mono*[5] campaign, which uses both physical characteristics of water off the coast of Iwaki (the meeting of warm and cold currents and the nutrient-rich environment this engenders) and the identities of those undertaking the fishing ('serious' and 'professional' yet local fishers and brokers) to demonstrate the quality and safety of Iwaki marine produce (Iwaki City 2015). Based on this understanding, we offer three related overarching observations on the implications of this research for how we think about epistemologies when it comes to governance of water – interplay with land, complexity, and scale.

First, post-disaster Iwaki fisheries demonstrate that the issue at hand is not always purely about 'water'. Rather, we may also be interested in things we put into water, things we take out of water and where water meets people. In Iwaki, the primary focus is on seafood that has been taken out of the water for human consumption – and how radioactive particles released from land find their way into seawater (either accidentally at the time of the nuclear accident or subsequently as part of deliberate controlled releases) and into fish. Indeed, it is on land that the effects on humans from marine radiation are ultimately felt. Produce caught at sea is, after all, screened, sold and consumed on land. And it is the economies and societies of communities on the coast that feel the effects of not being allowed to partake in the water-based practice of fishing.

The different ways of knowing about marine radioactivity post-Fukushima, then, draw on several other epistemologies. Some of these are epistemologies of 'water' in a narrow sense – physical science enquiries into ocean circulation that help understand distribution of radioactive matter (e.g. Bailly du Bois *et al.*

2014), or building a nuanced sense through 'fisheries science' of how recovery may proceed differently depending on techniques and species (e.g. Wada *et al.* 2013). Equally, however, constructing knowledge about marine radioactive contamination in Iwaki also entails drawing on epistemologies that are either land-based or not exclusive to water. These include the debates on how well the internal health effects of consuming contaminated produce are understood (summarised by Morris-Suzuki 2014), and the economic dimensions of how well post-disaster fish sell when taken to markets on land.

A key question this research throws up, then, is how water challenges or modifies some of these 'land-based' epistemologies. Sections 3 and 4 discussed how many of the issues around understanding sea water in Iwaki and Fukushima are bound up with issues taking place on the coast – access rights and technologies, perceptions of actors, visibility of monitoring – where the uncertainties and indeterminacies engendered by water bring additional complexities. This illustrates the importance, when considering epistemologies of 'water', to keep a check on the interplay between water and land, and to pay attention to how epistemologies of water interplay with land-based ways of knowing.

This leads into our second point – complexity. Boyes and Elliot (2014) hold that the number of activities taking place in marine environments, the national, regional and international entities that may have interests in a body of water, and the range of specialisations required to manage marine environments all engender extremely complex governance processes and structures. Boyes and Elliot make this argument from the perspective of policy and legislation; however, we would take this further and argue that the 'complexity' to which they refer goes far beyond policymaking (and indeed marine environments) to encompass the whole business of producing knowledge about and in water. Moreover, as argued by Hill (this volume), we would add that very complex governance structures and relationships exist not only at the national level upwards, but also within communities. This is demonstrated by the wide range of actors *within* Iwaki City involved in managing trial fisheries, and reinforces Hill's warning against viewing those that use water (or in our case those who catch and consume the produce within it) as passive end recipients of decisions made by external actors at larger spatial scales.

A significant source of this complexity in Iwaki fisheries is the myriad of data at play. This data may represent different facets of 'marine radioactivity' – fish meat, sea bed sediment, seaweed, even seawater itself – and may be collected in a range of ways including fishing with rods off piers, sampling in front of the FDNPP or taking sediment, water and fish from far out at sea by boat. The ways in which this data is collected depends on the finances, equipment and rights to which the different monitoring institutions have access. As discussed in Section 3, results are disseminated through various pathways depending on the organisation in question. Complexity in 'knowing' water also comes from the heterogeneity of experiences water engenders. Even prior to the 2011 disaster, physical factors such as the size and nature of ports and the way ocean currents carried different species around meant that within 'Iwaki' fisheries – let alone 'Fukushima' fisheries – hugely divergent experiences and knowledges existed over relatively short spatial distances. The fact water and fish do not move in straight lines serves only to add to this. Factors such as how far offshore fish are caught (and through which technique), and how rivers carrying radioactive material down from mountains interplay with the sea, means it is not necessarily the case that fish landed in ports further away from FDNPP are 'safer'.

The movement and interaction of water with land, therefore, gives great variability in what can/ought to be scientifically sampled, and questions how transferable knowledge produced from data collected at one place at one point in time can really be. Such complexity in data and experience is of course true on land as well, but what sea water adds to this is an extra variable, one difficult to comprehend fully due to limitations on what can be monitored and tracked. To borrow the Riesch (2012) typology, water moves us towards higher 'levels' of uncertainty where clear-cut responses are more difficult because our understanding of processes become ever less and assumptions increase.

A crucial challenge, therefore, is to reflect on how different epistemologies of water deal with uncertainty and complexity – and what happens if different ways of knowing water engender competing responses to complexity and uncertainty. It is fortunate in the case of post-disaster Iwaki fisheries that there appears to be good consensus that the various organisations' data is consistent, or at least agreement that continuous monitoring and the incremental restart of fishing is an appropriate response to uncertainty. However, in situations where different actors' epistemologies do not complement each other so well, it is vital to reflect on whose way of understanding uncertainty and complexity wins out, why and to what effect. The very nature of water – where risks and negative effects may not be distributed in a linear fashion across space and the greatest injustices may be felt by those physically distant from the issue at stake – makes it even more important to pay close attention to epistemic (Fricker 2007), procedural (McLaren 2012) and distributional (Shrader-Frechette 2002) injustices that may arise from the governance of uncertainty in water.

Nonetheless, despite the challenges water raises for making policy anchored to space, decisions about fisheries in Fukushima – and indeed, the governance of water bodies more generally – do sometimes have to be taken even if under conditions of uncertainty or indeterminacy. This gives rise to our third and final point – the question of how to address the scale of governance in water. Much like Casson (this volume), what we can see in the case of Iwaki – and Fukushima – fisheries is that there are different epistemologies of water operating at different spatial scales, each feeding into a governance system with its own intended outcome. At the national level, there is the centrally-set 100 Bq/kg limit and the restriction on coastal fisheries in Fukushima Prefecture, supported by data collected by the Fisheries Agency of Japan and also TEPCO. The critical debate on the Japanese government's desire to downplay the impacts of the nuclear disaster and expedite remediation (but see, for example Perrow 2013) lies beyond the scope of this chapter.

But it does not take much to imagine that a utility company like TEPCO with significant nuclear power assets would seek to prove its competence to facilitate reactor restarts elsewhere, or that the national government would at base wish to 'reconstruct' and rehabilitate disaster-affected areas. At regional level, prefectures carry out their own monitoring activities – this is where the Fukushima Prefecture 'government' finds itself in a strange position. On one hand, the empirical data for this research showed some suspicion of high-level prefectural government given its perceived economic desire to reju-venate tourism and stimulate sales of regional produce. On the other, local government employees 'on the ground' were viewed much more positively as citizens with a personal stake in the outcomes of fisheries radiation monitor-ing not afraid to get physically involved in collecting data. Locally, fisheries monitoring is actually more stringent than national standards, adopting lower radiation thresholds, training up new scientists at the fisheries cooperatives and independent citizens' groups supplanting 'official' monitoring with their own sampling and analysis. Local governance too aims to restart fisheries, but doing so at a speed and manner where addressing the concerns of local fishers and citizens plays a more central role.

Casson (this volume) holds that integrating information across a range of scales is crucial for developing a richer understanding on which to base envi-ronment-related policy. Casson makes this argument in the context of the Australian–Indonesian monsoon, however, it is equally applicable to fisheries in Fukushima and Iwaki where data is generated by actors operating at the local, prefectural, national and international scales. To this, we would add that looking across scales is necessary not only to understand the physical and envi-ronmental aspects of post-disaster Fukushima fisheries, but also to make sense of the economic, social and political dimensions of governing marine radio-activity. For instance, the concerns local actors have about ongoing leaks of contaminated water into the sea from FDNPP relate to a national-level actor in TEPCO and whether or not the claims to safety made by TEPCO are trusted and on what grounds. Fishers unable to catch and sell fish fully depend on compensation from TEPCO, the level of which is determined in negotiation with central government and as such is tied to central government perceptions of the extent to which it is safe to restart Fukushima fisheries. Likewise, the uptake of fish caught through trial fisheries and sold at markets across north-east Japan is contingent on how buyers and brokers from outside Iwaki and Fukushima perceive the risk and safety of Iwaki fish, itself a product of the extent to which they trust the screening methodologies and claims to safety made by the local fisheries cooperatives. Yet, just as Tsing (2005) holds that influence between the 'global' and 'local' works both ways, so it is also the case that decisions taken at a very local level by cooperatives, brokers and individual fishers as to what *they* feel constitutes an acceptable risk will influence when and in what quantity Iwaki fish are sent for sale across the rest of Japan.

Again, these issues are of course not unique to water, but the uncertainty and indeterminacy engendered by water means that trust in and perception of other actors' epistemologies becomes even more pronounced. If one cannot verify the safety of Fukushima fish oneself due to the nature of water, one must scrutinise the methods and motivations of those who do claim to be able to produce knowledge (after Wynne 1992). By necessity, water and its governance transcends scales due to its fluid nature, and also the additional movement of things like fish and radioactive matter across space before they enter and/or after they are removed from water. Linking back to the second challenge on complexity, of particular importance here is building a sense of *why* knowledge from one particular scale may dominate, and to what effect. In turn, there is also a need to imagine how to develop and enact new water governance strategies that are able to balance the different epistemologies, values and expectations working at different scales, and how to make visible the differences that may exist at these varying scales.

6 Conclusion

In this chapter, we have characterised some dimensions of water epistemology in Iwaki fisheries and suggested areas warranting further theoretical or empirical attention. Many of the issues we have raised are not unique to water. However, because water forces us to think more explicitly about what cannot be known and what the implications of this are for decision making, it may be that looking to sites of water such as Iwaki fisheries gives us greater analytical purchase on how to govern environments under conditions of uncertainty. At the same time, it is perhaps important to think more deeply about how exactly water changes land-based epistemologies beyond tropes of movement and flows. In this regard, we suggest the pluralities of ways in which water interacts with its margins (e.g. coastal areas) may be what makes governance based on spatial demarcation of knowledge more challenging.

Finally, whilst there is good evidence to suggest that marine radiation in Fukushima is subsiding and that knowledge of the nature of contamination is increasing, the Iwaki and Fukushima fisheries situation will remain difficult for some time. What happens next is not only a question of techno-scientific understanding of the safety of marine produce. Rather it is a much wider issue about how decisions based on certain understandings of water affect communities whose economies, cultures and identities are intrinsically bound up with fishing viewed as safe and trustworthy.

Acknowledgements

The fieldwork on which this chapter is based was possible due to a Japan Foundation Fellowship the lead author received, and partly by KAKEN 22310029 funding the second author received. The authors are grateful to Fukushima Prefectural Fisheries Development Office, Fukushima Prefectural Fisheries Research Station, Iwaki City Fisheries Cooperative, and Onahama

Danish Trawl Seines Fisheries Cooperative for support in arranging the field work in Iwaki; and to all members of the Iwaki fishing communities and citizens of Fukushima Prefecture who participated in this research. We are also grateful to the editors and reviewer for their comments on the chapter.

Notes

1 The Becquerel is an internationally-recognised measure of radioactivity. The number of Becquerels per kilogram (Bq/kg) is the unit of choice for measuring radioactivity in fish (and other foodstuffs) because it gives a basic physical measure of the amount of ionising radiation released by radioactive decay. The Sievert, by contrast, measures biological damage to the human body and the risk to health – which is dependent on the nature and extent of dose as well as the physical amount of radiation emitted (International Commission on Radiological Protection 2007).
2 'City' is the direct translation for an administrative district and does not necessarily mean the area is urbanised – although Iwaki does have an urban core inland.
3 For more on the fisheries monitoring process, we direct the reader to the English-language periodical reports produced by the Fisheries Agency of Japan, available at http://www.jfa.maff.go.jp/e/inspection/index.html
4 Kasumigaseki is the area in Tokyo where the Fisheries Agency is located.
5 This translates as 'Joban Produce'. 'Joban' is an old name for the Iwaki area, which interestingly combines parts of the Japanese characters for Iwaki and Hitachi in neighbouring Ibaraki Prefecture and hence transcends the boundary of Fukushima Prefecture.

References

Bailly du Bois, P., Garreau, P., Laguionie, P. and Korsakissok, I. (2014) 'Comparison between modelling and measurement of marine dispersion, environmental half-time and 137Cs inventories after the Fukushima Dai-ichi accident', *Ocean Dynamics*, vol. 64, no. 3, pp. 361–383.
Boyes, S. and Elliot, M. (2014) 'Marine legislation – The ultimate "horrendogram": International law, European directives and national implementation', *Marine Pollution Bulletin*, vol. 86, pp. 39–47.
Buesseler, K. (2012) 'Fishing for Answers off Fukushima', *Science*, vol. 338, pp. 480–482.
Casson, S. (2017) 'Trans-disciplinary analysis of Australian–Indonesian monsoon epistemologies and their implications on climate change adaptation strategies' in R. Baghel, L. Stepan and J. Hill (eds) *Water, Knowledge and the Environment in Asia: Epistemologies, Practices and Locales*, Routledge, London, pp. 36–51.
Choy, T. (2011) *Ecologies of Comparison: An Ethnography of Endangerment in Hong Kong*, Duke University Press, Durham, NC.
Dove, M. R. and Kammen, D. M. (2015) *Science, Society and the Environment: Applying Anthropology and Physics to Sustainability*, Routledge, London.
Fisheries Agency of Japan. (2015) *Results of the Monitoring on Radioactivity Level in Fisheries Products*, Fisheries Agency of Japan, Tokyo, http://www.jfa.maff.go.jp/e/inspection/ accessed 25 May 2016.
Fricker, M. (2007) *Epistemic Injustice: Power and the Ethics of Knowing*, Oxford University Press, New York.
Fukushima Prefecture Federation of Fisheries Cooperative Associations. (2014) *Inspection System* (in Japanese), FPFFCA, Iwaki, http://www.fsgyoren.jf-net.ne.jp/siso/buhin/kensa20140827.pdf accessed 25 May 2016.

34   *Leslie Mabon and Midori Kawabe*

Gray, T., Haggett, C. and Bell, D. (2005) 'Offshore wind farms and commercial fisheries in the UK: a study in stakeholder consultation', *Ethics, Place and Environment*, vol. 8, no. 2, pp. 127–140.

Hill, J. (2017) 'Development projects undermining farmer managed irrigation systems in Asia's high mountain valleys: Science as friend or foe?' in R. Baghel, L. Stepan and J. Hill (eds) *Water, Knowledge and the Environment in Asia. Epistemologies, Practices and Locales*, Routledge, London, pp. 70–86.

International Committee on Radiological Protection. (2007) *Annals of the ICRP Publication 103L, The 2007 Recommendations of the International Commission on Radiological Protection*, Elsevier, New York.

Irwin, A. (1995) *Citizen Science: A Study of People, Expertise and Sustainable Development*, Routledge, London.

Iwaki City. (2015) *What is* Joban-Mono? (in Japanese), Iwaki City Government, Iwaki, http://misemasu-iwaki.jp/joban/item/A5guidebook-201510.pdf accessed 25 May 2016

Iwaki City. (2016) *Iwaki City. Disaster Office Weekly Update* (in Japanese), Iwaki City Government, Iwaki, http://www.city.iwaki.lg.jp/www/contents/1450768683538/simple/zhigai20160518.pdf accessed 25 May 2016

Kyodo News. (2015) 'Fukushima Prefecture Fisheries Association: "The relationship of trust has collapsed". Angry voices at discharge of contaminated rainwater' (in Japanese), *Kyodo News*, 25 February, Online Edition, http://www.47news.jp/CN/201502/CN2015022501001290.html accessed 30 June 2015.

Mabon, L. and Kawabe, M. (2015) 'Fisheries in Iwaki after the Fukushima Dai'ichi nuclear accident: lessons for coastal management under conditions of high uncertainty?' *Coastal Management*, vol. 43, no. 5, pp. 498–518.

McKechnie, R. (1996) 'Insiders and outsiders: identifying experts on home ground', in A. Irwin and B. Wynne (eds) *Misunderstanding Science? The Public Reconstruction of Science and Technology*, Cambridge University Press, Cambridge, UK, pp. 126–151.

McLaren, D. (2012) 'Procedural justice in carbon capture and storage: a review', *Energy and Environment*, vol. 23, no. 2–3, pp. 345–365.

Matthews, A. S. (2011) *Instituting Nature: Authority, Expertise and Power in Mexican Forests*, MIT Press, Cambridge, MA.

Michael, M. (1992) 'Lay Discourses of Science: Science-in-General, Science-in-Particular, and Self', *Science, Technology & Human Values*, vol. 17, pp. 313–333.

Morris-Suzuki, T. (2014) 'Touching the Grass: Science, Uncertainty and Everyday Life from Chernobyl to Fukushima', *Science, Technology and Society*, vol. 19, no. 3, pp. 331–362.

Oughton, D. (2013) 'Social and ethical issues in environmental remediation projects', *Journal of Environmental Radioactivity*, vol. 119, pp. 21–25.

Perrow, C. (2013) 'Nuclear denial: From Hiroshima to Fukushima', *Bulletin of the Atomic Scientists*, vol. 69, no. 5, pp. 56–67.

Riesch, H. (2012) 'Levels of uncertainty', in S. Roeser, R. Hillerbrand, P. Sandin and M. Peterson (eds) *Essentials of Risk Theory*, Springer, New York, pp. 29–56.

Roberts, T. and Jones, P. J. S. (2013) 'North East Kent European marine site: Overcoming barriers to conservation through community engagement', *Marine Policy*, vol. 41, pp. 33–40.

Shrader-Frechette, K. (2002) *Environmental Justice: Creating Equality, Reclaiming Democracy*, Oxford University Press, New York.

Tsing, A. (2005) *Friction: An Ethnography of Global Connection*, Princeton University Press, Princeton.

Wada, T., Nemoto, Y., Shimamura, S., Fujita, T., Mizuno, T., Sohtome, T., Kamiyama, K., Morita, T. and Igarashi, S. (2013) 'Effects of the nuclear disaster on marine products in Fukushima', *Journal of Environmental Radioactivity*, vol. 124, pp. 246–254.

Wynne, B. (1992) 'Misunderstood misunderstanding: social identities and public uptake of science', *Public Understanding of Science*, vol. 1, no. 3, pp. 281–304.

Yoshida, N. and Kanda, J. (2012) 'Tracking the Fukushima Radionuclides', *Science*, vol. 336, pp. 1115–1116.

3 Trans-disciplinary analysis of Australian–Indonesian monsoon epistemologies and their implications on climate change adaptation strategies

Sarah Casson

Introduction

Monsoons are highly variable, complex climate patterns that directly affect the lives of half of the world. They cover landmasses in Asia, Africa, North and South America. As defined in the most recently published 2014 Intergovernmental Panel on Climate Change's report, 'a monsoon is a tropical and subtropical seasonal reversal in both the surface winds and associated precipitation, caused by differential heating between a continental-scale land mass and the adjacent ocean' (IPCC 2014, p. 1458). Beyond being a heavily studied climate phenomena, monsoons also reflect deep cultural notions across the globe. As Fein and Stephens (1987, p. ix) argue, monsoons are not only rain and wind patterns but also are deeply connected to every aspect of culture.

Monsoons provide the water that irrigates much of the world's agriculture, making them an important topic of study for both global trade as well as local communities (Naylor and Mastrandrea 2010). The Australian–Indonesian monsoon directly affects the lives and agricultural production of millions in Australia and Indonesia and, to some extent, those in Oceania and other Southeast Asian countries. Climate change increasingly exposes communities in impoverished parts of the world to more extreme weather patterns, ranging from mega-floods to mega-droughts instead of gentle, annual monsoon rains. The future of these communities – in which lives, cultures, and resources are intertwined with the monsoon – depends upon how resilient they are to these coming climate changes. With a population of over 240 million and much of its gross domestic product (GDP) coming from agriculture, Indonesia must create resilience strategies both to manage current unpredictability and to withstand future monsoon shifts. Such resilience strategies must be founded on a clear understanding of the Australian–Indonesian monsoon (Wheeler and McBride 2005, p. 125).

This chapter asks, whose knowledges and epistemologies define the problems of changing monsoons? Whose knowledge informs the solutions addressing the applied importance of agriculture and peoples' livelihoods that depend upon the regularity of the Australian-Indonesian monsoon? The primary objective of this chapter is to compare and contrast different scales of data that predict and describe the Australian–Indonesian monsoon. Each scale of analysis represents a different epistemology. The objective is reached through the review of modelling literature, examination of local meteorological dataset of 25 years, and analysis of ethnographic interviews. Most climate change literature prioritises the knowledges produces at two scales (global and regional) and undervalues the epistemologies used at two scales (local and household). This undervalued knowledge presents information not known at the prioritised scales of knowledge. This chapter concludes by suggesting that vastly different scales and data types need to be brought together in conversation to better understand the Australian-Indonesian monsoon. As climate change threatens to disrupt the monsoon patterns that many peoples in the world depend upon for water and agriculture, a better understanding of the Australian-Indonesian monsoon becomes critical. Such an improved understanding must be informed by multiple knowledges.

Scales at which the Australian–Indonesian monsoon is observed

Comparing and contrasting the scales at which the Australian–Indonesian monsoon is observed illuminates the ways in which the monsoon is currently understood. These four scales each provide tools to analyse the monsoon. The global scale provides an overall picture that situates the monsoon in context with other climate phenomena. The regional scale produces ways of understanding the exact behaviour of the monsoon itself. The local scale gives observed data that provide exact occurrences in a singular location. The individual scale shows the human experience of the monsoon. One scale is not more complete or better than the others. Instead, these four scales are complementary to one another.

Global scale

The global scale is drawn from coupled atmosphere–ocean circulation models of the global environment (Colman Moise and Hanson 2011; Hsu *et al.* 2013; Kim, Wang and Ding 2008; Meehl and Arblaster 2003; Schewe, Levermann and Meinhausen 2010; Wang, Kang and Lee 2004; Wang *et al.* 2008; Wang *et al.* 2014; Zhou *et al.* 2009). These models have been analysed within the Couple Model Intercomparison Project Phase 5 (CMIP5) run by the World Climate Research Programme. These models are seen as essential in understanding future behaviour of the Australian–Indonesian monsoon by the Intergovernmental Panel on Climate Change's Fifth Assessment Report on the Physical Science Basis for Climate Change (2014). Such models not only project possible changes in climate phenomena but also model the current and past climate behaviour (Kim, Wang and Ding 2008; Hsu, Li and Wang 2011).

The timescales that can be represented within models at the global scale are quite long: from the prehistoric to the far future.

These models provide a global context within which the Australian–Indonesian monsoon is situated among other climate phenomena, like the El Niño Southern Oscillation and the Madden–Julian Oscillation (Wang, Yang, Zhou and Wang 2008; Zhou, Wu and Wang 2009; Meehl *et al.* 2012; Wang *et al.* 2014). Several climate models simulations compare well to historical data on the Australian–Indonesian monsoon (Wang *et al.* 2008). The global scale is important because while the Australian–Indonesian monsoon has a geographic boundary, the exact limits are unknown and the monsoon's behaviour is likely influenced by climate phenomena outside the boundary limits (Hsu *et al.* 2013).

Yet, drawbacks are inherent to such a wide-reaching scale. Although climate models are able to output a few monsoon variables, they are not able to represent the entire system (Wang *et al.* 2008). Precipitation is a difficult variable to simulate within models because there is much uncertainty in the dynamics of cloud–aerosol interaction (Zhou, Wu and Wang 2009). Global climate models at the global scale simulate the Australian–Indonesian monsoon to predict the monsoon's spatial temporal behaviour and rely heavily on peer-reviewed assumptions because of a lack of data depth on the monsoon (Schewe, Levermann and Meinshausen 2010; Wang *et al.* 2014). To adjust for this, modellers focus on specific variables, like rainfall volumes and onset dates, to represent the entire Australian–Indonesian monsoon phenomena. Other monsoons, like the East Asian and the South Asian monsoon, are better understood at the global scale because more data is collected about them (Xu *et al.* 2006; Stowasser, Annamalai and Hafner 2009; Zhu *et al.* 2012; Annamalai *et al.* 2013).

Regional scale

To account for the global scale's inability to accurately model monsoons at a more regional scale, many monsoon modellers have begun to analyse monsoons at a regional scale (Mittal, Mishra and Singh 2013; Vera *et al.* 2011). Such models can provide projections of monsoon behaviour that is limited by spatial–temporal parameters (Moron *et al.* 2010). Models created at the regional scale can give better analysis of the behaviour of individual monsoons. The geographic context of the monsoon on a regional scale is more reliably integrated into the models as is the influence of other climate phenomena. Regional scale models of the Australian–Indonesian monsoon scale provide better analysis of the main influences of the monsoon's behaviour than can be understood at the global scale (Aldrian and Susanto 2003; Wheeler and McBride 2005; D'Arrigo *et al.* 2006a; D'Arrigo *et al.* 2006b; Qian *et al.* 2010; Moron *et al.* 2010). Examining the influences on monsoon behaviour, like El Niño Southern Oscillation, sea surface temperature and land topography, creates better predictions of both time and space. In doing so, this regional focus provides a stronger understanding of how a particular monsoon might shift with climate change than can be provided at the global scale (Naylor and Mastrandrea 2010; Robertson, Moron and Swarinoto 2009).

Monsoon researchers are currently working to downscale models of Australian–Indonesian monsoon to the regional scale in coordination with physical observations and statistical relationships. Models at the regional scale are created through dynamical or empirical downscaling of global climate models (Vimont *et al.* 2010; Moron, Camberlin and Robertson 2013; D'Arrigo *et al.* 2006a). Both downscaling methods extend the scientific assumptions used at the global scale to the regional scale. Dynamical downscaling relies on the output of a global scale climate model to create a higher resolution model of the Australian–Indonesian monsoon's spatial behaviour. Empirical downscaling relies upon the statistical relationship between global scale climate phenomena variables. These statistical relationships are replicated at the regional scale.

Dynamical downscaling is limited by the wealth of regional scale data for monsoon variables. Often there is a lack of the quality and quantity of data available to create strong dynamically downscaled models of the Australian–Indonesian monsoon at the regional scale (Naylor *et al.* 2007). Paleoclimatological data and historic rainfall patterns are heavily relied upon within models created at the regional scale with empirical downscaling (D'Arrigo 2006; Qian *et al.* 2010). Such reliance assumes future relationships among climate phenomena and monsoon influencers will be the same as historic relationships (Naylor *et al.* 2007). The Australian–Indonesian monsoon, however, is known to be a highly chaotic, non-linear system and cannot be assumed to replicate past relationships in the future, especially within future climate change scenarios.

Local scale

The Australian–Indonesian monsoon can exhibit highly localised behaviour (Aldrian and Susanto 2003; Vimont *et al.* 2010; Qian *et al.* 2010). To better understand the monsoon's localised behaviour, the Indonesia meteorological department created a 25-year dataset from weather observation in East Flores, Indonesia. Collected since 1989 but never analysed, this dataset provides an in-depth perspective on nine key variables deemed essential to understanding the monsoon by the East Flores government. The variables studied are: Rainfall Amount (mm), Rainfall days, Sunshine Percentage, Humidity Percentage, Average Windspeed (knot), Average Windspeed Direction, Maximum Windspeed (knot) and Maximum Windspeed Direction. Weather observations provide data on actual occurrences: exact historical records are known for the 25-year period. However, the limited temporal and geographic range of the dataset only allows for descriptive statistical analysis of the variables.

Household scale

The household scale provides a further in-depth understanding of the monsoon that often draws upon local and traditional knowledge. It is provided by the author's 42 ethnographic interviews with farmers in East Flores, Indonesia. East Flores, Indonesia is especially vulnerable to monsoon shifts. It is a rural region populated by farmers who have based much of their lives and livelihoods around the monsoon rains. While many remain their entire lives in this small region, some family members emigrate to Kalimantan, Sumatra or Malaysia for employment, returning only every couple years. This region has rain-fed agriculture, vast biodiversity and strong community structures, all of which are potentially threatened by shifting monsoons.

A household scale is a way to show a different understanding of the monsoon than the three other scales: everyday understandings of an essential climate phenomenon. The household scale shows not just what is happening daily but also what the variables mean within a community. Ethnoclimatology has proven to be a tremendously useful field within climate science (Orlove *et al.* 2000; Orlove *et al.* 2002; Robertson and McGee 2003; Schlacher *et al.* 2010). Using it for further understanding of the Australian–Indonesian monsoon should be encouraged.

Analysis of key variables

Investigating how the four scales of data approach the six variables is essential in understanding how these epistemologies conceptualise the monsoon and its behavioural implications for agriculture. The variables of monsoon onset and monsoon duration have a close relationship to the timing of planting and harvesting, respectively. The variable of monsoon onset is best understood as the starting date of the monsoon rainfall during the wet season. Duration is the temporal length of the monsoon rainfall and winds. Onset is much more easily measured than duration because the rains often come abruptly and leave gradually. Both onset and duration are incredibly important variables to agricultural production and are critical for farmers in knowing when to plant and harvest. Correct timing of the start and end of a monsoon wet season helps to ensure crop production (Fein and Stephens 1987). Crops are planted soon after a monsoon's onset and are harvested when the monsoon has fulfilled its duration of rainfall and wind.

Rainfall intensity and rainfall volume directly influence agricultural production. Crops not only need the proper rainfall timing, they also require consistent rainfall amounts in both time and space. Rainfall intensity measures the consistency of rainfall events over the monsoon wet season duration. Rainfall volume measures the volume of rainfall per month. Both are essential to agricultural production. Sporadic rainfall events or too much/little rainfall destroys fields and the crops growing therein.

Average windspeed and maximum windspeed also contribute to the overall production of agriculture harvested in a monsoon climate. Monsoon winds often increase at the end of a growing season. Understanding these winds is critical to agriculture production. The average windspeed variable represents the mean monthly windspeed. The maximum windspeed variable represents the extreme windspeed in a given month. Both are important and complementary variables for providing full understanding of a monsoon's wind. The average windspeed provides the monthly context and the maximum shows the extremes to which the windspeed varied from the average. Yet, maximum windspeed often draws much more attention because often it has far more dramatic repercussions, like broken stalks and large waves, than does the average variable. Both are hard variables to measure, from all four scales, and are hard to attribute solely to the Australian–Indonesian monsoon.

Monsoon onset

At the global scale, analyses in the modelling literature agree that the monsoon onset is complicated to model (Colman, Moise and Hanson 2011; Kim *et al.* 2011; Annamalai *et al.* 2013; Hsu *et al.* 2013). While it is difficult to model monsoon onset well, all literature at the global scale pays close attention to this variable (Kitaoh and Uchiyama 2006; Kim, Wang and Ding 2008). Most of the literature focuses upon creating a clear distinction between the onset of the Australian-Indonesian monsoon from other climate phenomena and from other monsoons (Colman, Moise and Hanson 2011; Hsu *et al.* 2013).

A regional scale is an incredibly difficult position from which to view the monsoon (Reid *et al.* 2013). It lacks both the on-the-ground daily realities of a local scale and the generalities of a global scale to view the Australian–Indonesian monsoon (Aldrian and Susanto 2003; D'Arrigo *et al.* 2006b; Haylock and McBride 2001; Moron 2010; Zhang *et al.* 2013). Most monsoon variables are incredibly hard to distinctly analyse at the regional scale. Monsoon onset is the best understood variable at the regional scale, but even this is still hard to analyse well. An example can be seen in Vimont *et al.* (2010) assessing the potential for downscaling Indonesian rainfall patterns from the large-scale to the regional. They found that downscaling is most accurate over the southern islands of Indonesia during the monsoon onset. Similarly, in their data analysis, Moron *et al.* (2009) found Australian-Indonesian monsoon dates to be potentially predictable. They view the monsoon onset as progressively coming to different sub-regions as it sweeps from the northwest to the southeast. They state that, in general, the monsoon onset begins in late August in Sumatra and continues across to Timor, where during mid-December, the latest monsoon onset of Indonesia occurs (Moron *et al.* 2009, p. 840).

The Government of East Flores, Indonesia, views onset as one of the most important variables to measure. Government officials understand onset as the month when monsoon rains begin to fall. As one agricultural officer said: 'The main trouble to agriculture in the region is rain and its unpredictability. The agricultural office hopes that farmers will be ready in September or October to plant right. So that whenever the rain does come, they can directly plant and are not late in planting'. She elaborated to say that the monsoon onset was shifting over time due to climate change, and that this timing shift caused many problems for the government and farmers:

Farmers have historically planted in September but now there is not always rain in September. But certain farmers continue to plant in September without waiting for the rains to come, which might not be until October or November. That is when there is major trouble because the crops will not grow and the seeds will die.

This statement, however, contradicts the dataset of rainfall patterns within East Flores, Indonesia: the 25-year dataset collected by the meteorology department shows no statistical significance in the change of onset over time. This discrepancy could be due to anecdotal evidence misleading the government, improper rainfall measurement methodology, the inadequate temporal length of the dataset or hyper-regional differences in rainfall patterns between the villages aided by the government and the singular location of data collected.

At the household scale, data collected from ethnographic interviews supports the view of the government: Monsoon onset is changing and changing drastically. One elder, a rice and corn farmer, summed up the problem: 'There have been changes in the rain. Today sometimes there is too much rain and then sometimes there is not enough rain. So, it is hard to tell when you should plant. Before, I always planted in October or November. Now it is difficult to predict when I should plant'. As one farmer stated, monsoon onset is a drastically important variable to understand: 'If the rains do not come or are late, all crop fields dry up and die. This is a big problem'. Delayed monsoon onset poses tremendous threat to the region's agriculture. Farmers depend upon a consistent onset date. A drastic change in that date – perceived or actual – could disrupt crop production and create social disruptions. Knowing perceptions at the household scale is as important as understanding how other climate phenomena influence the monsoon's onset at the global scale. Each provides critical data accessible only at its respective scale.

Monsoon duration

Duration, as a variable, is analysed and approached quite differently at the four scales. Much of the modelling literature that focuses on the global scale does not prioritise the monsoon duration variable of the Australian–Indonesian monsoon (Colman, Moise and Hanson 2011). Most view duration as a variable difficult to attribute to a singular monsoon and even harder to model any shifts due to climate change because of the influence of other variables (Kim *et al.* 2011; Kim, Wang and Ding 2008). When attempting to distinguish monsoons to analyse duration, most of the literature focuses on the East Asian monsoon or the South Asian monsoon (Hsu *et al.* 2013; Kitaoh and Uchiyama 2006).

At the regional scale, there is a large focus on internal timescale. The duration variable is understood as one of the key factors necessary to study in monsoons (Wheeler and McBride 2005; Zhang *et al.* 2013). Duration's direct relevance to farmers is obvious to most regional scale modellers and is used as a study rationale. Yet, the current data is not strong enough for good analysis of duration.

The dataset representing the local scale does not directly provide information on monsoon duration. It is not a specific variable recorded by the East Flores government but can be drawn out from other recorded variables, like monthly rainfall amounts. Using this methodology, no statistical significance is found in the changes of monsoon rainfall duration over time. Duration over the past 25 years was about three months in length during the same season per year.

Duration is a variable highly represented at the household scale because of the variable's direct influence on agricultural production. Most interviewees cited three months as the ideal rainy season and stated that a wet season duration that lasts longer than that could potentially kill all crops. Many saw duration as a set variable of three months that began after the monsoon onset, even with a delayed onset.

Rainfall intensity

Rainfall intensity is not well represented as a variable at the global scale. It is a difficult variable to model separately from climate influencers and other climate phenomena (Hsu *et al.* 2011; Lee and Wang 2014; Liu *et al.* 2009; Meehl *et al.* 2012; Wang *et al.* 2008; Zhou, Wu and Wang 2009). The modelling literature at the global scale, therefore, currently focuses on finding associated climate phenomena. Literature at the global scale suggests that the Australian–Indonesian monsoon does not correlate well with certain climate phenomena, like the El Niño Southern Oscillation, that often correlate well with the five other monsoon variables analysed herein (Colman, Moise, and Hanson 2011). Recent studies at the global scale have searched for other explanatory climate phenomena, like the Madden–Julian Oscillation, that might provide explanations for monsoon behaviour (Meehl *et al.* 2012; Wang *et al.* 2014; Wang *et al.* 2012; Zhou *et al.* 2009).

Modelling the variable of rainfall intensity at the regional scale is presented with many of the same complications as is faced at the global scale. Isolating rainfall intensity produced by the Australian–Indonesian monsoon from other climate influencers or phenomena proves difficult (Aldrian and Susanto 2003; Deni *et al.* 2010; Hendon 2002; Qian *et al.* 2013). Most of the literature at the regional scale argues that more data must be known about rainfall intensity (Deni *et al.* 2010; Moron *et al.* 2013). Some studies suggest that more data on the relationship between the El Niño Southern Oscillation and the Australian–Indonesian monsoon may prove quite useful (Qian *et al.* 2010; Moron *et al.* 2013).

The East Flores government does not collect data specifically on the rainfall intensity variable. By analysing the volume of rainfall per month and rainfall days per month, one can surmise the rainfall intensity. Analysis at the local scale shows that there has been a statistically significant increase over time in rainfall intensity in East Flores over the last 25 years.

Rainfall intensity is a well-represented variable at the household scale because of the variable's close connection to agricultural productivity. A heavy rainfall event or a period of unexpected drought in the wet season can quickly destroy a field of healthy crops. To many farmers interviewed, rainfall intensity represents the changing monsoon pattern in eastern Indonesia. Many remember a consistent rainfall pattern in years past but now see the monsoon as a period with increasing rainfall intensity in which the climate phenomena are becoming undependable.

Rainfall volume

From a global scale, it is incredibly hard to measure and model the Australian–Indonesian monsoon's rainfall volume as a variable (Schewe, Levermann and Meinshausen 2010; Wang *et al.* 2008). It is a poorly understood variable because of a lack of clear data, not its importance. Some models show increases over time, but the data is still inconclusive, vague and not statistically significant (Zhou, Zhang and Li 2008; Liu *et al.* 2009). As with other variables analysed from the global scale, studies suggest analysis of the relationship between the monsoon and the El Niño Southern Oscillation better represent the rainfall volume variable (Wang *et al.* 2012; Wang *et al.* 2008).

Models that approach the Australian–Indonesian monsoon at the regional scale often lack necessary clarity. An assumption from the global scale that is carried over to the regional scale is that rainfall volume can only be known after the onset and duration variables are understood (Qian, Robertson, and Moron 2013). As such, most of the literature at the regional scale focuses on modelling the Australian-Indonesian monsoon through other climate patterns in downscaled global models (Aldrian and Susanto 2003; Naylor *et al.* 2007; Qian, Robertson, and Moron 2013; Qian Robertson and Moron 2010). Most view two climate influencers – the El Niño Southern Oscillation and Sea Surface Temperature – to greatly influence the timing of the Australian–Indonesian monsoon.

At the local scale, rainfall volume is an important variable measured by multiple East Flores government departments, including the meteorology department. Lack of information sharing between departments aside, this variable is understood as important because to the East Flores government it represents the monsoon. To many officials in the government, the main monsoon tenet is rain, which is understood to be monsoon onset and volume of rainfall. Rainfall volume is also easily measured by an under-funded local government, making it easier to emphasise as important. The dataset produced by the meteorology department shows a statistically significant increase in rainfall volume over time.

Rainfall volume is not a well-represented variable at the household scale. While it is a variable easily measured by a government office, rainfall volume is not the most tangible variable to farmers when compared to the other five variables analysed herein. Attention to rainfall amounts at the household scale typically focuses on the season's biggest rain events and periods of inconsistent rainfall (measured by the rainfall intensity variable) rather than a month's total amount of rainfall (measured by the rainfall volume variable).

Average wind speed

Average windspeed is not well represented at the global scale. At such a scale, it is hard to visualise, and therefore model, the variable (Li, Tham and Chang 2001; Xu *et al.* 2006; Meehl and Arblaster 2011). It is understood as an important but complicated variable that has multiple characterizations (Colman, Moise and Hanson 2011). The geographic area covered by the global scale requires that the average wind variable not be understood as a singular variable but rather as a cluster of variables that include seasonal variation, lower- and upper-scale winds, and temperature (Colman, Moise, and Hanson 2011). Modelling litera-ture at the global scale suggests that further research into the variable be con-ducted with a more regional scale to distinguish monsoon-only influences on wind from other climate phenomena, like the Walker Circulation and El Niño Southern Oscillation (Colman, Moise, and Hanson 2011; Kim 2011; Li 2001).

Models at the regional scale cannot currently model windspeed (Haylock and McBride 2001; Qian 2013). Few studies concerning the regional scale focus on modelling average windspeed. Those that do, do so by justifying a concern about the variable by stating its influence on rainfall events because wind direction influences rainfall events (Naylor *et al.* 2007; Qian, Robertson and Moron 2010; Qian, Robertson and Moron 2013). At a regional scale, there is more of a concern about the average windspeed's cardinal direction than the speed at which the wind blows (Hendon 2002; Qian, Robertson and Moron 2010; Qian, Robertson and Moron 2013). As with the global scale on the average windspeed variable, studies at the regional scale focus on distinguish-ing the winds created by the monsoon from winds produced by other climate phenomena, like El Niño Southern Oscillation or sea surface temperature.

Average windspeed is a well-represented variable at the local scale. It is perhaps so well represented, however, because it is a variable easily measured at the local scale. Analysis of the 25-year dataset provided by the Indonesian meteorology department shows that average windspeed has increased in a statistically significantly manner over the last 25 years.

At the household scale, average windspeed is a well-represented variable. Most interviewees suggested that average windspeed is increasing over time. However, as with the global scale, the focus within the wind variables tended to be about the biggest wind events – the maximum windspeed – rather than the everyday average wind. Multiple elders suggested that the average wind-speeds of today have increased so much that what would have once been considered a maximum, singular event windspeed is now the norm of average windspeed.

Maximum windspeed

At the global scale, maximum windspeed is not a well-represented vari-able. It is a difficult variable to quantify and separate from climate influencers (Li, Tham and Chang 2001; Meehl and Arblaster 2011). Similarly, at the regional scale, maximum windspeed is not well represented because of the difficultly of isolating it as an individual variable and modelling it (Naylor *et al.* 2007; Qian, Robertson and Moron 2010; Qian, Robertson and Moron 2013). Extremes in both rain and wind are important characterisations to understand within global climate models, so it is likely that soon maximum windspeed will be a highly-studied variable through both the global and regional scales.

At the local scale, maximum windspeed is well represented. The East Flores government officials view maximum windspeed as an important variable to measure. They cite destruction to farmers' fields as rationale for the study of the variable. As such, the local scale is concerned both with the maximum monsoon windspeed scales and timing of those maximum events. One government official said, 'Nowadays not only in rainy season there is a problem with the strongest winds. Even in dry season, the scales of the strongest winds are increasing. It is a problem every month of the year, not just in the wet season'. Analysis of the dataset representing the local scale shows a statistically significant increase of maximum windspeed over the last 25 years.

Maximum windspeed is well represented at the household scale because of the variable's connection to agricultural productivity. Often the variable is referred to at the local scale as 'big winds'. These big winds are particularly important for farmers, both in speed and date. Many farmers said they not only could recall the periods of intense, maximum windspeed events but also knew how to tell when such events were coming. As one farmer said:

> Big waves and strong winds come in the rainy season. During the rainy season, the big winds start in the early morning and then continue for 3 days or 5 or 7 or 9. They always come in these amounts of days before it will stop. Usually they come in periods of 7 days.

Maximum windspeed is of great concern at the household scale because of its ability to quickly destroy a field of healthy crops.

Discussion

Current and future projections of the Australian–Indonesian monsoon at the global scale shows that this scale's epistemology recognises the importance of all six variables but that rich data at the global scale only exists for monsoon onset. Other variables are much harder to model at the global scale. Monsoon duration, while understood within the literature as important, is not well simulated within the global climate models because of a lack of clarity in the exactness of the variable (Wheeler and McBride 2005; Zhang *et al.* 2013). Rainfall intensity and volume are similarly lacking in depth of data richness within the literature analysing the global scale. Both are poorly modelled at the global scale and most literature focuses on distinguishing variability in rainfall intensity and volume produced by the Australian–Indonesian monsoon from other climate phenomena. Likewise, windspeed averages and maximums are often represented in the literature as variables too complex to be well represented at a global scale because of the difficulty of distinguishing the monsoon's influence on these two variables from the influence of other climate phenomena (Li, Tham and Chang 2001; Xu *et al.* 2006; Meehl and Arblaster 2011).

A review of the literature of the dynamically and empirically downscaled global climate models at the regional scale shows acknowledgement of the importance of all variables but variation in the ability to model all six. There is disagreement within the literature on the methodology for analysing the monsoon onset and monsoon duration variables (Wheeler and McBride 2005; Zhang *et al.* 2013). No literature at the regional scale can accurately model the variables of rainfall intensity and volume that are produced solely by the Australian–Indonesian monsoon. Instead the literature focuses on distinguishing variability produced by other climate phenomena, like El Niño Southern Oscillation, from variability produced by the monsoon (Aldrian and Susanto 2003; Deni *et al.* 2010; Hendon 2002; Qian *et al.* 2013). Using sub-seasonal and historical data, studies have advanced this knowledge and soon should be able to distinguish the monsoon from other climate phenomena at a regional scale (D'Arrigo 2006a; Moron *et al.* 2013). There is a severe lack of data at the regional scale for average windspeed and maximum windspeed (Haylock and McBride 2001; Qian 2013). As with rainfall intensity and volume at the regional scale, the literature focuses on distinguishing the effects of the Australian–Indonesian monsoon on average and maximum windspeeds from that of other climate phenomena (Naylor *et al.* 2007; Qian, Robertson and Moron 2010; Qian, Robertson and Moron 2013). Likely, this will be drastically improved in the coming years (Qian, Robertson and Moron 2013).

The analysis of the 25-year dataset of weather observations collected by the Indonesian meteorology department provides rich data on all six variables of the Australian–Indonesian monsoon. While spatially and temporally much more limited than the literature reviewed at the global and regional levels, the information provided by the local scale is quite rich. There is no statistical significance in the change over time of monsoon onset or monsoon duration variables. The analysis shows that there has been a statistically significant change over the last 25 years in rainfall intensity, rainfall volume, average windspeed and maximum windspeed. These data results provide an on-the-ground context of the Australian-Indonesian monsoon's behaviour at a singular location.

The household scale provides data on all six variables. Monsoon onset and monsoon duration have the richest data at the household scale because of the variables' direct influence on agricultural production. The data suggests that at the household level, both monsoon onset and duration are understood to be changing drastically over time in ways unfavourable to agriculture. Rainfall intensity is a variable with rich data at the local scale because of the threat it poses to crop production. Rainfall volume is understood as important but the data is not as rich as the other five variables. Similarly to rainfall intensity, the variables of average and maximum windspeed are well represented at the local scale because of the potential threat they pose to crop production. Like the local scale, the household scale is limited in the temporal and spatial depth of data but rich in the quality and quantity of data provided on all six variables.

Conclusion

None of the four scales provides a complete understanding of the Australian–Indonesian monsoon. Instead each scale contributes important information on the monsoon. Each prioritises certain characterisations of the monsoon over others out of necessity. One scale cannot provide the necessary depth and breadth for each variable of the monsoon. The study of monsoons – and water in general – demands trans-disciplinary research.

For example, monsoon onset, the best variable across all four scales, is represented quite differently at each scale. How each scale quantifies and qualifies onset varies greatly but can be quite complementary to one another. Knowing the context of monsoon onset against other climate phenomena (provided by the global scale), understanding the generalised annual patterns of onset (provided by the regional scale), quantifying the exact historical dates of a specific city's yearly monsoon onset (provided by the local scale), and analysing the human experiences and perceptions of the monsoon onset (provided by the household scale) gives a full picture of the variable. Each scale provides valuable information about the monsoon's onset that could not be found at any other scale. Looking at onset across scales, rather than analysing the variable only at one scale, provides a much richer understanding of the variable and the Australian–Indonesian monsoon.

A richer understanding of the monsoon is critical for any agriculture or climate change policy created for the region. Naylor et al. (2007) argue that the biggest risk Indonesia faces with climate change is the effect of a shifting monsoon on the country's agriculture. Predictions of future monsoons are not simply academic quandaries, they are tangible realities of great importance to all those who live in monsoonal areas (Naylor 2002). Such predictions would be strengthened if they incorporated information generated from all four scales of data rather than just relying upon the global scale or the regional scale. Through the combination of all scales a differentiated view is generated that is not achieved by analysing the study parameters exclusively at a single scale (Shackley and Wynee 1995; Hall and Sanders 2015). This new understanding puts into question the binary of expert and non-expert (discussed further in Mabon and Kawabe, this volume) in the pursuit of knowing water. One scale should not be deemed more expert than others. A hyper-valuing of some scales as expert and others as non-expert can only result in imperfect climate adaptation strategies.

One major reason academics study climate change's effect on monsoons is to help farmers adapt to the upcoming disruptions. While monsoon shifts may present great theoretical quandaries to dissect, they also pose tangible threats to the livelihoods of a third of the global population. Past shifts in monsoons severely disrupted societies (Cook et al. 2010). More recent disruptions in the

Australian–Indonesian monsoon have led to severe droughts, causing immense difficulties for Indonesia, a country heavily dependent on rice and corn production. A disrupted monsoon threatens crop production from a multitude of angles. How those threats are perceived within the academic community directs how adaptation strategies are developed. Prioritising one scale over the other three, as is often done in monsoon studies, risks understanding only certain aspects of the monsoon and risks forgetting the many farmers dependent upon the Australian–Indonesian monsoon's rains. As one such farmer said: 'There is nothing that can be done about bad rain. The only thing is to surrender to the rain. To pray. And just hope that the rain becomes steady again. This is the risk farmers face everyday.'

Acknowledgements

The author would like to thank Michael Dove, Karen Hebert, Ruth Barnes and Robert Barnes for their advice on this research and to thank Wiss Kedang for his translation work. The author is grateful for funding from Tropical Resources Institute, Council on Southeast Asia Studies, Carpenter-Sperry Research Fund and Charles Kao Research Fund.

References

Aldrian, E. and Susanto, R. D. (2003) 'Identification of three dominant rainfall regions within Indonesia and their relationship to sea surface temperature', *International Journal of Climatology*, 23(12), pp. 1435–1452.

Annamalai, H. *et al.* (2013) 'Global warming shifts the monsoon circulation, drying South Asia', *Journal of Climate*, 26(9), pp. 2701–2718.

Colman, R. A. *et al.* (2011) 'Tropical Australian climate and the Australian monsoon as simulated by 23 CMIP3 models', *Journal of Geophysical Research: Atmospheres (1984–2012)*, 116(D10).

Cook, E. R. *et al.* (2010) 'Asian monsoon failure and megadrought during the last millennium', *Science*, 328(5977), pp. 486–489.

Cook, E. R. *et al.* (2010) 'Asian monsoon failure and megadrought during the last millennium', *Science*, 328(5977), pp. 486–489.

D'Arrigo, R. *et al.* (2006a) 'Monsoon drought over Java, Indonesia, during the past two centuries', *Geophysical Research Letters*, 33.

D'Arrigo, R. *et al.* (2006b) 'The Reconstructed Indonesian warm pool sea surface temperatures from tree rings and corals: Linkages to Asian monsoon drought and El Niño-Southern Oscillation', *Paleoceanography*, 21(3), p. PA3005.

Deni, S. M. *et al.* (2010) 'Spatial trends of dry spells over Peninsular Malaysia during monsoon seasons', *Theoretical and applied climatology*, 99(3–4), pp. 357–371.

Fein, J. S. and Pamela L. Stephens (ed.) (1987) *Monsoons*. New York: Wiley.

Hall, E. F., and Sanders, T. (2015). 'Accountability and the academy: producing knowledge about the human dimensions of climate change', *Journal of the Royal Anthropological Institute*, 21(2), 438–461.

Haylock, M. and McBride, J. (2001) 'Spatial Coherence and Predictability of Indonesian Wet Season Rainfall', *Journal of Climate*, 14(18), pp. 3882–3887.

Hendon, H. (2002) 'Indonesian Rainfall variability: Impacts of ENSO and local air-sea interaction', *Journal of Climate*, 16, pp. 1776–1790.

Hsu, P. *et al.* (2011) 'Trends in global monsoon area and precipitation over the past 30 years', *Geophysical Research Letters*, 38(8).

Hsu, P. *et al.* (2013) 'Future change of the global monsoon revealed from 19 CMIP5 models', *Journal of Geophysical Research: Atmospheres*, 118(3), pp. 1247–1260.

IPCC (2014) *Climate Change 2013: The Physical Science Basis: Working Group I Contribution to the Intergovernmental Panel on Climate Change (IPCC) Fifth Assessment Report*. Cambridge: Cambridge University Press.

Kim, H. *et al.* (2008) 'The Global Monsoon Variability Simulated by CMIP3 Coupled Climate Models', Journal of Climate, 21(20), pp. 5271–5294.

Kim, H. *et al.* (2011) 'Global monsoon, El Niño, and their interannual linkage simulated by MIROC5 and the CMIP3 CGCMs', Journal of Climate, 24(21), pp. 5604–5618.

Lee, J. and Wang, B. (2014) 'Future change of global monsoon in the CMIP5', Climate Dynamics, 42(1–2), pp. 101–119.

Li, T. M. (2001) 'Masyarakat Adat, Difference, and the Limits of Recognition in Indonesia's Forest Zone', Modern Asian Studies, 35(3), pp. 645–676.

Liu, J. *et al.* (2009) 'Centennial variations of the global monsoon precipitation in the last millennium: results from ECHO-G model', *Journal of Climate*, 22(9), pp. 2356–2371.

Meehl, G. A. *et al.* (2012) 'Monsoon regimes and processes in CCSM4. Part I: The Asian-Australian monsoon', *Journal of Climate*, 25(8), pp. 2583–2608.

Mittal, N. *et al.* (2013) 'Combining climatological and participatory approaches for assessing changes in extreme climatic indices at regional scale', *Climatic Change*, 119(3–4), pp. 603–615.

Moron, V. *et al.* (2013) 'Extracting Subseasonal Scenarios: An Alternative Method to Analyze Seasonal Predictability of Regional-Scale Tropical Rainfall', *Journal of Climate*, 26, pp. 2580–2600.

Moron, V. *et al.* (2009) 'Spatial Coherence and Seasonal Predictability of Monsoon Onset over Indonesia', *Journal of Climate*, 22, pp. 840–850.

Moron, V. *et al.* (2010) 'Local versus regional-scale characteristics of monsoon onset and post-onset rainfall over Indonesia', *Climate Dynamics*, 34(2-3), pp. 281–299.

Naylor, R. *et al.* (2002) 'Using El Niño Southern Oscillation climate data to improve food policy planning in Indonesia', *Bulletin of Indonesian Economic Studies*, 38(75–91).

Naylor, R. *et al.* (2007) 'Assessing risks of climate variability and climate change for Indonesian rice agriculture', PNAS, 104(19), pp. 7752–7757.

Naylor, R. and Mastrandrea, M. (2010) 'Coping with Climate Risks in Indonesian Rice Agriculture', in Filar, J. A. and Haurie, A. (eds.) *Uncertainty and Environmental Decision Making*, New York: Springer Science+Business Media.

Orlove, B. S. *et al.* (2000) 'Forecasting Andean rainfall and crop yield from the influence of El Niño on Pleiades visibility', *Nature*, 403, pp. 69–71.

Orlove, B. S. *et al.* (2002) 'Ethnoclimatology in the Andes: a cross-disciplinary study uncovers a scientific basis for the scheme Andean potato farmers traditionally use to predict the coming rain', *American Scientist*, 90, pp. 428–435.

Qian, J. H. *et al.* (2013) 'Diurnal Cycle in Different Weather Regimes and Rainfall Variability over Borneo Associated with ENSO', *Journal of Climate*, 26(5), pp. 1772–1790.

Qian, J. *et al.* (2010) 'Interactions among ENSO, the Monsoon, and Diurnal Cycle in Rainfall Variability over Java, Indonesia', *Journal of the Atmospheric Sciences*, 67.

Reid, J. S. *et al.* (2013) 'Observing and understanding the Southeast Asian aerosol system by remote sensing: An initial review and analysis for the Seven Southeast Asian Studies (7SEAS) program', *Atmospheric Research*, 122, pp. 403–468.

Robertson, A. W. *et al.* (2009) 'Seasonal predictability of daily rainfall statistics over Indramayu district, Indonesia', *International Journal of Climatology*, 29(10), pp. 1449–1462.

Robertson, H. A. and McGee, T. K. (2003) 'Applying local knowledge: the contribution of real history to wetland rehabilitation at Kanyapella Basin, Australia', *Journal of Environmental Management*, 69, pp. 275–287.

Schewe, J. *et al.* (2010) 'Climate change under a scenario near 1.5 C of global warming: monsoon intensification, ocean warming and steric sea level rise', *Earth System Dynamics Discussions*, 1(1), pp. 297–324.

Schlacher, T. A. *et al.* (2010) 'Use of local ecological knowledge in the management of algal blooms. Environmental Conservation', *Environmental Conservation*, 37, pp. 210–221.

Shackley, S., and Wynne, B. (1995). Global climate change: the mutual construction of an emergent science-policy domain. Science and Public Policy, 22(4), pp. 218–230.

Stowasser, M. *et al.* (2009) 'Response of the South Asian Summer Monsoon to Global Warming: Mean and Synoptic Systems★', *Journal of Climate*, 22(4), pp. 1014–1036.

Vera, C. *et al.* (2011) 'Understanding and Predicting Climate Variability and Change at Regional Scales (Community Paper of WCRP/OSC Parallel Session B5)'.

Vimont, D. J. *et al.* (2010) 'Downscaling Indonesian precipitation using large-scale meteorological fields', International Journal of Climatology, 30(11), pp. 1706–1722.

Wang, B. *et al.* (2004) 'Ensemble simulations of Asian-Australian monsoon variability by 11 AGCMs', Journal of Climate, 17(4), pp. 803–818.

Wang, B. *et al.* (2008a) 'How accurately do coupled climate models predict the leading modes of Asian-Australian monsoon interannual variability?', *Climate Dynamics*, 30(6), pp. 605–619.

Wang, B. *et al.* (2008b) 'Interdecadal Changes in the Major Modes of Asian-Australian Monsoon Variability: Strengthening Relationship with ENSO since the Late 1970s★', *Journal of Climate*, 21(8), pp. 1771–1789.

Wang, B. *et al.* (2012) 'Recent change of the global monsoon precipitation (1979–2008)', Climate dynamics, 39(5), pp. 1123–1135.

Wang, B. *et al.* (2014) 'Future change of Asian-Australian monsoon under RCP 4.5 anthropogenic warming scenario', *Climate Dynamics*, 42(1–2), pp. 83–100.

Wheeler, M. C. and McBride, J. L. (2005) 'Intraseasonal variability of the Australian–Indonesian monsoon region', in Lau, W. and Walliser, D. E. (eds.) *Intraseasonal Variability of the Atmosphere–Ocean System*, New York: Praxis.

Xu, M. *et al.* (2006) 'Steady decline of east Asian monsoon winds, 1969–2000: Evidence from direct ground measurements of wind speed', *Journal of Geophysical Research: Atmospheres* (1984–2012), 111(D24).

Zhang, H. Q. *et al.* (2013) 'The response of summer monsoon onset/retreat in Sumatra-Java and tropical Australia region to global warming in CMIP3 models', *Climate Dynamics*, 40(1–2), pp. 377–399.

Zhou, T. *et al.* (2009) 'The CLIVAR C20C project: which components of the Asian–Australian monsoon circulation variations are forced and reproducible?', *Climate Dynamics*, 33(7–8), pp. 1051–1068.

Zhou, T. *et al.* (2009) 'How well do atmospheric general circulation models capture the leading modes of the interannual variability of the Asian-Australian monsoon?', *Journal of Climate*, 22(5), pp. 1159–1173.

Zhou, T. *et al.* (2008) 'Changes in global land monsoon area and total rainfall accumulation over the last half century', *Geophysical Research Letters*, 35(16).

Zhu, C. *et al.* (2012) 'Recent weakening of northern East Asian summer monsoon: A possible response to global warming', *Geophysical Research Letters*, 39(9).

4 An epistemological re-visioning of hybridity

Water/lands[1]

Kuntala Lahiri-Dutt

Introducing wet theory

Land has been understood and explained by excluding water. The common idea of land is best expressed in the Oxford Dictionary's definition: 'land' is something that exists in opposition to water; that is, 'the part of the earth's surface that is not covered by water',[2] meaning that land excludes swamps, estuaries, tidal areas, lakes, ponds and streams. Historically, geography has conventionally been the leading discipline propagating this epistemic divide, but other disciplines have equally contributed to deepening the chasm. Sauer in 1925 described land as 'a unit of geography' (Sauer [1925] 1963, p. 321). Equivalent terms, 'area' and 'region', gave rise to the *place facts of geography*; thus, landscape became a recognizable entity with definite limits leaving the waters beyond them. Even before Sauer, the instructions of classical geomorphologists such as Davis (1900) on the 'content of geography underlined the complete separation of land from water. Hartshorne's (1939, p. 150) ideas of landscape (as the 'appearance of a land as we perceive it') privileged land.[3] Geographical metaphors are generally associated with land: territory, field, place, horizon, soil. Like the traditional categories offered by political geographers – frontiers, boundaries and borders, rim lands, and peripheries. Geographical conceptualisations of hybridity are embodied in consideration of various hybrid 'landscapes'. Land and water have definitely and absolutely been established as two completely separate epistemic categories.

Moreover, land is valorised in environmental studies as an area of ground, the basis for all human activities; what Leopold and Schwartz (1989) describe as the 'Abrahamic conception of land', laying for humans a vast expanse of nature to exploit for themselves. The post-enlightenment rational need for removing ambiguities sealed the dichotomy but also associated the categories with usefulness or the lack of it; Domonoske (2012, p. 4) considers this 'Water-Land associated with Nothing-Something' as expressing a division between enchanted pre-modernity and the dis-enchanting Enlightenment, putting land, man and rationality against water and its peoples. The very process of 'reclamation' of land became one where human beings converted the useless watery land into solid ground. The implied dichotomy or a blurred partition also disguised a conflict between water and land – in particular, the struggle between silt, or fine soil, and water because silt is perceived to pollute the clear and clean water, and impede the flow of rivers.[4]

This chapter challenges this separation by extending the idea of hybrid waterscapes beyond not just the nature-culture binary but beyond the deeply entrenched water-land dichotomy, thereby reconceiving both as environments that have the potential to critique standard cultural dichotomies. It engages with a kind of hybrid environment not fully dealt with yet in water epistemology: where not only the binary division of nature and culture breaks down, but where land and water are also inseparable, giving rise to a nebulous, ever-changing and fluid environment. Thinking of hybridity in this manner turns it more fully into what Appadurai and Breckenridge (2009) outline as 'wet theory'; that is, a flexible theory that is able to accommodate messiness and contextual variations. Thinking of hybridity as fluid and transient can help to relinquish the notion of permanence in land (and landscapes) and bring to the fore the constant negotiations between the land and waters – the seas, rivers, and lakes that geographers have long constituted as *lying outside the terra firma.* Wet theory does not need to rely on a 'hard edge' – that is, the clean division in the sciences between land and water created in order for accurate measurement, planning and control – and can allow us to rethink lands as spongy and aqueous, and as uncertain and fluid. The examples I provide demonstrate that hybrid water/lands are not only coproduced by nature and culture, but also constitute a blend of water and land where the two are merged with each other imperceptibly and changeably, posing a difficulty of defining them as 'either-or'.

This chapter takes the readers to the Bengal[5] floodplains, flying over the waterscapes of which during the monsoon months could visually convince the reader of my argument. Conventional environmental wisdom would suggest that this is what floodplains look like, but does not explain why in Bengal, still described in most books as 'a riverine land', rivers and their floods came to be constructed as an enemy. The sight of a land soaked in water would be a starting point in questioning the boundaries between land and waters that conventional environmental training might have created. Hybridity in this paper constitutes an attempt to elucidate the waterscapes in the lower Gangetic delta, a densely peopled riverine plain that is part land, part water, but is neither in its entirety. In thinking about such waterscapes, the theoretical lens of hybridity is useful in that it allows one to question the water-land binary. Scale is significant, and Bengal is illustrated as one big aqueous terrain of rivers, with contextual reference to the Sundarbans (the mangrove delta), then zooming onto the *haors*[6] – the vast bowl-like depressions in eastern Bangladesh; and the *chars* – river islands, which are also called '*charbhumi*' or *char-bada jami* (*bhumi* and *jami* are both 'land' in Bangla). At all these geographical scales – from the breadth of the delta mouth to the microcosmic worlds of chars that lie within the riverbeds – this paper shows how this aqueous land is more than a product of fluvial action, and how colonial interventions – including changes in land tenure and revenue collection – and postcolonial dam-building coproduced this hybrid environment.

Neither the Sundarbans, nor the *haors* and the *chars*, can be described as land or water, and certainly not a combination of the two, which is the way 'wetlands' have been considered in mainstream environmental science. Water/lands are not necessarily 'wetlands' as identified by environmental experts by the presence of certain, predefined, amount of waters in certain times in the annual water cycle. The flux and fluidity of water/lands, their ever-changing and uncertain and often subjective) material qualities, and above all their co-constitution as socio-natures make this analytical category quite different from wetlands. The paper argues for precisely this epistemological re-visioning to open up the space within water/river studies for thinking about hybridity in a new way that is robust enough to take us beyond thinking of land and water as two rigid and indissoluble categories.

These malleable environments present empirical examples of new meta-phorical terrains where environmental historians and cultural geographers might make more sense to one another, and 'make it possible to imagine nature as both a real actor *and* a socially constructed object without reducing it to a single pole of nature-culture dualism' (Demeritt 1994, p. 163). These environments are also integral parts of the 'vague, diffuse or unspecific, slippery, ephemeral, elusive or indistinct' world, a world that 'changes like a kaleido-scope, or doesn't really have much of a pattern at all' and is impossible to be 'distorted into clarity' (Law 2004, p. 2). To explain the messy world of these environments, one must avoid the 'methodological moralizing' of individual disciplines and categorising (Law 2007, p. 599), and invoke transdisciplinarity to displace monolithic interpretation. This is where – to explain lands that are soaked in water – ideas of hybridity can open up the possibilities of wet theory, a new water epistemology.

Hybrid environments so far: Wading towards a wet theory of socio-natures

In recent years, a number of scholars have highlighted that nature and society are coupled together, offering an understanding of environments as historically shaped and culturally constituted. Consequently, one can trace the sources of the critique of positivist binary division between nature and culture in the con-versation across several disciplinary borders. A remarkable boundary-crossing, uncommon in the history of social sciences, reflects, as Inglis and Bone (2006) suggest, an increasing interest by social scientists on human manipulation of both biological life and so-called 'natural' environmental forces and phenomena.

Arguably, the greatest contributions have been made by environmental historians who have illuminated the changing relationship between human-ity and nature (see Radkau 2008 for a comprehensive overview). Their early work attempted to emphasize the 'social construction' of nature, such as forests (see Jeffrey 1998). From this, studies moved on to what Agrawal and Sivaramakrishnan (2001) describe as 'social nature', where the artificiality of boundaries between arable lands, forests and pastures are demolished, problematizing the ways in which certain modes of livelihoods are commonly associated with each of these categories.

Geography and related fields claim a territorial right over the complex domain of nature, society, territory and scale. Geographers, primarily Anglo-American, have also contributed to the ongoing debate of what is commonly described as 'socio-nature'. For them, the concept of hybridity has been crucial to break down the schizophrenic division between nature and culture that has ailed geography almost since the inception of the discipline. Drawing together notions of relational dialectics and hybridity, they have offered a rethinking of the nature-culture divide. Two major arguments are relevant for this paper. First is that the nature-culture divide is neither static nor diametrically cor-related, but is constituted by traits in both categories that are indistinguish-able from one another (Hinchcliffe 2007; Head and Muir 2007; Barnes 2008).[7] Secondly, those following this paradigm shift have also challenged, and in some cases rejected, the once common belief in the equilibrium idea in nature, a condition of constant balance, by showing that the idea of equilibrium is a myth that environmentalists have long perpetuated (Bracken and Wainwright 2006, p. 167). Instead, a large number of cornerstone ecological processes – floods in rivers for example – are now recognized as non-equilibrium dynam-ics, long-term shifts and historical conditionalities, such as path dependencies and trajectories (Zimmerer 2000, 2007). These shifts from a normative model of 'nature in equilibrium' have great relevance in rethinking floodplains and the functions of rivers (Lahiri-Dutt 2015). The renewed emphasis on flux rede-fines nature/society hybrids, and is in stark contrast to environmental principles rooted in the belief of nature as tending towards equilibrium, and has great sig-nificance for environment management practices (Eden and Bear 2011, p. 394).

Within this overall rethinking about new water epistemologies, geographi-cal studies have built close links with environmental history. For example, in studying the essential relations between water and society, analyses of both the history of water and how the idea of water articulates with its material and rep-resentative forms to produce this history have been presented by Linton (2010). A crucial common ground between environmental historians and geographers has been in thinking about 'the production of nature'. A purely physical envi-ronment, unaffected by human enterprise, hardly exists today; what is seen as *the environment* is a product of human interaction and modification over years.

Grove and Damodaran (2006), however, refer to early geographical contri-butions such as those from Gordon East to show how contemporary colonial anxieties were expressed in the work of academic geographers who had begun to understand the extent and consequences of human interventions in nature, that is, on 'man's role in changing the face of the Earth'. Problematically, East (1938, p. 11) was noted for his environmental determinism: 'If only by its more dramatic interventions, a relentless nature makes us painfully aware of the uneasy terms on which human groups occupy and utilize the earth'. One can say that the anxiety of environmental historians is rooted in the unpredictable and recurrent nature of dramatic and disastrous natural events in the contemporary world. The difficulty is that such views may lead to what is known in geography as neodeterminism.

Today, the role of time in shaping nature is interpreted differently: nature is no longer seen as a historical actor 'exist[ing] apart from our understanding of it' (Cronon 1992, p. 40), and the key actor in causing change in human society. Such an idea accepts the dichotomy of nature and culture as *fait accompli*; nature is understood either as an 'immaculate linguistic conception' (i.e. a mental or social construction) or an object knowable through absolutist knowledge of real world entities and processes that are separate from human intervention (i.e. unadulterated physical environment; Whatmore 2002, p. 2). In envisioning a new epistemology for water, it is important to remember the recent contributions in environmental research that highlight the complex relationships and hybrid nature of landscapes, and that are not always in full agreement with the perspectives of environmental historians.

The critique of the dangers posed by neo-environmental determinism has two major strands. One emerges from postmodernists, who equate nature with a text whose meaning depends on the reading of it, thus denying its material basis. Although this perspective has been valuable in denaturalizing hegemonic ways of perceiving the environment, Demeritt (1994, p. 164) explains that the world cannot be 'denatured', and that too sharp a focus on human ways of seeing renders nature illusory. The other strand of critique derives from cultural ecologists, who have contested the nature-culture binary and highlighted the correlations between nature and society (Inglis and Bone 2006). Interestingly, both rely on *terra firma* to explain their views; the just as the idea of 'nature', the idea of unmoving land too remains *the* basic building block of socio-natures.

The obsession with stable land (and hence permanence) remains unquestioned in thinking about hybridity. The hegemony of land as the basis of landscapes has also created an unbridgeable chasm – that between water and land – within the discipline. Such an unquestioned acceptance of land-based physical environment as the essence of landscape assemblages belies the fluidity inherent in the conceptualisation of hybridity. More importantly, it undermines the postcolonial origins of the idea of hybridity (see Bhabha 1994; Nandy 1983) and its association with anti-essentialism. In other words, the invocation of socio-natural 'assemblages' (Braun 2006, p. 645) might risk hybridity being conceptualized in absolute (eventually leading to essentialist) ways, leaning upon a mixing of two or more elements. The term hybridity, as Canclini (1995) shows in his study of traditional and modern cultures in Latin America, involves the movement of people in and out of ways of being rather than moving from one to another in a linear fashion. This conceptualisation is closer to Bhabha's (1990, p. 211) third space of hybridity, which is not an identity but rather an *identification*, a fluid and uncertain process of identifying with and through another object.

Thinking through the uses of the term hybridity demonstrates the extent to which meaning has travelled from the original, complex, intended sense. The challenge that arises is how to close this gap, theoretically and with robust empirical examples of environments that are in motion, unstable, uncertain, and unpredictable – environments that can morph from one into another, and can fuse into each other. If hybridity challenges conventional categories, how could a more foundational dichotomy between water and land be critiqued? This is where Appadurai and Breckenridge's (2009) provocation to think of aqueous uncertainties through the lens of wet theory may soften the many hard edges and capture the fluidity inherent in the idea of hybridity. Wet theory, they argue, offers an explanation that can accommodate flux, flow, change and other boundary-blurring phenomena within the core of historical and physical occurrences, and not regard them as exceptional or outliers. Wet theory is also flexible enough to accommodate and absorb new contextual information without pretending to be universal, or breaking on the weight of its own rigidity. Finally, wet theory does not pretend to be certain, and bows down to the empirical facts of local contexts. Clearly, wet theory has far-reaching significance for critical water studies wanting to take the concept of hybridity further to ensure we do not lose sight of complexities, do not renounce uncertainty, and do not give up mud and silt in favour of either land or water.

The next section takes the readers to the muddy lands of Bengal and show how the spongy environments of Bengal – environments that combine and confuse water with land, not just comprise a synthesis of culture and nature as lived-in landscapes – have been produced not merely by nature, but also by colonial and postcolonial interventions in river control. As Iqbal (2010) notes, an historical perspective is central to this section. The aim is to show that Bengal's character as a fluid landscape of shifting river courses, inundated irrigation, and river-based life changed during British colonial rule. Legal and engineering interventions sought to stabilize land and water and create permanent boundaries that privileged land and protected it from inundation by annual floodwater; within one hundred years of the introduction of three key colonial legal interventions, Bengal had become a land-based peasant milieu. A modernist view of the environment firmly believed in a watertight divide of water and lands, robbing the rivers of their histories and extracting them from their social contexts of human experience.

Bengal: The soft edges of a soaked land

The exact outlines of the two geographical entities, the geologic unit Bengal basin and the geomorphic unit Bengal delta, are difficult to pinpoint. Broadly, one can say that the Bengal delta is part of the Bengal basin, which is a bowl-like composite formation created by the stretch and sag of the eastern part of Gondwanaland and described as 'a large subsurface sedimentary province filled up by sediments of pre-trappean and post-trappean age'[8] (Dasgupta 2010, p. 198). Since then, the sea has retreated and the delta has prograded further southward. The combined flows of three major river systems, the Ganga, Brahmaputra and Meghna, bring enormous quantities of silt from the surrounding hills and the Himalayan mountain range to make the Bengal delta. Experts are divided about the boundaries of the delta: Bagchi (1944, pp. 8–19) believed that 'The region between the Ganges and the Brahmaputra and that between the Brahmaputra and the Meghna ... [has] no doubt been built up by the materials brought down by the rivers'. The massive weight of the mud, brought down from the surrounding hills and mountains over millennia, means that silt is in places up to five kilometres deep, pushing down the bedrock and creating a near-flat surface.[9] The flat delta, only about 10–15 meters above sea level in the northern parts, and much lower closer to the sea , is a dynamic, constantly changing environment, and is actively being built in the eastern part even today, innumerable rivers flow, frequently changing their channels as they act not just as conduits of water but also silt and mud. A tremendous amount of biological life flows along with the fecund rivers' life-giving silt. Together, the rivers carve out new lands and create new horizons for human communities, whose lives and experiences are framed by the rivers. Together, the water channels intersecting each other create a moist and unstable maze that has nurtured riverine ways of life and culture for hundreds of years.

Other waterscapes of Bengal

The water/lands of Bengal are not only manifested in its shifting rivers. Unmissable is the Sundarbans, literally 'the beautiful forests',[10] located at the mouth of the Bengal delta, and housing one of the most fascinating mangrove forests (Jalais 2010). This is where the innumerable distributaries of silt-laden rivers dump their sediments to create one of the largest, most ecologically complex and densely inhabited waterscapes in the world. The mangrove forests of the Sundarbans alternate between saline and fresh water, and between littoral and fluvial environments (Richards and Flint 1990). Land and water are intermingled throughout these mangrove forests as they are in the Bengal basin.[11]

Closer to the delta mouth are large chunks of moving silt flowing as compacted balls in the waters, the 'shoals' within the river channels that are floating, sometimes above the water but often below it. These shoals are formed at the very lowest reaches of the Ganga by the sediments that various tributaries pour into the river. These shoals often wrecked the seafaring vessels that used Calcutta port (O'Malley 1914). The tidal nature of the area means the rivers can carry out the sediments for only about 12 hours a day, and if the storms and tidal surges coincide during the monsoons when the rivers bring down colossal amounts of silty waters, the rivers spill over across the banks. It is due to these tides that the water tastes saline closer to the sea and brackish well inland, and floods, therefore, lead also to increased salinity of the soil. The United Nations Environment Programme estimates that 60 per cent of this 10,000 square kilometre biosphere heritage that is known as the Sundarbans is within the political boundaries of Bangladesh. Here, the land is constantly being altered and reshaped by tidal as well as human action, making the Dampier–Hodges Line[12] – supposed to mark the boundary of the forested part – a mere fiction. People have lived on and used these lands and waters for a very long time, but in recent years the numbers have soared due to migration and also natural growth (Kabir and Hossain 2008).

Other waterscapes include the 500 or so *haors* located in the north-eastern corner of the Bengal basin, in Sylhet (Bangladesh), just to the south of the Meghalaya hills of the Eastern Himalayas, which receive on average over 12,000 millimetres of rainfall annually. The waters drain off to the plains, and stay there for about six months of the year, reaching depths of up to seven metres (Rashid, 1991). In its entirety, the *haor* waterscapes cover about 6,000 square kilometres of saucer-shaped, bowl-like land, surrounded by the numerous tributaries of the Surma and Kushiyara Rivers. The hydraulic rhythm of life adjusts to the rise and fall of the waters; rice crops are harvested in May just before the entire area is again flooded, and as the farmers retreat, the fishermen move in (Duyne-Barenstein 2008).

Chars are sandy, silty pieces of land that rise from the shallow riverbeds in the lower Gangetic plains of deltaic Bengal (Lahiri-Dutt and Samanta 2007). In North Bihar and Eastern Uttar Pradesh flats, the coarse sands create uncultivable islands, called *diaras* locally, but farther down the river in Bengal, the finer alluvium has built more expansive and flatter chars. Within a few years of the emergence of chars above water, the humid climate assists the growth of coarse catkin grass and reeds, leading to a slow process of organic breakdown that facilitates the growth of flora. The chars are generally temporary in nature but can also be permanent; some chars are known to have existed for tens of years and to have become more stabilized as the river flow changes. Yet, chars are always susceptible to riverbank erosion and an entire char can vanish overnight. Lying within the riverbank, the chars are also heavily inhabited, often by the poorest and the most vulnerable people, the 'river-gypsies' who make a living off this environment. Lahiri-Dutt and Samanta (2013) show that these sandy masses not only constitute moveable bodies of land and water in varying proportions, but offer a novel way of thinking about the hybrid environment.

These waterscapes are also metaphors for an ungovernable and borderless state of the environment. Within them, the very idea of border – environmentally between the land and water, and politically between two administrative units – loses its usefulness.

Living in waterscapes

Deltas are essentially dynamic entities where silt is stored and re-assorted by rivers at their own free will. In Bengal rivers neither flow along a certain route, nor is the land fixed and permanent. The rivers of Bengal are seasonal: during the monsoon months, they are in spate, revealing a tremendous fierceness, unleashing a brutality in destructive powers through their floods.[13] Consequently, and over years of coping with floods, village communities had developed various means of tuning their ways of life to the excesses of water, giving rise to a rhythm of life that is adjusted to the rise and fall of rivers. *Pulbandis* – low-lying, non-extensive and poorly maintained embankments – lined rivers in certain places to allow spillovers into fields. Houses in rural areas were built on raised plinths that withstood the onslaught of the worst flooding. Crops chosen for cultivation actually thrived in floodwaters, growing taller as the floodwaters rose during the monsoon months. The indigenous rice varieties grew rapidly ahead of the floodwaters, and the cropping calendar was also suited to phases of inundation (Brammer 1990; Hofer and Messerli 1997). In his seminal lectures at Calcutta University in 1930, William Willcocks (1930, pp. 9–12) described this flood dependence as 'overflow irrigation'. Broad and shallow canals carried the fine clay and humus-rich crest waters of the floods into the fields, and frequent cuts on the banks of the canals – spill channels, called *kanwa*s in Bhagalpur and *hana*s in the lower parts of Damodar valley – inundated the fields, fertilizing the soil and helped to check the spread of malaria, turning rural Bengal into the most productive part of India. Even standing water had its use: jute crops were retted in the stagnating water of the swamplands (Chapman and Rudra 1995). An intricate network of ponds, aqueducts and water tanks provided seasonal storage of water as well as drainage.

The water/land gave rise to a highly sophisticated and rich artisan economy that grew on water-based prosperity that led to an artisanal mode of secondary production. The historian of Bengal delta, Willem van Schendel (1991), quotes the eighteenth-century traveller Orme: 'in the province of Bengal … it is difficult to find a village in which every man, woman and child is not employed in making a piece of cloth.' Until the early colonial period, even British travellers, such as Buchanan (1798, p. 7), commented that many inhabitants of densely populated parts of Bengal treated agriculture as a subsidiary occupation: '[I]n this part of the country, there is hardly such a thing as a farmer.' Today, Bengal is still primarily defined by its rural character, but one that has farming at its base. This fundamental transformation from a sophisticated artisan economy to a peasant farming one was accomplished during the colonial period, and was associated with a conversion of the watery and soft land into a hard and dry land. The following section outlines that transformation.

Colonial separation of land from rivers

The British Empire marked an exceptional ecological moment in the history of South Asia, integral to which was 'the relentless transformation of environments and landscapes' (Kumar *et al.* 2011, p. 1). The colonial period is the watershed that shaped the ecology (and economy) of Bengal, marking the moment when land and waters became firmly separated both materially and epistemically. Two overlapping strands in environmental historians' debate on the colonial transformation stand out as relevant for Bengal: first is Whitehead's (2010, p. 84) argument that the colonialists envisaged lands in the newly acquired tropics as remaining underused or lying waste, in need of being put to better use;

second is the not unrelated argument of D'Souza (2009) that the accumulation of capital from selected parts of the environment was the driving force of this transformation. Colonial intervention in the landscape, particularly the privileging of land through the introduction of a new land tenure system, was the most crucial development in greater Bengal (D'Souza 2007; Hill 2008; Mishra 2003; Klingensmith 2007). As Bengal became 'the great environmental laboratory' (Hill 1997) to test European theories on the purpose, use and control of nature in all its manifestations, the annual floods became more disastrous and riverbank erosion increased (Chapman and Rudra 2007; Lahiri-Dutt 2008; Rudra 1996, 2004). The two strands of argument meet in Bengal, the strange water/land that needed to be stabilized to firm up a systematic process of revenue collection.

Experiments in creating a land-based economy began soon after 1760, when Bengal and its ceded territories came under British rule.[14] This land-based polity regarded rivers as incidental because it was land that was to yield revenue.[15] One of the first tasks was to isolate the rivers by constructing embankments along their courses. The control was not always physical and ideological, rivers and lands needed to be separated. This was accomplished, according to Swamy (2011, p. 138) by the 'transmission' of ideas and institutions from England.[16] Two *land laws* played crucial roles in this regard, the Permanent Settlement Act introduced in 1793, soon after the British conquest of Bengal, and the *Bengal* Alluvion and Diluvion Act of 1825, pertaining specifically to river islands or chars.

At the heart of the application of English property law in India was the idea of establishing *absolute* and *permanent* land tenure. The Permanent Settlement Act gave land away to the *zamindars* or 'local landlords in perpetuity'[17] and reduced the complexities of revenue collection. This Act ignited debate as to what constitutes land, and what are its productive and unproductive uses (this deliberation is outlined in Mookerjee 1919). A category of 'wasteland (*baze zameen*) needed to be defined as the opposite of useful land that is farmed. A sub-category was that of *khas mahals*, lands that were not included within the area of any permanently settled area (Mitra 1898). Whitehead (2010) describes such a division as not only a 'constructed different landscape of value', but also a landscape loaded with the social subjectivities of groups inhabiting these territories. For wastelands to become the subliminal *other* to private land or state-appropriated property, the waters needed to be separated from land completely. The water/land of chars and the *haors* and, according to Wescoat (1990), the Sundarbans, provided an answer to this search.

Land in Bengal thus became 'the estate' for the British (Phillips 1886, p. 41) and in opposition, the rivers became problematic (Sherwill 1858). The Act had the ecological effect of stabilizing the lands (and waters) and changing the meaning of property: the cultivators began to lose the right to occupy the land that they had enjoyed since ancient times because British colonizers had enumerated the characteristics of the zamindari property as an absolute right of proprietorship in the soil, subject to the payment of a fixed amount of revenue to the government. This absolute right needed an absolute boundary between the land and the rivers, and altered the meanings that these elements of nature held to local residents. Consequently, higher embankments marked the boundaries on rivers to protect revenue-yielding land. As the embankments rose in height to completely segregate the rivers in places, silt accumulation on riverbeds caused further decay to rivers and the traditional overflow irrigation systems, causing the crumbling fluvial systems to radically increase the shifting of their courses.[18] As such movements became more unpredictable, thus those settlements that began with the blessing of the river now needed protection from it.

The other law is the Bengal Alluvion and Diluvion Act (BADA) created in 1825 and is one that still rules the oversight of ownership rights of char lands.[19] The need to make the delta more productive led colonial administrators to contemplate the problematic issue of legalizing the shifting river courses and moving lands produced by unending accretion and erosion (Eaton 1990). Massive survey operations were also initiated to produce cadastral maps to obtain more accurate knowledge of what was deemed as 'resource', and of assessing the capacity and potential of these lands (Cederlöf 2014). BADA established a set of rules to guide the courts to determine the claims to land 'gained by alluvion' or accretion, and the resurfaced land previously lost by diluvion or erosion.[20]

Although chars exist within riverbanks, the difficulty remains that when a piece of land is lost to riverbank erosion, it may not arise in exactly the same location or arise at all within the foreseeable future. This means the owner has no certainty that they will get it back if and when it resurfaces, or when another char rises nearby. BADA considers two main categories of chars rising within the riverbanks: those rising *in situ* and new accretions. For the right to land that once existed, but was diluviated, and subsequently resurfaced in the old site, BADA considers that right to be incidental to one's title to a tangible property, derived from the principle of justice and equity. The right to property is not affected only because that property has been submerged under water, and the owner is deemed to be in 'constructive possession' of the land during the time of its submergence and it can be claimed back when it reappears out of water and can be identified as land. For this, however, the owner must continue to pay rent for the diluviated land. BADA ensures that when new land rises within a river, it should be considered as 'an increment to the tenure of the person' to whose land it is contiguous, subject to the payments of revenues assessed by the state. Thus, the key to establishing land rights in the court of law was the payment of rent, even on diluviated land, obviously benefiting the richer classes.

Changes in social relations of production in agrarian Bengal in response to the laws have been well documented (see for example Guha 1963; Field 1883); less documented is the transformation of Bengal's ecology along with changing values of land and water. The laws unleashed a cycle of change in which one thing led to another – interventions on land and water changed production relations and exacerbated power inequalities within communities. The absolute right over land, Roy (2010) argues, allowed the colonial British to accumulate primary capital through land taxation, determine agrarian relations, and alter the meanings attached to the elements of land and water. This had unforeseen impacts in shaping the meaning of waterscapes of Bengal. Bengal was transformed from a riverine into a land-based community (Urch 2008).

Legal control of rivers went hand-in-hand with engineering controls. Once the idea of land was entrenched as being in need of protection from unpredictable waterways, it was easy to wall-in the latter with embankments and dikes, encouraging them to remain within fixed courses, thus rendering land even more permanent (Lahiri-Dutt 2000). Walled-in and bounded by embankments, the rivers were further controlled by the construction of dams and barrages upstream (Lahiri-Dutt 2008). In the process, the fluid worlds of waterscapes became less visible.

Conclusion: Critical environmental studies reframing ideas of hybridity

The water/lands of the Bengal delta are hybrid environments not only because they are natural and lived-in worlds, but because they are products of colonial reconstruction of land and rivers that was followed in postcolonial phase of water resource development (that has been critiqued by Iyer 2015, among others). The water/lands of Bengal illustrate diverse and fluid worlds, and encourage us to reinterpret hybridity not just as either one environment or the other, and not even as the mixture of two environments, but as *sometimes a given environment, sometimes another, sometimes both and sometimes neither.* The water/lands are not just a mixture but present an epistemological challenge to our understanding of water.

Yet, there is a need to ask the 'so-what' question, and consider the implications of water/land in advancing critical theory. In the contemporary world where sea levels are rising, deserts spreading and conflicts over resources looming, and where sustainability science experts are forecasting collapse and doom all around, is it reckless and nihilist to talk about wet theory? Can wet theory make itself legible to mainstream environmental sustainability science and resource management experts who are looking for 'fixes' to prescribe to deal with change? Can wet theory converse with (and enrich) our understandings of uncertainty and risk? Can wet theory avoid being embraced by dark and deep ecology proposing uncivilizational returns? Finally, how can the 'liminal spaces' of hybrid water/lands be re-presented 'not [only as] lines of separation but zones of interaction ... transformation, transgression and possibility' (Howitt 2001)? These important challenges are yet to be dealt with by the proponents of hybridity. A flexible conceptualization of hybrid environments inherent in wet theory would enliven the methodological debates on water.

In laying out the characteristics of wet theory, Appadurai and Breckenridge (2009, p. ix) say '[this theory] recognizes its own uncertain footing, that is humble to the ruthless tyranny of context, and that is always ready to negotiate with the facts'. For academic subalterns, conventionally treated as 'sources of data' in international academia, it thus opens up an opportunity to flaunt their 'view of nature from below'. Empirical evidence will undoubtedly play a crucial role, lest in critiquing empty concepts we resort to empty rhetoric. Muddy, soggy Bengal, with its ill-defined legal and geographical boundaries, provides us with those contextual facts to push the conceptualisation of hybridity into domains not yet well trodden by Anglo-American theorists of water.

Mainstream environmental studies, based on the water-land dichotomy, turned water/lands invisible to both social and environmental scientists by focussing on wetlands as a simplified category. Yet, thinking of Bengal through the lens of hybridity invokes reconsideration of the way water (and land) has been considered. Bengal as water/land offers us the rare opportunity to dredge beyond the conventional perception of water and land as two rigid and indissoluble categories, and to transcend the idea of permanency and hard edges – both natural and political – instead recognizing dynamic, constantly changing environments, people, communities and the cultures deeply embedded in them.

Notes

1   This chapter is a revised version of my paper, published in 2014 as Beyond the water-land binary in geography: Water/lands of Bengal re-visioning hybridity, *ACME: International Journal for Critical Geography*, Volume 13, issue 3, pp 505–529.
2   See http://www.oxforddictionaries.com/definition/english/land (accessed on 11 August, 2014).
3   Hartshorne (1939, p. 152) further pointed out that the appropriate meaning of the term would be 'the section of the earth surface and sky that lies in our field of vision as seen in perspective from a particular point'. However, he insisted that the term 'landscape' could not be given a clearly fixed and defined meaning (Olwig 2003, p. 875).
4   Reynard (2013, p. 41) thinks that environmental historians, on the contrary, have put water at the centre of their debates: 'Environmental historians interested in the impact of natural forces, be they, to human eyes, catastrophic or simply erratic, have often turned to contexts where water is at work.'
5   A point of note is my treatment of Bengal as one geographical unit regardless of its political boundaries. Bengal was partitioned in 1947 to comprise two countries, India and East Pakistan (later in 1971 Bangladesh). Despite this division, Bengal remains physically and culturally one entity, and I refer to this entity unless specifically mentioned otherwise.
6   The term *haor* is a local Sylheti dialect of the Bengali word *sagor* (literally, 'the sea'), and indeed they are localised in Sylhet. Besides *haors*, similar waterscapes of Bengal include *beels* and *baors*. A *beel* is usually a depression produced by erosion, is generally smaller, and spread throughout the area of moribund rivers. A *beel* can also be the remains of a river channel that has changed its course. A *baor* is an ox-bow lake and is found in the moribund parts of the Bengal delta. Many of these waterscapes are dry during the summer but expand into broad and shallow sheets of water during the monsoon months.
7   Some even suggest that we are living in a post-natural condition in which human-fabricated phenomena pervade all parts of the biosphere and where environmental threats are invariably human-induced.
8   The Bengal basin occupies an area of about 90,000 square kilometers, extending southwards well into the offshore regions of the Bay of Bengal. Geographically, the basin includes West Bengal and Bangladesh.

9 The Ganges and the Brahmaputra together carry more sediment than any other river system: double that of the Amazon, four times as much as the Nile. During the monsoons, 13 million tonnes of silt are carried to the delta every day. Most of this continues into the Bay of Bengal, but some gets deposited on the riverbeds along the way (Nicholson 2007, p. 122).

10 There are different interpretations of the name in English; Yule and Burnell (1903) believe the etymology has derived from *sunderbunds*, implying that '*ban*' is a derivative of the term *bund* or the embankment. Danda (2007, p. 28) notes that the notion of 'beautiful forest' is most likely a retroformation, representing an ecological romanticism of valuable forests.

11 W.W. Hunter, a colonial chronicler, notes it as a vast alluvial plain in which the process of land formation was ongoing, describing it as 'a sort of drowned land, covered with jungle, smitten by malaria, and infested by wild beasts' (1875, p. xiii). He portrays the Sundarbans as an area 'intersected by a thousand river channels and maritime backwaters, but gradually dotted, as the traveller recedes from the seaboard, with clearings and patches of rice land' (ibid., p. 287).

12 Between 1829 and 1830, the forested littoral was surveyed by Commissioner Dampier and Lieutenant Hodges. The line identified by them today is roughly the border of South 24 Parganas district.

13 W. W. Hunter observed that '[m]any decayed or ruined cities attest the alterations in riverbeds within historic times' (1882, p. 30) and thought that the history of this part of India is linked to the changing courses of its rivers. More recently, Rudra observed that 'the layers of silt of the Ganges delta hide history' (2008, p. 3).

14 Wescoat (1990) quotes Ascoli, a revenue collector of the Sundarbans, to explain the neglect of water resources in the colonial administration of mangrove forests in the Bengal delta. Revenues were based on 'land', i.e. spatially delimited areas of economic access and control; water was regarded as merely one of many resources attached to the land. Revenues were derived from commodities that laborers produced: timber, food, fish, and fibre. However, revenue collection was organized through systems of entitlement to land.

15 Ascoli (1921, p. 156) noted 'it is merely the fact that revenue is more concerned with land than with water that has tended in this book to hide the importance of rivers'.

16 Chaudhuri (1927, p. 18) noted that Lord Cornwallis asked Jagannath Tarkapanchanan to write 'a digest of Hindu law' and suggested that under the ancient laws and customs of India, landed property was/is? Vested in the peasant. Mitra (1898, p. 2) supports this view by noting Manusmriti ('A field, says Manu, is his who clears it of jungle') and the rule laid by the Prophet who was followed by the Muslim rulers ('whoever cultivates waste lands does thereby acquire property in them').

17 The (mistaken) understanding was that the land was ultimately owned by the state, and hence zamindari property offers an absolute right of proprietorship in the soil subject to the payment of a fixed amount of revenue to the government (Field [2010/1883] on colonial land regulations).

18 One of the effects of river control was rampant malaria. By the time C.A. Bentley, the colonial director of public health in Bengal, wrote his 1925 treatise, the soaring mosquito population had started to result in the decline of agricultural fertility and deterioration of public health. Klingensmith supports this cause-effect explanation (between mosquitoes caused by a choked drainage system and malaria): 'Out of a provincial population of around 45 million people in 1911, Bentley estimated that 30 million in Bengal were afflicted by malaria, more than 10 million severely so' (2007, p. 34).

19 It is most likely that BADA added a word to Bengali dictionary; the popular folk term *badajami* means charlands.

20 Donaldson (2011, p. 159) offers an historical account of ideas of accretion and avulsion that caused difficulties in firmly drawing boundaries between territories.

References

Agrawal, A. and Sivaramakrishnan, K. (2001) *Social Nature: Resources, Representations and Rule in India*, Oxford University Press, New Delhi, India.

Appadurai, A. and Breckenridge, C. A. (2009) 'Foreword', in A. Mathur and D. Da Cunha (eds) *Soak: Mumbai in an Estuary*, Rupa & Co., New Delhi, India, pp. 1-3.

Ascoli, F. D. (1921) *A Revenue History of the Sundarbans from 1870 to 1920*, Bengal Secretariat Book Depot, Calcutta, India.

Bagchi, K. G. (1944) *The Ganges Delta*, Calcutta University Press, Calcutta, India.

Barnes, T. J. (2008) 'History and philosophy of geography: Life and death 2005–2007', *Progress in Human Geography*, vol. 32, no. 5, pp. 650–658.

Bhabha, H. (1990) 'The Third Space, interview with Homi Bhabha', in J. Rutherford (ed.) *Identity, Community, Culture, Difference*, Lawrence and Wishart, London, pp. 202–221.

Bhabha, H. (1994) *The Location of Culture*, Routledge, London.

Bracken, L. J. and Wainwright, J. (2006) 'Geomorphological equilibrium: Myth or metaphor?', *Transactions of the Institute of British Geographers NS*, vol. 31, no. 2, pp. 167–178.

Brammer, H. (1990) 'Floods in Bangladesh II: Flood mitigation and environmental aspects', *The Geographical Journal*, vol. 56, no. 2, pp. 158–165.

Braun, B. (2006) 'Environmental issues: Global Natures in the space of assemblage', *Progress in Human Geography*, vol. 30, no. 5, pp. 644–654.

Buchanan, F. H. (1798) *An Account of the District of Purnea in 1809–1810*, Bihar and Orissa Research Society, Patna, India.

Canclini, N. G. (1995) *Hybrid Cultures: Strategies for Entering and Leaving Modernity*, University of Minnesota Press, Minneapolis, MN.

Cederlöf, G. (2014) *Founding an Empire on India's North-Eastern Frontiers, 1790–1840*, Oxford University Press, New Delhi, India.

Chapman, G. and Rudra, K. (1995) *Water and the Quest for Sustainable Development in the Ganges Valley*, Mansell, London.

Chapman, G. and Rudra, K. (2007) 'Water as friend, water as foe: Lessons from Bengal's millennium flood', *Journal of South Asian Development*, vol. 2, no. 1, pp. 19–49.

Chaudhuri, K.C. (1927) *The History and Economics of the Land System in Bengal*, The Book Company, Calcutta, India.

Cronon, W. (1992) 'A place for stories: Nature, history, and narrative', *Journal of American History*, vol. 78, pp. 1347–1376.

Danda, A. A. (2007) 'Surviving in the Sundarbans: Threats and responses, an analytical description of life in an Indian riparian commons', PhD thesis, University of Twente, Enschede, The Netherlands.

Dasgupta, A. (2010) *Phanerozoic Stratigraphy of India*, World Press, Kolkata, India.

Davis, W. M. (1900) 'The physical geography of the lands', *Popular Science Monthly*, vol. 57, pp. 157–170.

Demeritt, D. (1994) 'The nature of metaphors in cultural geography and environmental history', *Progress in Human Geography*, vol. 18, no. 2, 163–185.

Donaldson, J. W. (2011) 'Paradox of the moving boundary: Legal heredity of river accretion and avulsion', *Water Alternatives*, vol. 2, no. 2, pp. 155–170.

Domonoske, C. (2012) 'Ecology, language, and water/land: Ambiguous fenlands and challenged dichotomies of enlightenment in Graham Swift's Waterland', *Papers and Publications: Interdisciplinary Journal of Undergraduate Research*, vol. 1, no. 1, Article 7, http://digitalcommons.northgeorgia.edu/papersandpubs/vol1/iss1/7, accessed 21 March 2014.

D'Souza, R. (2007) *Drowned and the Dammed: Colonial Capitalism and Flood Control in Eastern India*, Oxford University Press, New Delhi, India.

D'Souza, R. (2009) 'River as resource and land to own: The Great Hydraulic Transition in Eastern India', Paper presented at Asian Environments Shaping the World: Conceptions of Nature and Environmental Practices, Singapore, 20–21 March 2009.

Duyne-Barenstein, J. (2008) 'Endogamous water resources management in North-east Bangladesh: Lessons from the Haor Basin', in K. Lahiri-Dutt and Robert J. Wasson (eds) *Water First: Issues and Challenges for Nations and Communities in South Asia*, Sage Publications, New Delhi, India, pp. 349–371.

East, G. (1938) *The Geography Behind History*, T. Nelson, London.

Eaton, R. (1990) 'Human settlement and colonization in the Sundarbans, 1200–1750', *Agriculture and Human Values*, vol. 7, no. 2, pp. 12–31.

Eden, S. and Bear, C. (2011) 'Models of equilibrium, natural agency and environmental change: Lay ecologies in UK recreational angling', *Transactions of the Institute of British Geographers NS*, vol. 36, no. 3, pp. 393–407.

Field, C. D. (1883) *Landholding and the Relation of Landlords and Tenants in Various Countries, Reprinted 2010*, Gale, Farmington Hills, MI.

Grove, R. and Damodaran, V. (2006) 'Imperialism, intellectual networks, and environmental change: Origins and evolution of global environmental history, 1676-2000', Part 1, *Economic and Political Weekly*, pp. 4345–4358.

Guha, R. (1963) *A Rule of Property for Bengal: An Essay on the Idea of Permanent Settlement*, Mouton & Co, Paris, France.

Hartshorne, R. (1939) *The Nature of Geography, Association of American Geographers*, Lancaster, PA.

Head, L. and Muir, P. (2007) *Backyard: Nature and Culture in Suburban Australia*, University of Wollongong Press, Wollongong, NSW, Australia.

Hill, C. V. (1997) *River of Sorrow: Environment and Social Control in Riparian North India, 1770-1994*, Association for Asian Studies, Ann Arbor, NY.

Hill, C. V. (2008) *South Asia: An Environmental History*, University of California Press, Santa Barbara, CA.

Hinchcliffe, S. (2007) *Geographies of Nature*, Sage, London.

Hofer, T. and Messerli, B. (1997) *Floods in Bangladesh: Process Understanding and Development Strategies*, Institute of Geography, Swiss Agency for Development and Cooperation and the United Nations University, Berne, Switzerland.

Howitt, R. (2001) 'Frontiers, borders, edges: Liminal challenges to the hegemony of exclusion', *Australian Geographical Studies*, vol. 39, no. 2, pp. 233–245.

Hunter, W. W. (1875) *A Statistical Account of Bengal, Vol. I, District of 24 Parganas and Sundarbans*, Trubner and Company, London.

Hunter, W. W. (1882) *The Indian Empire: Its Peoples, History and Products*, Trubner and Company, London.

Inglis, D. and Bone, J. (2006) 'Boundary maintenance, border crossing and the nature/culture divide', *European Journal of Social Theory*, vol. 9, no. 2, pp. 272–287.

Iqbal, I. (2010) *The Bengal Delta: Ecology, State and Social Change, 1840-1943*, Palgrave Macmillan, Basingstoke, UK.

Iyer, R. (2015) *Living Rivers, Dying Rivers*, Oxford University Press, New Delhi, India.

Jeffrey, R. (ed.) (1998) *The Social Construction of Indian Forests*, Centre for South Asian Studies, Edinburgh, Scotland.

Kabir, D. M. H. and Hossain, J. (2008) *Resuscitating the Sundarbans: Customary Use of Biodiversity and Traditional Cultural Practices in Bangladesh*, Bangladesh Unnayan Onneshan - The Innovators, Dhaka, Bangladesh.

Klingensmith, D. (2007) *One Valley and a Thousand: Dams, Nationalism, and Development*, Oxford University Press, Oxford, UK.

Kumar, D., Damodaran, V. and D'Souza, R. (2011) 'Introduction' in D. Kumar, V. Damodaran, and R. D'Souza (eds) *The British Empire and the Natural World: Environmental Encounters in South Asia*, Oxford University Press, New Delhi, India, pp. 1–16.

Lahiri-Dutt, K. (2000) 'Imagining rivers', *Economic and Political Weekly*, vol. 35, no. 27, pp. 2395–2400.

Lahiri-Dutt, K. (2008) 'Negotiating water management in the Damodar Valley: Kalikata hearing and the DVC', in K. Lahiri-Dutt and R. Wasson (eds) *Water First: Issues and Challenges for Nations and Communities in South Asia*, Sage Publications, New Delhi, India, pp. 316–348.

Lahiri-Dutt, K. (2015) 'Towards a more comprehensive understanding of rivers', in R. Iyer (ed.) *Living Rivers, Dying Rivers*, Oxford University Press, New Delhi, India, pp. 421–434.

Lahiri-Dutt, K. and Samanta, G. (2007) 'Like the drifting grains of sand: vulnerability, security and adjustment by communities in the Charlands of the Damodar Delta', *South Asia, Journal of the South Asian Studies Association*, vol. 32, no. 2, pp. 320–357.

Lahiri-Dutt, K. and Samanta, G. (2013) *Dancing with the River: People and Lives on the Chars in South Asia*, Yale University Press, New Haven, CT.

Law, J. (2004) *After Method: Mess in Social Science Research*, Routledge, London and New York.

Law, J. (2007) 'Making a mess with method', in W. Outhwaite and S. P. Turner (eds) *The Sage Handbook of Social Science Methodology*, Sage Publications, Thousand Oaks, pp. 595–606.

Leopold, A. and Schwartz, C. W. (1989) 'Introduction', in R. Finch, *The Delights and Dilemmas of a Sand County Almanac*, 2nd Edition, Oxford University Press, New York, p. viii.

Linton, J. (2010) *What is Water: The History of a Modern Abstraction*, University of British Columbia Press, Vancouver, Canada.

Mishra, D. K. (2003) 'Life within the Kosi embankments', *Water Nepal*, vol. 10, no. 1, pp. 277–301.

Mitra, S. C. (1898) *The Land-Law of Bengal*, Thacker, Spink & Co, Calcutta, India.

Mookerjee, R. (1919) *Occupancy Right: Its History and Incidents Together with an Introduction Dealing with Land Tenure in Ancient India*, University of Calcutta, Calcutta, India.

Nandy, A. (1983) *Intimate Enemy: Loss and Recovery of Self under Colonialism*, Oxford University Press, New Delhi, India.

Nicholson, J. (2007) *Ganges*, BBC Books, London.

Olwig, K. (2003) 'Landscape: The Lowenthal Legacy', *Annals of the Association of American Geographers*, vol. 93, no. 4, pp. 871–877.

O'Malley, L. S. S. (1914) *Bengal District Gazetteers: 24 Parganas*, Bengal Secretariat Press, Calcutta, India.

Phillips, H. A. D. (1886) *Our Administration of India: Being a Complete Account of the Revenue and Administration in All Departments with Special Reference to the Work and Duties of a District Officer in Bengal*, W. Thacker & Co, Bombay, India.

Radkau, J. (2008) *Nature and Power: A Global History of the Environment*, Cambridge University Press, Cambridge, UK.

Rashid, H. E. (1991) *Geography of Bangladesh*, University Press Limited, Dhaka, Bangladesh.

Reynard, P. C. (2013) 'Explaining an unstable landscape: Claiming the islands of the early-modern Rhône', *Environment and History*, vol. 19, no. 1, pp. 39–61.

Richards, J. F. and Flint, E. P. (1990) 'Long-term transformations in the Sundarbans wetlands forests of Bengal', *Agriculture and Human Values*, vol. 7, no. 2, pp. 17–33.

Roy, T. (2010) 'Rethinking the origins of British India: State formation and military-fiscal undertakings in an eighteenth century world region', *Working Papers No. 142/10*, London School of Economics, London.

Rudra, K. (2008) *Banglar Nadikatha* ['The Story of the Rivers of Bengal', in Bangla], Sahitya Sansad, Kolkata, India.

Sauer, C. (1925). 'The morphology of landscape', *University of California Publications in Geography*, vol. 2, pp. 19–54. Reprinted in J. Leighly (ed.) (1963) *Land and Life: A Selection from the Writings of Carl Sauer*, University of California Press, Berkeley and Los Angeles, CA, pp. 316–350.

Sherwill, W. S. (1858) 'Report on the Rivers of Bengal and Papers of 1856, 1857, 1858 on the Damoodah Embankments etc.', *Records of the Bengal Government's Selection No. 29*, G. A. Savielle Printing and Publishing Co. Ltd, Calcutta, India.

Swamy, A. V. (2011) 'Land and law in colonial India', in D. Ma and J. L. van Zanden (eds) *Law and Long-Term Economic Change: A Eurasian Perspective*, Stanford University Press, Stanford, CA, pp. 138–157.

Urch, D. (2008) *Crescent and the Delta: The Bangladesh Story*, The University of Buckingham Press, Buckingham, UK.

van Schendel, W. (1991) *Three Deltas: Accumulation and Poverty in Rural Burma, Bengal and South India*, Sage Publications, New Delhi, India.

Wescoat, J. L. (1990) 'Common law, common property, and common enemy: Notes on the political geography of water resources management for the Sundarbans area of Bangladesh', *Agriculture and Human Values*, vol. 7, no. 2, pp. 73–87.

Whatmore, S. (2002) *Hybrid Geographies: Natures, Cultures, Spaces*, Sage Publications, London.

Willcocks, W. (1930) *Lectures on the Ancient System of Irrigation in Bengal and Its Application to Modern Problems*, University of Calcutta, Calcutta, India.

Yule, H. and Burnell, A. C. (1903) *Hobson-Jobson: A Glossary of Colloquial Anglo-Indian Words and Phrases, and of Kindred Terms, Etymological, Historical, Geographical and Discursive*, Asian Educational Services, J. Murray, London.

Zimmerer, K. S. (2000) 'The reworking of conservation geographies: Non-equilibrium landscapes and nature-society hybrids', *Annals of the Association of American Geographers*, vol. 90, no 2, pp. 356–369.

Zimmerer, K. S. (2007) 'Cultural ecology (and political ecology) in the "Environmental Borderlands": Exploring the expanded connectivities within geography', *Progress in Human Geography*, vol. 31, no. 2, pp. 227–244.

5 Science as friend or foe?

Development projects undermining farmer managed irrigation systems in Asia's high mountain valleys

Joseph K. W. Hill

'*Dhobi ka kutta, na ghar ka na ghaat ka*'
'The washerman's dog belongs neither to his house nor to the washing place'
(Comment by a villager on government
irrigation projects)

Introduction

Comprehension of different and disparate knowledge systems or bodies of knowledge at differing scales is key to the successful management and development of natural resources, water for agriculture (irrigation) included. Scientific knowledge is known to be fragmented, a result of the disciplinary and departmental organisation of higher education (Norgaard and Baer 2005). Such knowledge is applied either by scientists working to formulate or evaluate policy (systematic research) or by the staff of line agencies (government or non-government) working in development programmes or projects, for example, technical engineers.

Ahlborg and Nightingale (2012), focusing on Nepalese forestry, use the concept of scales of knowledge to foreground the fact that scientists' choice of observational scale, for example, temporal, spatial, as well as epistemological position ranging from realist to constructionist perspectives has policy implications, and is thus political. Mosse (2005) convincingly shows that development practice is rarely steered by science-based policy rather it is driven by agencies' exigencies and their need to maintain relationships. Much of this is known tacitly by rural folk, including those who use and maintain small-scale irrigation systems. Often their knowledge, also known as local or traditional knowledge, is side-lined in development projects. Thus, four bodies of knowledge relevant to development projects can be delineated: systematic 'scientific' research, management experience ('practice'), political judgement (Head 2008) and local knowledge. There is a pressing need, if the great sums of money invested by donors and agencies to develop irrigation facilities are to be used wisely, for decision-makers to fully comprehend, acknowledge and respect each of these various bodies of knowledge.

This chapter posits that the misapplication of scientific knowledge (both technical and management-related) is undermining farmer managed irrigation systems (FMIS) in the high mountain valleys of Asia. This argument is based on research conducted over a four-year period (2011–2015) in India's trans-Himalaya (Ladakh's Kargil), Pakistan's Karakorum (Baltistan) and Tajikistan's Pamir (Gorno-Badakhshan). Field research took place at the irrigation system- and village-levels, not at district- or regional-levels, which was a deliberate scalar choice due to the researcher's academic training as well as the dearth of comparable studies on FMIS across the wider region. In these vast, high altitude desert landscapes, generations of villagers have built and maintained gravity-flow irrigation systems to convey snowmelt, spring or river water to their farmland. The management of such systems has for centuries been an integral part of everyday life for local populations. Similar irrigation institutions (for maintenance, water distribution, etc.) are observed in farmer-managed irrigation systems across northern India, northern Pakistan and eastern Tajikistan's mountain valleys. Yet, communities have experienced hugely differing forms of development intervention over the last century or so, ever since the onset of the geopolitical struggle between the British and Russian Empires.

The division of the region into South Asia and Central Asia continues to greatly influence the way development interventions are formulated, implemented and transformed on the ground. Tajikistan's incorporation into the Russian Empire and later the Soviet Union profoundly altered its society. To this day, its population's second language is Russian and migrants from almost every household work in Russia and remit money home. The Soviet Union (SU) heavily invested in Gorno-Badakhshan's FMIS. Since its collapse there has been an influx of Western non-government agencies (NGOs) staffed by professionals, most of whom have knowledge of Russian and management experience in former SU countries. By contrast, and keeping in mind the similarities in ecology, livelihood and FMIS, the communities residing in the mountain valleys of Pakistan's Baltistan and India's Kargil speak Urdu as a second language, migrate to their respective mainlands, and have experienced state-led development only in recent decades (though in northern Pakistan, NGOs have also implemented FMIS development projects). The Line of Control, created by India and Pakistan, not only curbs movement of people and goods but also restricts the sharing of development experiences. Thus, geopolitics determines the types of agencies, for example, government or NGO, present in any given locale and their agendas, as well as the production and sharing of knowledge regarding FMIS, including how best to design and implement interventions to support FMIS.

This chapter presents a case study from India's Kargil (located in Ladakh) where FMIS interventions are decided upon by bureaucrats and elected politicians, designed and overseen by government engineers, and implemented by politically connected contractors; without the aid of systematic 'scientific' research and with the local knowledge and participation of water users almost entirely sidelined.

Fragmented scientific knowledge, policy and practice

Today's world is characterised by an unprecedented fragmentation and specialization of knowledge. Scientific knowledge is disjointed and more often than not, scientists do not work together across disciplines to gain a scientific understanding of a given problem (Norgaard and Baer 2005). Little in the way of guidelines to support interdisciplinary work exists in methodology texts, while the understanding of how scientists collectively learn and know is not formally incorporated into the interfaces between science and policy, or science and democracy (ibid.). Projects and the policies behind them are conceived by outside actors – national-level government actors and water/NGO professionals – and actually serve their 'immediate interests', in other words, to bring institutional reality into line with their policy prescriptions (Mosse 2005). As Allan (2006, p. 38) states, water users are rarely a part of the policy process; rather they are regarded as the passive recipients of projects:

> All policy-making discourse is partial in that it is made by coalitions which reflect those who can best construct and deliver the most persuasive arguments … For example, governments rarely confront large farming communities existing on low incomes … Policy arguments are driven by immediate interests, rather than by high-minded notions of long-term collective action based on social equity, economic efficiency or environmental considerations.

Allan clarifies that policy is not made on the basis of rational science, although science can play a role if its messages are effectively constructed and included in the policy-making process. He argues that local discourses over water have been impacted by 'Northern' (i.e. Western) ideas through a sequence of international water management paradigms. This Kuhnian perspective[1] (Kuhn 1970) is a useful framework for understanding the quagmire of FMIS in mountainous regions. Allan identifies five water management paradigms:

1   pre-modern communities, having limited technical and organisational capacity;
2   industrial modernity and the hydraulic mission of the twentieth century (led by Europe, North America and the former Soviet Union);
3   environmental awareness, from the 1980s to the present;
4   economic valuation of water, from the 1990s to the present;
5   and supplementing the third and the fourth, 'Integrated Water Resources Management' (IWRM), based upon the notion that water allocation and management are political processes. (Allan 2006, pp. 46–68)

For the mountain communities encountered in this research, there has been no linear progression through these five outlined paradigms. The communities and irrigation systems of the first paradigm are partially subjected to the second (engineering) and fourth (management) paradigms although they've been bypassed by the third (environmental), while the fifth (IWRM) is yet to be introduced.

More precisely, the communities residing in Tajikistan were subjected to the second paradigm during the Soviet period, and are now faced with NGOs pushing the fourth paradigm; whereas the communities in India and Pakistan's mountains were only subjected (in a major way[2]) to the engineering paradigm as late as the end of the twentieth century, while the third, fourth and fifth paradigms are yet to come into play. From this perspective, it is clear that agencies working on either side of the Central Asia–South Asia geopolitical divide, could learn a great deal from the experiences of FMIS on the other side.

Yet, such learning does not take place, for several epistemological reasons. First, the division of the larger region into the two geopolitical spheres of Central Asia and South Asia led to the creation of two, broader academic areas of specialization, but also to very different forms of culture, government and society. Thus experts, professionals and officials from one region rarely travel to the other. Second, mainstream science has led us to believe that so-called *'pre-modern communities'* require outside knowledge to improve their condition, and not vice versa. Thus 'experts' believe their knowledge is superior to locals'. Third, the 'sanctioned discourse' currently in vogue in Central Asia is that of the fourth paradigm, 'economic efficiency'. For example, international donors and agencies, using as justification the widely-held belief that farmers in Tajikistan, Kyrgyzstan and other Central Asian countries were deskilled during the Soviet period, impose the globally promoted 'water users association' (WUA) model upon farmers as a conditionality to funding. Over twenty years ago, it was noted that the replication of standard organisational formats are often inappropriate to local farming systems (Vincent 1994); yet donors and agencies in Central Asia appear to be unaware of the earlier, unsuccessful attempts to impose such committees in South Asia, and thus continue to insist that such committees be formed.

Water policy reforms are a political process, shaped by the discourses that precede their formulation, the outcomes reflecting the interests of the participants as well as the absence of the interests excluded (Allan 2006; Molle 2008). While it is often assumed that experts have the required knowledge, this is not always the case. For example, FMIS are built by local water user communities and managed collectively according to their own norms and organisational constraints. Having persisted for centuries, these systems are nowadays struggling to adapt to wide, societal and structural changes. Following Beccar *et al.* (2002), it is the combination of several elements that make irrigation systems 'work': physical elements (e.g. water source, infrastructure), normative elements (rules, rights, obligations), organisational elements (for operation, maintenance) and agro-productive elements (technology, capital, knowledge, etc.). Thus, irrigation projects impact upon the normative and organisational basis of irrigation, for example, by altering the amount of water available for distribution or by modifying infrastructure and thus maintenance requirements

(Beccar *et al.* 2002, p. 10). This is because the technology *itself* contains the norms included by its designers (Boelens 1998, p. 93). Therefore, due to inadequate or faulty understanding by outside agencies:

> many irrigation projects, despite the obvious relationship of existing water rights with the future modification of the infrastructure and the increased flow rate, do not consider this issue important or explicitly address it. They fail to understand that the irrigation infrastructure is not just a concrete structure carrying water, but a collective project in which users can, by investing, create individual rights when they build it, and consolidate them when they maintain it.
>
> (Beccar *et al.* 2002, p. 11)

Such misunderstandings are a direct consequence of the fragmented (scientific) knowledge held by the staff of government and NGOs, as well as the influence of (geo-) politics upon the implementation of policies (Mosse 2005). Thus, irrigation projects can – in the medium to long-term – undermine the very systems they seek to support.

The next section demonstrates how using local knowledge, farming communities in Asia's high mountain valleys have integrated into their irrigation systems all the paradigms of Allan's schematic – engineering, management, environmental and 'integrated management'. Allan's labelling of '*pre-modern communities* with limited technical or organisational capacity' (Allan 2006, p. 46), though appearing Eurocentric, alludes to local communities' inability to organise effectively beyond their immediate vicinity.

Farmer managed irrigation systems in Asia's mountain valleys

In Asia's high mountain valleys, most villages are perched precariously on river terraces, hillsides or on the sides of alluvial fans. The farming systems in India's Ladakh, Pakistan's Gilgit-Baltistan and Tajikistan's Gorno-Badakhshan are similar (see Herbers 2001, p. 275). Irrigation is central to agricultural production and indeed human life because of a severe deficit of rainfall. For example, in the Karakorum's valley bottoms, Schmidt (2009, p. 21) and Vander Velde (1989, p. 7) estimate rainfall to be less than 150 mm per annum. It's understood that gravity-flow irrigation systems were developed hundreds of years ago, to utilize river water, spring water or melt water from glaciers and snow fields. Raunig (as cited in Bliss 2006, p. 125) states that archaeological data from the Pyandsh valley in the Pamir prove that irrigation of wheat and barley can be attested for as early as 1550 to 1200 BC. Depending on the local topography and water sources, irrigation channels range in length from less than 1 km to 20 km, their discharge varying accordingly.

The institutions developed by communities for managing irrigation systems share many commonalities. When constructing a new channel, the traditional method for determining the slope of a channel is to use water as a level. Water is allowed to flow along the channel as it is dug on a carefully estimated line. Elders are consulted for advice on past glacial movements, avalanche and mud flow paths, and stream flows from glacial and snowmelt or springs. To equitably allocate water, irrigation turns are taken according to an established roster during periods of water scarcity, notably between March and May. Under this system, each household in a command area takes its water on a specific day for a specified and equal period of time. Farmers whose turns are close to one another frequently informally trade or exchange water. Vander Velde (1989) observed that food crops are given priority over trees and fodder crops such as alfalfa, and that vegetables take priority over food grains. Springtime and occasionally mid-season maintenance of irrigation systems reflects the channels' common property origins. The common portion of the channel is maintained with an annual contribution, in the form of labour or produce (and more recently, cash) from all the farmers served. Some villages employ a watchman during the irrigation season to patrol the common portion of certain channels, to adjust the headwork, plug leaks, repair small breaches and monitor water supply. Where watchmen are not employed, villagers take regular turns to patrol and maintain irrigation channels.

An example from Karchay Khar village in India's Ladakh

Ladakh is the meeting point of the Tibetan–Buddhist and Persio–Islamic worlds. Due to this, the regional literature on farmer-managed irrigation is divided into two bodies, focussed on Buddhist villages in Indian Ladakh (e.g. Labbal 2000; Wacker 2007; Gutshow 1998), and Shia Muslim villages over the Line of Control in Pakistan's Gilgit-Baltistan (e.g. Schmidt 2004; Fazlur-Rahman 2007; Kreutzmann 2000; Schmid 2000). Closed to outsiders from Indian independence until the 1970s, India's Ladakh was carved into two districts in 1979, creating the predominantly Shia Muslim Kargil district and the largely Buddhist Leh district. Due to the inhospitable desert conditions, Kargil district's agricultural land comprises a total of just over 10,000 ha, of which 38 per cent is under grim (barley), 32 per cent alfalfa or grass (used as fodder), 13 per cent wheat, 7 per cent pulses, 5 per cent millet and 4 per cent vegetables (GoJK 2011, p. 24). Almost all the cultivated area (96.5 per cent) is recorded as irrigated by 'canals' (ibid., p. 34), which indicates the importance of FMIS to the region.

The political landscape of Kargil includes Members of the Legislative Assembly (MLAs), Members of Parliament (MPs), locally elected councillors to the autonomous hill council formed in 2003[3] and religious leaders known as *sheikhs*. The Suru valley, studied in-depth by Grist (1998), is one of Kargil's principal valleys, its population almost entirely Shia Muslim. Local *sheikhs* – divided in their loyalty to two Kargil-based Shia schools – take a prominent role in party politics and local village-level decision-making and dispute resolution. Thus, to the detriment of local resource management, party politics, hill council politics and Shia Muslim politics divide Kargil district's local population into factions from the village level upwards.

Located mid-way up the Suru valley, Karchay Khar Revenue Village, also a *gram panchayat*,[4] has about 120 households and comprises three villages, Karchay Khar, Stiankung and Stakbourik (see Figure 5.1). This case study focuses upon the village of Karchay Khar consisting of about 67 households. Survey data for Karchay Khar[5] show that all households have members working outside of the village. It shows that 85 per cent of households have migrant labourers, two-fifths of whom have travelled outside of Ladakh to work, 96 per cent of whom have worked as labourers within Ladakh. At the time of the survey, in spring 2014, only two-fifths of households had a member outside of the village working; however, only one of the 27 surveyed households had no adult male present in the village. This data indicates that seasonal migratory work is an important source of earnings, however, village-based livelihoods remain paramount to any household's functioning. Farming, silviculture, and livestock rearing are the basis of village livelihoods, all of which necessitate availability of (irrigation) water.

Karchay Khar is situated on the alluvial fans of two mountain streams, Chogo Lungma and Sheshabroq Lungma. Up until the time of Indian independence, the village depended on these two mountain streams for its water supply. Chogo Lungma provides water to the village's most fertile land – upon which barley, wheat and peas are grown – via irrigation channels that include Kharzong yurba and Richen yurba. The water of Sheshabroq Lungma supplies six irrigation channels that irrigate six areas of land known as *broq*, upon which trees, alfalfa and some cereals are grown. Another channel, Stakbourik yurba, draws water from the Barsoo River. It supplies water to some of Karchay Khar's

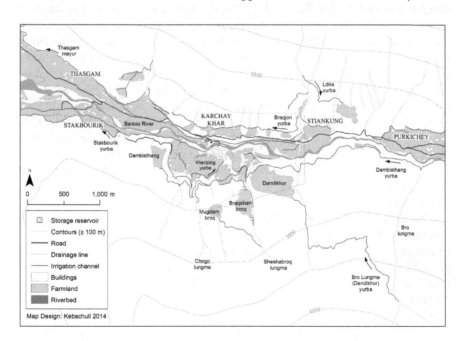

Figure 5.1 Karchay Khar gram panchayat, showing its villages and irrigation channels.

farmland located close to the riverside, and is the primary channel supplying Stakbourik village's farmland. Few if any of the villagers apply chemical fertiliser or pesticides to their farmland, and the concept of 'weed' does not exist. 'Weeding', as it is conventionally called, is mostly the job of women, who take the grasses home to feed their livestock. There is an intrinsic notion of environmentalism in local livelihoods.

Around 1950, Karchay Khar's community decided to build a new irrigation channel called Dambisthang yurba (see Figure 5.1). Construction of the channel took between 14 and 18 years. Work is said to have stopped for three years when the village's *nambardar* (headman) died, which is illustrative of the embeddedness of irrigation systems in the village social structure. The current village *nambardar*, Mohd Issa, said:

> Around 1947 my grandfather, who was the *nambardar*, and others like Rozee's grandfather, started construction of Dambis kuhl. No money, no food was provided in those days. The villagers made tools like the *kudali* from the wood of Apricot trees. Around 1961 they arrived at Dambisthang. So building the channel took 14 years of work.
>
> (Paraphrased excerpt from an interview conducted in 2014)

The father of Mohd Rozee confirmed the above, adding that the local government provided dynamite for blasting rock. The irrigation channel is a remarkable feat of engineering. It is designed so that in the springtime, when the snow melts, the 7 km long channel taps the water of the mountain stream Bro Lungma located several kilometres to the east of the village. When the stream dries up, around July, the channel draws water from the Barsoo River opposite Purkichey village. The channel crosses over the mountain streams, Sheshabroq Lungma and Chogo Lungma, and in places runs along sheer cliff faces. It's remarkable that without the use of modern engineering surveys, the villagers correctly determined the channel's slope. By contrast, a 2005 project to create a second branch to Stakbourik yurba, funded by a 100 thousand rupees donation by an MLA, failed because the slope of the short section of channel being carved from a cliff face was incorrectly determined. This perfectly illustrates the danger posed by the erosion or neglect of local knowledge.

Water-related disputes between communities are common. The villagers of Karchay Khar and Stiankung are engaged in a long-term dispute regarding usage rights to the water of Bro Lungma, because prior to Dambisthang yurba's construction the villagers of Stiankung used Bro Lungma's water to irrigate a small piece of land. After the case reached the local Tehsil office in Sankoo, an agreement was reached that allows Stiankung's villagers to draw water for their land from Dambisthang yurba without having to involve themselves in Dambisthang yurba's annual maintenance work. A separate dispute, which continues to be a sour point between Stakbourik and Karchay Khar's villagers, centres on the five-hectare piece of land at Dambisthang yurba's tail end (Figure 5.1). Stakbourik's community also have land there, and wanted to participate in the construction of

Dambisthang yurba to gain rights to irrigate their land, however, Karchay Khar's community didn't allow them to participate.

There are no government records of Dambisthang yurba or its water rights, yet the FMIS is managed extremely well. An example highlights its autonomy. In 2000, an agreement was made, in the presence of *sheikhs,* between all the users of Dambisthang yurba and the seven or so households residing in Richen hamlet (belonging to Karchay Khar village), positioned just below the channel where it supplies Dambisthang. The agreement stipulates that should the channel collapse and damage anyone's property, whoever's turn it is to irrigate at that moment will be liable to compensate those affected. The villagers have a strong incentive to maintain Dambisthang yurba because besides supplying the farmland at its tail end, the channel's water is used to irrigate almost all of the village's farmland (Figure 5.1). A water user, on his or her turn, can direct the channel's water to irrigate whichever piece of land he or she wishes to.

The system of irrigation turns (also known as 'water rights') for any given water source is complicated for outsiders to decipher. The system of turns may relate to a stream (as in the case of Sheshabroq Lungma and Chogo Lungma) or to an irrigation channel (as in the case of Dambisthang yurba or Dandikhor yurba).[6] The water rights of Sheshabroq Lungma and Chogo Lungma were recorded during the land revenue settlements of 1901 and their re-assessments in 1911,[7] which included a record of rights for each village. The record contains a section called the *Riwaj-i-Abpashi* (irrigation customs) that indicates the irrigation rights of water users to the channels (Coward 1990, p. 81). For example, the 1910 records for Karchay Khar (Singh 1910) state that in the first half of the agricultural year there are no water alloca-tion rules for Kharzong yurba and Richen yurba. However, later in the year, when the stream's water flow decreases, a system of water distribution (irriga-tion turns) is implemented. There are 24 households that share the stream's water over a 12-day cycle, two households per day. The household names were recorded in 1910 and this same system continued to be followed in 2013, though informally adjusted over time to reflect demographic changes, land sales, and so on (see Hill 2014a, p. 17).

The government has no records of the water rights of Dambisthang yurba because the FMIS was created after the last settlement. Thirty-four households created the channel, and the system of turns allows for three of these households to take water each day over a 12-day cycle, totalling shares for 36 households.[8] If water supply is less and crops are parched, the cycle switches to a nine-day cycle of four households per day. The rights to the water of Dambisthang yurba remain attached to the 34 households that originally built the channel. One or two generations from its construction, and due to the subdivision of households, nowadays some 70 households have rights to the channel's water. The example of Mohd Baqir and his two brothers, who each live in separate households though they descend from one right-holder to the channel's water, illustrates how this works in practice. The three brothers utilise their one share. In a given year, if 15 working days must be spent repairing the channel, then each of the brothers will participate for five days. By contrast, in a household that has not subdivided, a representative (most often male) would have to work the full 15 days.

The village's two main streams (Chogo and Sheshabroq Lungma) and two large irrigation systems (Dambisthang and Dandikhor yurba) each have an appointed official, known as a *helbie*, who oversees water allocation and maintenance work, making sure that every household provides the required labour, and if not, ensuring compensation is paid. The *helbie* is not paid for his work; rather the post is hereditary. For Dambisthang yurba and Dandikhor yurba, due to their lengths and treacherous paths, a watchman, known as a *chustrunpa*, is selected each year to manage the channel, for which he is remunerated. The watchmen used to be paid in kind – in grain (barley, also wheat) – by each household. In 2012, Dambisthang yurba's watchman received 20 kg of grain from each of the 33 households (not including the *helbie*). However, this remuneration has recently been monetarised, reflecting changing societal values. In 2013, the *chustrunpa* was paid 250 rupees by each of the 33 right-holders, totalling around 8,250 rupees. The village *nambardar* explained that four households descend from his grandfather who co-built Dambisthang yurba, so each household must contribute one-quarter of the annual payment to the channel's watchman. He further explained that should a household not attend maintenance duties, a fine of 100 rupees is levied. The *nambardar* lamented that nowadays *'paise ke baat chalte hai'* (money rules). This intricate system of management – that now incorporates monetarised payments – shows the sophisticated economic and management dimensions to FMIS. In the next section the role of external scientific knowledge applied during development interventions is discussed.

Irrigation projects to 'support' farming communities

The example of Karchay Khar village illustrates how high Asia's farmer managed irrigation systems integrate the first four paradigms in Allan's (2006) schematic, namely 'pre-modern' with limited capacities, engineering, environmental and management-economic. Research conducted in Tajikistan's Pamir (Hill 2013) and Pakistan's Karakorum (Hill 2014b) exemplifies the same. One could argue that FMIS are already 'integrated' in an IWRM sense (Allan's fifth paradigm), because the practices and discourses of water users show a clear recognition that 'water allocation and management are political processes'. However, IWRM is defined as a process that incorporates all stakeholders, and this is not yet the case with respect to FMIS – as this section shows – because nowadays the stakeholders involved in FMIS include non-local actors residing outside of the village. To illustrate this, I draw on the example of Dambisthang yurba, for which in 2011 the Ladakh Autonomous Hill Development Council, Kargil (LAHDC, K) earmarked a staggering 20 million rupees[9] to be spent on its renovation over a period of ten years.

This 20 million rupees project is overseen by the office of the executive engineer, public works department (PWD), Kargil. A junior engineer (JE), who works under the guidance of the assistant executive engineer, manages the project. During an interview in 2013 at his chamber in the Baroo office in Kargil, the JE was unwilling to show the project's file. However, opening Google Earth on his laptop, he discussed the ongoing project while pointing to sections of Dambisthang yurba: in the first year (2011) of the project 2.5 million rupees was spent on concrete lining and support treatment at the headwork, in the second year just over half a million rupees was spent on support work in three places, and in 2013 a culvert was under construction. Each year, the JE explained, he chooses a sub-project and prepares a tender. Several contractors apply by fronting 'insurance' money, and the selected applicant will complete the sub-project, balancing quality and cost, for his payment comes from whatever remains of the budget. In June 2013, during field research, three of Karchay Khar's men (a skilled labourer in charge, with two men working as labourers) were seen working on the culvert mentioned by the JE.

Later in June 2013, a section of Dambisthang yurba close to Chogo Lungma collapsed. The local councillor, elected to the LAHDC, K, paid a visit to the village and was taken to the site. Convinced by the villagers, he helped them to procure materials (wire mesh and piping) for its repair from Panchayat funds. Yet, without using these materials the villagers skilfully repaired the channel. It turned out that the water users used the opportunity provided by the councillor's visit to procure material that they could store to use later, elsewhere in the village. This illustrates how savvy the villagers can be when it comes to politics and the procurement of development resources. Two years later, in 2015, the repair site close to Chogo Lungma was found to be standing strong. A different, nearby part of the channel had collapsed and again the water users had repaired it without the aid of government funds.

In 2013, water users said that due to the project to renovate Dambisthang yurba, they now only spend between 5 and 15 days repairing the channel as compared to between 15 and 30 days in the past. While allegations of corruption abounded – it is common knowledge that a part of all project funds goes to various officials – it appeared that the project supported the FMIS's physical infrastructure without affecting its normative and organisational dimensions. However, in 2015 while visiting Dambisthang yurba's headwork where it draws water from Bro Lungma, a gang of Nepalese labourers were found concretising two sections of the channel. Not a single villager was present. The *sarpanch* (the elected leader of the *gram panchayat*) explained that nobody in the village had had the funds to front for the tender put out by the JE that year and thus a contractor from another village had applied and been granted the contract. In 2015 there was also unusually heavy rainfall that swelled the Barsoo River, washing away much of the concretised headwork created under the project in 2011.[10] This is problematic because the villagers can only repair such damage with further external support (both material and financial). Thus, a project that in 2013 and 2014 appeared quite benign looked quite different in 2015: the use of outside labourers and outside materials will most certainly upset the technical-normative-organisational unity of the FMIS.

How science is undermining farmer managed irrigation systems

That water has a multi-faceted epistemological status is keenly evident from the case study of FMIS in Asia's high mountain valleys. The majority of these systems were developed and designed by villagers who knew one another and their local resources and means (local knowledge). To such water users, an irrigation system is more than simply a technological infrastructure; it's an ongoing interaction between the physical infrastructure and organisational and normative elements (Boelens 1998), embedded in the local community and related to other resources and institutions. While the technology is an expression or materialisation of irrigation norms, its physical constraints influence the social organisation required for the system's management (ibid.; Mosse 2003, p. 4). For any particular irrigation technology to work effectively, particular social conditions have to be fulfilled (Boelens 1998). Thus, when farmers themselves develop and design their own technologies, the process maintains technical-normative-organisational unity, because the designers and water users share norms and assumptions regarding the technology. However, when any or several of these components – the technical, normative or organisational – are disturbed, the balance of the configuration changes. This is why both societal changes that alter relationships among water users, and interventions by outside agencies that alter the physical or organisational dimension to a system, can and do undermine FMIS.

In recent decades, government and non-government agency support to FMIS in both Ladakh and northern Pakistan (Hill 2014b) has increased significantly. Conversely in eastern Tajikistan, support to FMIS has dwindled, leaving farming communities dependent on the piecemeal support of mostly Western-funded donors (Hill 2013). Irrigation systems developed through government or non-government agencies' interventions involve the interaction of engineers and other stakeholders as well as the introduction of technologies and materials often beyond the means of local farmers. The choice of technical design is often imposed upon local farmers, and with it, preconceived notions of the purposes the introduced technologies should serve, and the institutional forms through which they should be achieved (Boelens 1998; Vincent 1995). When irrigation technology (e.g. concretisation) or organisational forms (such as the Water User Association model) are introduced through intervention, the technical-normative-organisational balance may be upset. Ambler (1994) cites evidence that putting in permanent headworks and concrete lining reduces labour needs but at the same time lowers productivity and equity. When the users of FMIS are not the planners, their own objectives and norms are not included in the design stage, and they later face social requirements and norms defined by other actors (Boelens 1998). This explains the faultiness of projects decided upon by bureaucrats, designed by engineers, given to non-local contractors able to front the required money, and undertaken by non-local labourers who are not the users of the FMIS.

The limitations of a specialised, dispersed scientific understanding of environmental problems is highlighted by Norgaard and Baer, who critique scientists' lack of discussion over the interdependency of science and democracy. Most of the time scientists do not even think of themselves as engaged in democratic discourse (Norgaard and Baer 2005, p. 956). However, 'science is itself framed by unstated social commitments' (Irwin and Wynne 1996, p. 2), and 'what counts as "science" may be shaped by social relations and institutional structures so that the very constitution of science will reflect wider social interests' (ibid., p. 8). The public application of science to the management of Dambisthang yurba, as detailed in the previous section, is a clear example of outside actors conceiving policies and projects that serve their, rather than water users', immediate interests, in other words, the project facilitates the channelling of funds to politically important personalities (politicians, bureaucrats or contractors). Water users are regarded as passive participants, sidelined to such a degree that ultimately, as seen in 2015, they do not renovate their own irrigation channels. This confirms Allan's claim that policy arguments 'are driven by immediate interests rather than by high-minded notions of long-term collective action' (2006, p. 38).

Allan's Kuhnian schematic of international water management paradigms was used to draw attention to two interlinked epistemological concerns surrounding FMIS. First, the farmers behind FMIS are not merely *pre-modern communities, having limited technical and organisational capacity*; rather they have engineering skills that can surpass modern engineering surveying techniques, have an environmental awareness that surpasses the modern agricultural scientists' obsession with high-chemical input and energy-intensive agricultural seed-fertiliser packages,[11] and have organisational skills that surpass the economic-efficiency paradigm's mania for formal committees and standardised 'irrigation service fees'. The farmers behind FMIS are practising IWRM, while it is the outside actors (the bureaucrats, the engineers) that need to learn this perspective. Second, the Kuhnian schematic presents a model as to how outside agencies interact with FMIS. In India's Ladakh the second paradigm – hydraulic mission of the twentieth century – is still the frenzy, whereas in the former Soviet Union international NGOs and donors narrowly conform to the fourth paradigm – economic valuation of water.

Conclusion

To farmers, irrigation systems are simultaneously hydrological, engineering, organisational and farming system entities, and thus irrigation management is positioned within and forms part of the logic of their communities' social environment and production systems. Yet, in these decades of rapid societal change, a multitude of knowledge systems have come to bear on FMIS through the interactions of outside actors. At the local level, both Head's (2008) delineation of three bodies of knowledge ('practice', 'political' and 'scientific'), and Ahlborg and Nightingale's (2012) concept of scales of knowledge are useful to comprehend the quagmire faced by local farming communities. Various local and regional actors (e.g. contractors, engineers, local politicians and bureaucrats) directly apply certain bodies of knowledge to FMIS. While such development practice is framed in a language of science-based policy (e.g. engineering), it takes place through political manoeuvring. Other knowledge systems extending from beyond the region are also influential, some of which are captured well in the paradigmatic model of Allan (2006).

The title of this chapter is purposively provocative. Can the role of science be questioned if, as stated by Irwin and Wynne (1996), the very constitution of science reflects wider interests? Chambers (1988) and Coward (1985) had three decades ago called for a framework to study irrigation that integrates both technical and social science perspectives; yet, this can hardly be said to have materialised. Fragmented knowledge, be it local versus expert, micro- versus macro-level, bureaucratic versus civil society, or engineering versus social science, is at the heart of the problems facing the effective management of water. Thus, bureaucrats are driven by abstract and macro-level concerns, for example, to raise crop yields by improving irrigation or to expand bureaucratic power and reach of their government. Engineers survey an irrigation channel from its headwork to tertiary channels, and privileging their form of technical knowledge plan the concretisation of sections deemed weak. Until professionals, most of whom have been trained at centres of higher education, acknowledge the interlinkages between science and democracy, to allow less visible groups (i.e. farmers) to express their views and preferences, can it be expected that irrigation interventions pertaining to support FMIS will succeed in doing so? Thus, this chapter questions whether applied science is 'friend or foe'?

Acknowledgements

The author would like to thank Jenny Kebschull for helping to produce the village map. This study was funded by the German Ministry for Education and Research (BMBF), under the research programme 'Crossroads Asia: Conflict–Migration–Development'.

Notes

1 In the sense that one paradigm replaces another when the members of a scientific or practitioner community understand why they held their previous beliefs and how these beliefs could have misled them (Erickson 2005, pp. 73–74).
2 Settlement reports of the early twentieth century show that the government, through its Public Works Department, funded a small number of irrigation projects (Singh 1913).
3 Following demands, in 1995 an act was passed allowing Ladakh to create its own autonomous hill council. Leh district formed its council in 1995 (van Beek 1999), and Kargil district's political leaders formed theirs, the Ladakh Autonomous Hill Development Council, Kargil (LAHDC, K) in 2003.
4 A *gram panchayat* is a cluster of villages, the lowest administrative unit. *Gram panchayat* elections took place in 1999 (Aggarwal 2004), but from 2004 to 2011 the system was moribund due to political difficulties in Kashmir.

5 Twenty-seven households were randomly selected from 59 households in Karchay Khar.
6 Without government support, Karchay Khar's villagers built the 5km long Dandikhor yurba (Bro Lungma yurba) in the 1980s. Fifty-five households use its water, sourced from snowmelt at its headwork at 3,800m, to grow alfalfa (fodder).
7 Wacker (2007) and Hill (2014b) show the relevance of these settlements to FMIS.
8 One of the extra 'shares' belongs to a household that forsook farmland that stood in the path of the channel; the other share is for the Mosque and Imambara's farmland.
9 In 2014, roughly €300,000.
10 Across Kargil, the headworks of run-of-the-river FMIS were washed away due to the heavy rains of 2015. I saw damage to FMIS in Stakbourik and Gyaling villages too.
11 During a PRA session in Karchay Khar village, five varieties of native wheat and one introduced high-yielding variety were compared. While the introduced, fertiliser-responsive seed was said to have a higher yield, it is also the most costly to produce and the least tasty and nutritious.

References

Aggarwal, R. (2004) *Beyond Lines of Control. Performance and Politics on the Disputed Borders of Ladakh, India*, Duke University Press, Durham and London.

Ahlborg, H. and Nightingale, A. J. (2012) 'Mismatch between scales of knowledge in Nepalese forestry: Epistemology, power, and policy implications', *Ecology and Society*, vol. 17, issue 4, no. 16.

Allan, J. A. (2006) 'IWRM: The new sanctioned discourse?', in P. P. Mollinga, A. Dixit and K. Athukorala (eds) *Integrated Water Resources Management. Global Theory, Emerging Practices and Local Needs*, Sage Publications, New Delhi, India.

Ambler, J. (1994) 'Small-scale surface irrigation in Asia. Technologies, institutions and emerging issues', *Land Use Policy*, vol. 11, no. 4, pp. 262–274.

Beccar, L., Boelens, R. and Hoogendam, P. (2002) 'Water rights and collective action in community irrigation', in R. Boelens and P. Hoogendam (eds) *Water Rights and Empowerment*, Koninklijke Van Gorcum, Assen, Netherlands.

Bliss, F. (2006) *Social and Economic Change in the Pamirs* (Gorno-Badakhshan, Tajikistan), Routledge, London, New York.

Boelens, R. (1998) 'Collective management and social construction of peasant irrigation systems: A conceptual introduction', in R. Boelens and G. Davila (eds) *Searching for Equity. Conceptions of Justice and Equity in Peasant Irrigation*, Koninklijke Van Gorcum, Assen, Netherlands.

Chambers, R. (1988) *Managing Canal Irrigation: Practical Analysis from South Asia*, Cambridge University Press, Cambridge, UK.

Coward, E. W., Jr. (1985) 'Technical and social change in currently irrigated regions: Rules, roles, and rehabilitation', in M. M. Cernea (ed.) *Putting People First: Sociological Variables in Rural Development*, Oxford University Press, New York.

Coward, E. W. J. (1990) 'Property rights and network order: The case of irrigation works in the Western Himalayas', *Human Organisation*, vol. 49, no. 1, pp. 78–88.

Erickson, M. (2005) *Science, Culture, and Society: Understanding Science in the Twenty-First Century*, Polity Press, Cambridge, UK.

Fazlur-Rahman (2007) 'The role of Aga Khan Rural Support Programme in rural development in the Karakorum, Hindu Kush and Himalayan region: Examples from the northern mountainous belt in Pakistan', *Journal of Mountain Science*, vol. 4, no. 4, pp. 331–343.

GoJK (2011) 'Statistical handbook (2010–2011)', Directorate of Economics and Statistics, Planning and Development Department, Government of Jammu and Kashmir (GoJK), issued by District Statistics and Evaluation Officer, Kargil, India.

Grist, N. (1998) 'Local politics in the Suru Valley of Northern India', PhD thesis, Goldsmiths' College, London University, UK.

Gutschow, K. (1998) 'Hydro-logic in the Northwest Himalaya: Several case studies from Zangskar', in I. Stellrecht (ed.) *Karakorum-Hindukush-Himalaya: Dynamics of Change Part 1*, Rüdiger Köppe Verlag, Köln, Germany.

Head, B. W. (2008) 'Three lenses of evidence-based policy', *The Australian Journal of Public Administration*, vol. 67, no. 1, pp. 1–11.

Herbers, H. (2001) 'Transformation in the Tajik Pamirs: Gornyi-Badakhshan-an example of successful restructuring?', *Central Asian Survey*, vol. 20, no. 3, pp. 367–381.

Hill, J. (2013) 'The role of authority in the collective management of hill irrigation systems in the Alai (Kyrgyzstan) and Pamir (Tajikistan)', *Mountain Research and Development*, vol. 33, no. 3, pp. 294–304.

Hill, J. (2014a) 'Farmer-managed irrigation in the Karakorum (Skardu) and trans-Himalaya (Kargil)', *Ladakh Studies (Journal of the International Association of Ladakh Studies)*, vol. 31, pp. 4–23.

Hill, J. (2014b) 'Irrigation practices, irrigation development interventions, and local politics: Re-thinking the role of place over time in a village in Baltistan, in the central Karakorum', *Crossroads Asia Working Paper*, no. 16.

Irwin, A. and Wynne, B. (1996) 'Introduction', in A. Irwin and B. Wynne (eds) *Misunderstanding Science? The Public Reconstruction of Science and Technology*, Cambridge University Press, Cambridge, UK.

Kreutzmann, H. (2000) 'Water management in mountain oases of the Karakoram', in H. Kreutzmann (ed.) *Sharing water. Irrigation and Water Management in the Hindukush-Karakoram – Himalaya*, Oxford University Press, Karachi, Pakistan.

Kuhn, T. S. (1970) *The Structure of Scientific Revolutions*, The University of Chicago Press, Chicago, IL.

Labbal, V. (2000) 'Traditional oases of Ladakh: A case study of equity in water management', in H. Kreutzmann (ed.) *Sharing water. Irrigation and Water Management in the Hindukush-Karakoram – Himalaya*, Oxford University Press, Karachi, Pakistan.

Molle, F. (2008) 'Nirvana concepts, narratives and policy models: Insights from the water sector', *Water Alternatives*, vol. 1, no. 1, pp. 131–156.

Mosse, D. (2003) *The Rule of Water: Statecraft, Ecology and Collective Action in South India*, Oxford University Press, New Delhi, India.

Mosse, D. (2005) *Cultivating Development: An Ethnography of Aid Policy and Practice*, Vistaar Publications, New Delhi, India.

Norgaard, R. B. and Baer, P. (2005) 'Collectively seeing complex systems: The nature of the problem', *Bio Science*, vol. 55, no. 11, pp. 953–960.

Schmid, A. T. (2000) 'Minority strategies to water access: The Dom in Hunza, Northern Areas of Pakistan', in H. Kreutzmann (ed.) *Sharing water. Irrigation and Water Management in the Hindukush-Karakoram – Himalaya*, Oxford University Press, Karachi, India.

Schmidt, M. (2004) 'Interdependencies and reciprocity of private and common property resources in the Central Karakorum', *Erdkunde*, vol. 58, no. 4, pp. 316–330.

Schmidt, M. (2009) 'About water towers and arid mountain valleys: Water management in the Karakoram', Tagungsband zum EGEA (European Geography Association for Students) Western Regional Congress, 23–28 March 2009, in Bingen, Mainz, Germany.

Singh, T. (1910) 'Kartse Khar riwaj-i-Abpashi. Jamabandi. Mouza Kartse Khar, Ilaka Kartse, Tehsil Kargil, Wazarat Ladakh', by Raja Thakur Singh, Mokhtamum (Administrator) of Bandobast, 1910 (1967 Bikrimi), place of publication unknown.

Singh, T. (1913) 'Assessment report of the Skardo Tahsil of the Ladakh District', by Thakar Singh, Settlement Officer, Baltistan, 1913 (1970 Bikrimi), The 'Civil and Military Gazette Press', Lahore.

Van Beek, M. (1999) 'Hill councils, development, and democracy: Assumptions and experiences from Ladakh', *Alternatives: Global, Local, Political*, vol. 24, no. 4, pp. 435–459.

Vander Velde, E. J. (1989) 'Irrigation management in Pakistan mountain environments', *Country Paper-Pakistan No. 3*, International Irrigation Management Institute, Colombo, Sri Lanka.

Vincent, L. (1994) 'Lost chances and new futures: Interventions and institutions in small-scale irrigation', *Land Use Policy*, vol. 11, no. 4, pp. 309–322.

Vincent, L. (1995) *Hill Irrigation. Water and Development in Mountain Agriculture*, Intermediate Technology Publications, London.

Wacker, C. (2007) 'Can irrigation systems disclose the history of the villages in Ladakh? The example of Tagmachig', in J. Bray and N. T. Shakspo (eds) *Recent Research on Ladakh*, J&K Academy of Art, Culture and Language, Leh, India.

6 Competing epistemologies of community-based groundwater recharge in semi-arid north Rajasthan

Progress and lessons for groundwater-dependent areas

Chad Staddon and Mark Everard

1 Introduction

Groundwater is an especially important resource in arid and semi-arid regions, where surface water is scarce and subject to high evapotranspiration losses. Groundwater supports over 85 per cent of India's rural domestic water requirements, 50 per cent of urban and industrial water needs and nearly 55 per cent of irrigation demand (Government of India 2007). In addition, 92 per cent of India's groundwater extraction is used for irrigation (Central Ground Water Board 2006). The area of groundwater-irrigated agricultural land more than doubled in the two decades to 2009 (Jha and Sinha 2009). The number of mechanized wells escalated in the last four decades of the twentieth century, from less than one million in 1960 to more than 19 million in the year 2000 (Jha and Sinha 2009). Across India, more than 22 million operational wells support the rural economy (Wani *et al.* 2009). Groundwater exploitation has also contributed significantly to poverty reduction in rural India, and to wider socio-economic development and the Indian economy in general. Small and marginal farmers comprise 20 per cent of the total agricultural area, yet control 38 per cent of the net area irrigated by wells (Jha and Sinha 2009).

This chapter focuses on groundwater management in a semi-arid region of the north-western Indian state of Rajasthan, the location of an ongoing civil society-led initiative in community-led groundwater recharge, which we argue constitutes nothing less than a new epistemology of water for the region. Conventional understandings of water as an abstract entity cohering with state and national scales of governance are being challenged in the Arvari (and elsewhere – see Lopez Gunn 2009, 2012) by an avowedly localist epistemology. This way of seeing water rematerialises water with reference to its links to local livelihoods, sustainability, community and – most importantly – a new hydropolitics. We use an empirical case study of local management of groundwater in the Arvari catchment of Rajasthan (see Figure 6.1) to argue that an Ostrom-type common pool resource (CPR) regime can not only operate effectively, but can constitute a model for adaptive management based on advanced groundwater science and democratic decision-making combined with a restoration of 'Gandhian' principles of interaction between competing users and with the natural world. This new epistemology of water explicitly rejects the 'cult of the expert' embedded in conventional state-led water management (what we will term 'Nehruvian' water management) but does not lapse into on a backward-looking romanticism (cf. Gupta 2011). The global movement away from locally-embedded water governance systems towards centrally mandated and increasingly neo-liberalised systems also complicates our attempt to examine successful *local* responses to the above challenges (cf. Budds and McGranahan 2003; Staddon 2010; Staddon, Langberg and Sarkozi 2015).

Our analysis is located at the juncture between critical political ecology (Forsyth 2002; Staddon 2009) and institutionalist perspectives on CPR management (Ostrom 1990, 1997). We suggest that the requirements outlined by Ostrom for a well-functioning CPR are present in Alwar District, including:

- clearly defined boundaries to the commons;
- consistent appropriation and provision rules;
- participatory collective-choice arrangements;
- effective monitoring by accountable parties;
- graduated sanctions;
- accessible conflict resolution mechanisms;
- minimal recognition by the state of rights to organise; and
- nested governance with local CPRs.

However, whilst we do see the Arvari programme for groundwater management as a sort of CPR, this insight is not in itself enough to explain either how the system came into being or how and why the Indian state (operating within a highly centralised Nehruvian epistemological frame) has come, albeit grudgingly, to accept it as the legitimate manager of water resources in Arvari. A political ecological perspective, focusing attention on the politics of relations between competing resource managers (e.g. the Indian state versus local Arvari citizens and the non-governmental organisation, NGO, Tarun Barat Sangh, TBS), is indispensable for showing how the present situation of uneasy accommodation has shaped both sides of the central-local rivalry. What's more, a political ecological perspective offers the prospect of, as Staddon (2009) notes, seeing the epistemological shifts attendant on the hearing of often unheard (or ignored) voices; of women, ethnic minorities, the aged, etc. – *and indeed of the 'environment' itself*. This sort of "critical political ecology" is more than mere stakeholder engagement as it is attuned to the ways in which previously marginalised voices can produce meaningful discourses about the natural world which subsequently result in material practices of management and exploitation of the natural world (cf. Forsyth 2002).

For us the TBS case study is interesting not just because it involves a successful challenge to the prevailing Nehruvian state-centric epistemology of water resource management, but because it demonstrates that success often relies on a new ontological and epistemological politics of community-environment relations. For other critical scholars of groundwater management, 'social capital' is a key aspect of local success (Lopez-Gunn 2012), and this is where TBS has been particularly successful. Specifically, TBS has sought to replace the centrally-mandated and highly bureaucratised groundwater management paradigm (which we term here 'Nehruvian' after Jawaharlal Nehru) with locally-rooted management structures more closely aligned with local structures of administration and legitimation (which we term 'Gandhian' after Mahatma Gandhi).[1]

The analysis presented in this paper is part of a long-running engagement with the Arvari region initiated by one of the authors (Everard) in the 1990s. We have made considerable use of the large, but mostly 'grey', literature on groundwater management in this part of Rajahstan (e.g. Agrawal 1996; Jayanti 2009; Rathore 2003). This literature has been cross-referenced with other literatures on groundwater management in developing areas and/or community-based approaches to water management, including especially the work of Birkenholtz (2009), Lopez-Gunn (2009) and Subramaniam (2014). Desk-based research has been augmented by direct observation and data collected by Everard during field visits to the TBS ashram (headquarters) and neighbouring villages in the Arvari catchment in 2013, 2015 and 2016. Interviewees included TBS staff (Rajendra Singh, Abhinav Agrawal[2] and Kanhaiya Lal) as well as village elders (particularly Rooparam, the headman of Hameerpur village, and a number of other local decision-makers). Ad hoc discussions took place with other local people, for which TBS staff acted as translators. The catchment visit was augmented by telephone and other technology-mediated discussions with various people with experience of ecosystem management in the district (e.g. Dharmendra Kandal, director of the NGO Tigerwatch) as well as others with international expertise in wetland and catchment management. Additionally, through a separate project we are attempting to appropriately classify and document the growing abundance of biodiversity (flora and fauna) created by successful groundwater recharge.

Section two introduces the challenges of groundwater management in the Arvari basin and in particular the inability of the Nehruvian paradigm to address them properly. In section three, we then discuss the history and achievements of a remarkable NGO, Tarun Bharat Sangh, which for more than 30 years, has promoted a Gandhian water paradigm as an alternative more likely to lead to successful community and environmental outcomes. Section four outlines the key empirical achievements (and challenges) of the TBS approach organised using the 'STEEP' (social, technological, economic, environmental, political) model, developed initially to assess a range of global environmental change issues (Morrison and Wilson 1996). In the case explored here, the STEEP framework provides a platform for exploring the higher order analytical issues raised by applications of CPR and political ecological approaches. We conclude the paper with a few key lessons that are transferrable to water management in other arid and semi-arid, groundwater-dependent regions, with suggestions for follow-on research.

2 Groundwater dependence in Rajasthan's Alwar district

The Arvari (or Arwari) River Basin, a predominantly rural catchment with a rapidly growing population lying within the Alwar District of north-eastern Rajasthan, is the geographical focus of this study (see Figure 6.1). The oldest of the historical Rajasthani kingdoms, Alwar has a land area of 7,832 km², representing about 2.5 per cent of the total area of the contemporary state of Rajasthan (Ramusack 2004). The 2011 Indian census put the total population of Alwar District at 3,671,999, a 23 per cent increase since the previous census in 2001 (Census 2011.co.in). Ethnically the region is quite diverse though in many Alwar District communities Meena or Gujjar castes/tribes predominant (Tiwari 2008). Unemployment is high in the district and many livelihoods depend on a mixed economy of subsistence and market garden agriculture with some animal husbandry.

Located southwest of Sariska National Park, the Arvari River catchment is semi-arid, occupying an area of 476 km² comprising 46 micro-watersheds through which the now-perennial 90 km river main stem flows on its way

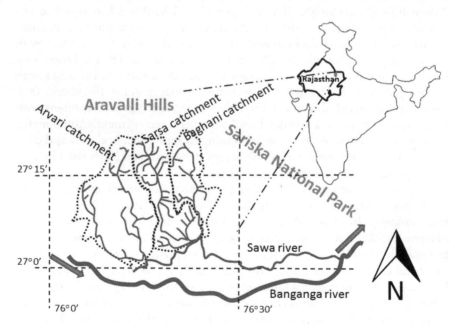

Figure 6.1 Location of Arvari and Adjacent catchments in Alwar District, Rajasthan.

to its confluence with the Sarsa and Baghani Rivers, before joining with the much larger Banganga River (Glendenning and Vervoort 2010; Rathore 2003). Long-term average annual rainfall for the district is 657.3 mm, though this rainfall is unevenly distributed both spatially and temporally, strongly affecting agricultural potential. Available renewable surface water per person is under 1400 m³/year, making Alwar District 'water stressed' according to the common metrics used by UNEP and other global water management bodies (Staddon 2010). The primary soil type is aridisol, meaning that background organic content is low as is water retention. Access to agricultural land is, for many, problematic due to caste/tribe power asymmetries and the state-sponsored land dispossession that created Sariska National Park in 1978.

Irrigated agriculture in Alwar district is a significant contributor to local and district economies, with irrigation mainly supplied from wells and tube wells supporting about 83 per cent of the cultivated area of 507,171 hectares (Kumar 2011). An estimated 35,470 electric motors and 66,502 diesel pump sets are currently used for irrigation purposes. A considerable proportion of the irrigated area is double-cropped with, in declining areal extent, of Bajra (pearl millet), jowar (sorghum), guar (cluster bean), maize, arhar (pigeon pea), cotton, ground nut, rice and pulses. Owing to the predominantly vegetarian diet of Alwar residents, rough grazing also occurs but is not intensive. Employment in agriculture is at 67 per cent, with 82 per cent of those workers tending small land holdings. The National Bank for Agriculture and Rural Development (NABARD) and other assistance agencies have run programmes in Alwar to help farmers diversify production, though the lack of reliable irrigation water has until recently been a serious constraint.

Rathore (2003) reports that the valleys of the Aravali Hills were well-vegetated up to the mid-1930s. At this point, timber rights were sold to private interests and within ten years ecological decline and associated increased incidences of both seasonal flooding and drought ensued. Sale of forest rights and the sub-lease of land for mining were reportedly instigated by a prince anticipating that a free India would diminish his prerogatives (Tarun Bharat Sangh 2013; Ramusack 2004). Around the same time, administrative reforms passed water management responsibilities from local to state and national government control. Ceding of water management responsibilities from local control led in turn to a shift in perception about the value of water, disengagement of local people with management of their own assets and responsibility for supply-side management, and hence widespread abandonment and consequent degradation of community water management structures (Rathore 2003). The emphasis shifted instead towards greater mechanical efficiency in resource exploitation, including a decline of bullock-operated wells with greater investment in energisation (diesel and electric pump sets) for extraction from ever-deeper wells. Management was also centralised, with local communities and land managers being relegated to the status of 'clients' of increasingly powerful water management bureaucracies based in faraway cities Boyce *et al.* 2007).

This situation led to substantial depletion of groundwater, prompting a range of adverse ecological, social and economic consequences. Government data reported by Rathore (2003) records that groundwater extraction in Alwar district was 66 per cent of the available resource in 1984, 110 per cent in 1988, 108 per cent in 1999, 119 per cent in 1995, 100 per cent in 1998 and 118 per cent in 2001. Continued overexploitation was clearly unsustainable, contributing to a cycle of linked ecological and social eco-degeneration. With few options, people began to migrate away from the villages of Alwar district to cities, mirroring depopulation trends across considerable areas of rural India subject to a similar cycle of centrally-mandated groundwater overuse. Rathore (2003) stated that 'Migration to urban and peri-urban areas is symptomatic of the deepening crisis in the farm and rural sectors', substantially driven by declining access to water.

Groundwater depletion has serious potential social and economic implications, constituting the key nexus in the negative spiral of interlinked ecological, social and economic degradation leading to community decline. Downing (2002), Seckler *et al.* (1999), Postel (1999), Vaidyanathan (1996) and others observe that the still largely unregulated pump irrigation revolution observed across much of South Asia, particularly since the 1970s, is leading to serious socio-ecological threats. Seckler *et al.* (1999) in particular warned that a quarter of India's food harvest was at risk if groundwater management was not improved, with Postel (1999) suggesting that 10 per cent of the world's food production depends on overdraft of groundwater with 50 per cent of this area located in Western India alone. Singh and Singh (2002) estimate that declining groundwater levels could reduce India's harvest by 25 per cent or more. Failing water resources represent a major threat to food security in India (Kumar 2003). There are also significant distributional impacts associated with groundwater over-extraction, as the relatively high costs of deepening the generally shallower wells owned by small and poorer farmers often excludes them from access to water long before wealthy farmers and other affluent users (Moench 1994). After 1947, successive waves of state-brokered 'accumulation by dispossession' (Harvey 2003) with respect to water and land resources (Subramaniam 2014) helped create a 'perfect storm' of environmental, economic and political decline.

3 The Tarun Baharat Sangh programme of water management in Alwar Region

Tarun Bharat Sangh (TBS, translating as 'Young India Association') is an activist organisation inspired by Gandhian philosophical principles, particularly those related to non-violent resistance (in this case against the neoliberal 'Nehruvian' Indian state) and self-reliance. Originally established in 1975 in Jaipur, TBS received a considerable boost in 1985 when community organiser Rajendra Singh (sometimes now referred to in the media as the 'waterman' or 'rivermaker') and several of his friends moved to the area (Tiwari 2008).

Against the backdrop of rural depopulation and economic and ecological decline discussed in the previous section of this paper, TBS founded its work on helping local communities rediscover old ways of managing land and water resources. TBS has been active in the simultaneous restoration of the supportive natural resources of soil, water and ecology and of human livelihoods, understood from their Gandhian perspective as interdependent and spiritually alive. The key educational emphasis of TBS changed in 1985 when Singh was told by a village elder that the primary issue in the region was not lack of education but lack of water (Jayanti 2009; Singh, personal communication). Many rivers in the region, including the Arvari, had by then stopped flowing or flowed only intermittently, and agricultural land productivity was at an historic low and still declining. Reversing this trend became a priority for the fledgling organisation, located in an ashram near the town of Thanaghazi in the Alwar district of Rajasthan.

The principal work of TBS has therefore revolved around locally and the practically-focussed measures. The approach of TBS is defined as 'Community self-reliance through natural resource conservation', embedding the Gandhian ethos of *Jal Swara* including participation, equity and decentralisation at the heart of water management (Jayanti 2009). In particular, this revolves around restoration or creation of small, localised water harvesting schemes (WHSs) that can recharge groundwater, restoring greenery and soil moisture to village hinterlands and rejuvenating local rivers.

Local people believed that rehabilitation of derelict *johads* (earthen dams designed to intercept rainwater long enough to permit infiltration into soil) would help them overcome drought. TBS, therefore, began the work of reconstructing *johads* drawing heavily on voluntary labour. Singh and his colleagues also took advice from a lower-caste older lady to build *talabs* (a kind of small pond created by shallow bunds) near the village of Mangu Meena to stop and store water during the concentrated monsoon period, allowing it time to subsequently percolate into the ground and recharge groundwater. The first such water capture structure was built by the villagers of Bhaonta-Koylala. Construction was considered risky, and results uncertain, but the dam functioned as anticipated, restoring soil moisture and ecology for improved food production, rejuvenating local grazing and other vegetation, and re-establishing some vitality to the local river (Singh 2009). There was also interest in restoring *anicuts* (weir structures that create deep, more permanent, water pools). Table 6.1 presents a brief summary of the different WHSs championed by TBS in Alwar District.

This initial innovation led to a widespread and rapid proliferation of interest from the many parched, depopulated villages in the vicinity. TBS activities correspondingly expanded to other villages as they attracted funds, mainly from international donors, for constructing WHSs. Construction of hundreds of *johads*, *talabs*, *anicuts*, and so on, was to follow, with TBS usually contributing 50 per cent of costs largely from international donor sources matching investment by beneficiaries in the villages of the Arvari watershed. Significant

Table 6.1 Characteristics of WHSs and associated water bodies in the Arvari, Sarsa and Baghani catchments visited by one of the authors

Arvari catchment	
JS	Jabar Sagar (27.207373°N,76.202331°E, 386 metres altitude) is an anicut on the Arvari river serving farmland around Harmeerpur. It was one of the earliest installed with support from TBS, in the latter half of 1980s. The anicut is the third between the source of the Arvari and Harmeerpur village, but is one of around 100 anicuts and johads within the catchment and 49 in the vicinity of the village. Wheat and gram crops, including a variety of ruderal weeds, are grown right up to the water's edge, which is at its maximum after monsoon rains in July-September after which water level retreats seasonally.
KA	Anicut near Kalid (Kaler) village (27.155427°N, 76.224163°E, 386 metres altitude), downstream of Harmeerpur. The large concrete anicut holds water in the Arvari river perennially. It is heavily used for grazing with consequently barren banks and bare drawdown zones. Fish of different species were clearly visible, and in 1996 were the subject of a conflict over fishing rights, discussed elsewhere in this paper.
Sarsa catchment	
BE	Beechkharaga (27.249158°N,76.30568°E, 403 metres altitude), a newly-completed johad (constructed 10 December 2014 to 27 February 2015) where mountain slope drops to valley edge, serving to retain run-off and recharge aquifer and adjacent open well. The johad is located at the head of a monsoon nala (drainage channel) near Jaitpur village, upper Sarsa catchment, serving land owned by 9 families. The cost of the scheme was 1.8 lakh Rupees, with 67% of costs routed by TBS from the 'Wells for India' fund and 33% provided by the village.
GP1	Gopalpura (27.268606°N, 76.30753°E, 407 metres altitude), a johad built in 1985 (the first constructed under the guidance of TBS) in the upper Sarsa catchment. The johad holds water all year, recharging groundwater and it is also extensively used for stock watering. There are now 17 water-harvesting structures in the vicinity of Gopalpura, serving 80 families (including the three Chabutra Wala anicuts in the adjacent shallow valley).
CW	Chabutra Wala (27.270369°N, 76.310328°E, 408 metres altitude) is a series of three anicuts with water level control sluices built across a shallow valley in the upper Sarsa catchment in 1985 by TBS. The anicut surveyed was the highest upstream of three anicuts. The three anicuts retain surface water until the land is ready for sowing, when water is released downstream by removing wooden stoppers from holes in the water control sluice. Farmed land upstream of the anicuts supports crops of wheat, gram, mustard, bindi, brinjal, potato and carrot. Chabutra Wala is one of 17 WHS in the vicinity of Gopalpura, serving 80 families.
GP2	Johad by road north of Gopalpura (27.276656°N, 76.302532°E, 411 metres altitude). The johad was at the time holding water, but appears to dry down in full summer once all water has seeped into aquifers or evaporated. There was evidence of extensive watering of and trampling by animals.
GK	Golakabass (27.10133°N, 76.321519°E, 339 metres altitude), a relatively new anicut across the Sarsa River, north-west of a road crossing downstream from which is a broken former dam. The Golakabass check dam spans the river approximately 10 km upstream from its confluence with the Sawa River.

Baghani catchment	
TI	Tilda (27.188411°N, 76.414071°E, 329 metres altitude) is a check-dam constructed across the Tilda River, upstream of its confluence with the Baghani River, forming a deep, clear-water pool ringed by patera (the local name for *Typha angustata*). A temple is located at the head of the impoundment with ghats (steps to the water's edge) around which fish shoal and swim with bathing children. At the downstream end, women were washing clothes on the concrete check dam, which is also used as a crossing place.
TE	Tehela (27.249588°N, 76.441471°E, 344 metres altitude), a check dam on the Jalumbragarh river (a tributary of the Baghani system). The Tehela check dam was installed around 2000, near the town of Tehela (population 5-6,000). It constitutes a shallow impoundment supporting extensive stock watering (water buffalo, sheep, goats) with wallowing buffalo present, activities which eliminate much marginal vegetation (except stands of invasive, tall and woody *Ipomea carnea* growing as an emergent close to the dam wall)
MAu	Mandalwass, upper impoundment (27.277571°N, 76.33273°E, 496 metres altitude), a large dam on the headwaters of the Baghani River built in 1993 immediately above smaller dam on the top of a high mountain ridge. The upper Mandalwass dam is deep (estimated at 18-20 feet in low summer weather) with the head heavily grazed, resulting in little vegetation and dense, greenish (assumed algal) water. This water condition is exacerbated by high fish stocks (reported but unknown species) in the impoundment, for which the village people allot contracts to commercial fishermen providing an annual income used to refurbish the upper and lower Mandalwass dams. The rocky margin of the impoundment was being used during the survey period for washing clothes.
MAl	Mandalwass, lower impoundment (27.279707°N, 76.333966°E, 496 metres altitude), the lower of two impoundments repaired at the time the larger, upper dam was built in 1993, the impounded water much shallower, clearer and well-established and densely vegetated, located on the top of a high mountain ridge. There are some houses adjacent to the slope to the south of the impoundment, with extensive grazing by buffalo in the riparian zone.

catchment-scale outcomes ensued, leading Singh to remark that 'We never realised that we were recharging a river. Our effort was just to catch and allow water to percolate underground' (Down to Earth 1999a). Demand for building WHSs remains high in villages; but TBS is only able to build or renovate around 300 structures annually due to funding limitations.

Construction and management of WHSs is central to TBS activities, in many cases resurrecting traditional technologies and knowledge and the social infrastructure necessary to operate them sustainably. The importance of strengthening social infrastructure cannot be overstated because without it the physical structures are vulnerable to neglect and abandonment (a pattern often following from the all-too common passive donor client relationship). Therefore, TBS only responds to demand from village groups, and works to improve local management and decision-making processes along Gandhian principles. Consequently, the process of building and maintaining WHSs has been coincident with the resurrection of traditional village institutions in many villages (Kumar and Kandpal 2003). Prominent among these are *Gram Sabha*, traditional village decision-making bodies, which traditionally included discussion and decision-making about water management (Jayanti 2009). Whilst some *Gram Sabha* became dormant after construction of WHSs, many remained active where support from TBS continued, building social capital with some village institutions then progressing to tackle other related issues including protecting forests, building schools and other developmental works (Kumar and Kandpal 2003). *Gram Sabhas* decide on village land and water management issues, serve as liaisons with TBS and bring in local indigenous knowledge and skills into the water structures (whether *johads*, *anicuts* or other designs). In considering ongoing wise stewardship of water resources, *Gram Sabha* are frequently also active in zoning of land-use and regulating uses of pastureland to avoid ecological degradation with broader social and economic consequences (Singh, personal communication; Tiwari 2008).

Responding only to demand from villages (rather than individuals) has the potential weakness of fragmenting action across the landscape (Kumar and Kandpal 2003). To achieve a more integrated approach, TBS promoted formation of an Arvari Water Parliament (*Arvari Sansad*), which since its inception in December 1998 meets twice a year to determine water sharing and management issues, including dispute resolution, across the catchment (Rathore 2003). The parliament has identified measures such as reforestation of the formerly forested but subsequently degraded Aravali Hill range to improve hydrological health in the larger catchment (Jayanti 2009). At local scale, village development committees (VDCs) have also been established throughout the catchment, instigated by village elders to promote local collaboration in water management.

By 2010, TBS was working with more than 700 villages in Rajasthan and had completed more than 200 WHSs and related works (Subramaniam 2014), although the numbers of WHSs reported in different literature ranges from 366 (Glendenning and Vervoort 2010), to 375 Jayanti (2009), to 650 (Down to Earth 1999b). Narain and Agarwal (in Boyce *et al.* 2007), citing a United Nations report, say that TBS helped to build almost 2,500 WHSs in almost 500 villages. The figure of 3,200 reported by Kumar and Kandpal (2003) probably includes structures outside of Arvari District and may also include repeat visits.

Though remarkable, TBS experience is not entirely unique. Wani *et al.* (2009) report significant groundwater rises in 'treated areas' where community-based participatory methods have been developed at benchmark sites in several other Indian states/provinces, as well as in Thailand, Vietnam and China. These initiatives, which bring together institutions from scientific, non-government, government and farming sectors, have been found to improve productivity by up to 250 per cent, as well as to restore groundwater levels and to reverse the degradation of natural resources (Wani and Ramakrishna 2005; Wani *et al.* 2006). The chapters in Boyce *et al.* (2007) explore similar processes around the world.

As we shall see, the experiences of TBS-brokered social-ecological infrastructure in Arvari has not been uniformly positive, smooth or entirely free of challenge, particularly from the state and central governments, whose perspectives on water management are very different and whose departments wield far more institutional and legal power than TBS and local *Gram Sabha* or the *Arvari Sansad*. In times of crisis or challenge from the central state, however, the TBS community development methodology often gives needed robustness and resilience to local communities.

4 Review of outcomes in the Arvari catchment

In this section, we present the outcomes of empirical work in Rajasthan using the STEEP framework to thematically structure our analysis. We will consider the theoretical implications of these outcomes in the concluding section of the chapter.

4.1 Social outcomes

Though effects are uneven locality to locality, ecological and hydrological recovery through TBS-brokered WHSs has reversed the trend of abandonment of villages in the Arvari catchment. Children and young people are now more likely to remain in their villages, and schools are thriving as more people anticipate viable livelihoods. Also, *johads* and other WHSs build on existing cultural traditions of the area, and are a 'seed crystal' for revitalisation of the traditional governance arrangements such as *Gram Sabha* that are central to their continued operation. These are fully social-ecological systems inasmuch as each requires the other for continued smooth functioning.

Significant amongst the successes of TBS and related community initiatives around integrated water management, is the empowerment of women who, freed from the drudgery of traditional roles foraging for water, fodder and fuel in a water-stressed environment, have more time to participate and learn following ecosystem restoration (Jayanti 2009). Before TBS started promoting community-based WHSs in 1985 women would typically spend six to seven hours daily searching for and carrying water but, as a result of rising water tables and access to water through installation of hand pumps and wells close to housing, this now takes much less time (Singh, personal communication; Tiwari 2008). TBS has been further active in empowering women through enabling democratic engagement and improving educational programmes including Ayurvedic (traditional herbal) medicine. Declining drudgery across all sectors of society, particularly women, contributes to greater potential for engagement in village-scale decision-making and other productive activities.

However, social challenges remain. Major disconnections and social differentials remain in terms of resource ownership and capacity to access water. This is not helped by the established framework of top-down regulation and economic incentives emanating from the central Indian state. Cochran and Ray (2009), Terry *et al.* (2015) and others consider 'equity in process' as well as 'equity in outcome' as central to community-based development efforts, in particular rainwater harvesting programmes. The symbolic capital of contributing to project design and development is important to 'community understandings of equity as the distribution of benefits from the project', serving to retain community heterogeneity through a very open and broad-minded approach to costs and benefits. Rathore (2003) found that 58 per cent of surveyed households in the Arvari catchment preferred community ownership of water resources in forest areas, though 42 per cent feared capture of community governance and land assets by the richer members of their own communities. According to Subramaniam (2014) and Tiwari (2008), this is already happening in some villages, because larger landholders can better afford to drill wells, allowing them to exploit restored groundwater flows.

4.2 Technological outcomes

Kumar and Kandpal (2003) conducted a review of TBS projects between 1994/5 and 2003, cumulatively representing a total investment of 16.2 million Indian rupees (approximately US$400,000 at 2003 exchange rates). They found that '[t]he scale of work adopted by TBS is staggering…' and had 'shown that rejuvenation of traditional water harvesting structures on a wide scale is indeed possible'. Kumar and Kandpal (2003) further observed that work promoted by TBS had a positive impact on water availability in the region, both in agriculture-dominant and animal husbandry-dominant villages, resulting in significant economic gains, greater protection against drought and a marked reduction in distress-related migration. Soil erosion was reduced significantly by measures such as voluntary field bunding, and farmers were also able to diversify into cash crops as well as livestock due to better assured water availability. Whereas the Arvari River had been dry outside of the monsoon season in the late 1980s, by 2009 it ran perennially as a result of successful groundwater rehabilitation. In the wider Alwar District, more than 10,000 WHSs were constructed between 1994 and 2008, restoring perennial flows to five formerly seasonal rivers – the Bhagani-Teldehe, Arvari, Jahajwali, Sarsa and Ruparel – benefitting 250 villages and their local natural environments (Jayanti 2009). In some cases, as Tiwari (2008) points out, *Gram Sabhas* imposed restrictions on the range of crops or husbandry activities possible, on the grounds of soil and water management.

Glendenning and Vervoort (2010) evaluated over 366 WHSs built in the Arvari catchment since 1985. They also estimated potential recharge from seven WHSs, across three different structure types and in six landscape positions based on monitoring of water level fluctuations in 29 dug wells. The average daily potential recharge from WHSs varied between 12 and 52 mm/day, while estimated actual recharge reaching the groundwater ranged from 3 to 7 mm/day. The large difference between recharge estimates could be explained through soil storage, local groundwater mounding beneath structures, and lateral transmissivity in the aquifer. Overall, approximately 7 per cent of rainfall recharged groundwater via WHSs in the catchment during both the comparatively wet and dry years of field analyses, with key differences between WHSs due to engineering design and location. These results indicate that recharge from WHSs positively affects the local groundwater table, but also has the potential to move laterally and impact surrounding areas.

A further study testing the approach and outcomes achieved by TBS drew upon 'hard' physical science and engineering data as well as the narratives of local people from 500 families in 36 affected villages (Agrawal 1996; Tiwari 2008). It was observed that 90 per cent of the effort and financial resources routed through TBS were directed at water harvesting and conservation, including linked soil and forest conservation. Agrawal (1996) noted that no hydrological calculations were completed to assess volumes of storm run-off, flood flow and the amount of water needed by local people, all of which would have been required as inputs to a more structured engineering approach to the design of WHSs. Selection and specification of design of WHSs was instead based on instinct, deliberation and consensus within village *Gram Sabhas*. Using rainfall data, addressing both amount and timing, and assuming a run-off coefficient, Agrawal (1996) determined an ideal storage capacity of 1000 m^3/hectare of catchment area to capture flow and promote infiltration, but also noted that 'the optimal *johad* storage for these areas would be 1000–1500 m^3/hectare which would raise annual average groundwater table by 20ft'. Comparing this 'ideal' with 166 community-engineered *johads* in the Arvari catchment, Agrawal (1996) considered that 35 were too small, 49 were small, 61 were optimal (800–1200 m^3/hectare), 16 were too large and only 5 were excessive, providing strong evidence that traditional knowledge routed through traditional consensual processes, though not articulated through engineering discourse, produced robust and appropriate designs.

4.3 Environmental outcomes

Rising water tables have had profound implications for Arvari catchment ecosystems. Amongst the most visual successes during field visits was restoration of perennial flows to the Arvari River which, as reported by Singh (2009) and Jayanti (2009), is one of several rivers now flowing perennially in Alwar district that had formerly run only during monsoon rains. During one late summer visit by one of the authors, the Arvari contained significant areas of open water supporting livestock watering and wetland biota. Below the village of Hameerpur (also Hamirpur), wildlife noted in Table 6.2 was observed (without the benefit of collecting equipment or local keys hence tentative identifications) in large bodies of water held back by check dams constructed for the dual purpose of promoting groundwater infiltration and providing watering of livestock. The extensive beds of hydrophytes and the presence of fish, frogs and Odonata and other obligate aquatic organisms demonstrate the ecological health of the water body. Singh (personal communication) and Rooparam

Table 6.2 Wetland species observed during site visit near dammed Arvari river downstream of Harmeerpur, with generic description and tentative identification

Hydrophytes	
	Potamogeton crispus? (curly-leaved submerged)
	Potamogeton spp. (floating-leaved)
	Potamogeton pectinatus? (filiform-leaved submerged)
	Cyperus spp. (emergent marginal)
	Extensive beds of submerged, rooted *Elodea*-like plants
	Oxalis spp. (floating-leaved and emergent forms)
Wetland birds	
	Red-wattled lapwing (*Vanellus indicus*)
	Geese (flock of five unidentified small black and white geese)
	Moorhens (*Gallinula chloropus*)
	Common sandpiper (*Tringa hypoleucos*)
	Little egret (*Egretta garzetta*)
	Little cormorant
	Black-winged stilt (*Himantopus himantopus*)
Odonata	
	Damselflies (unidentified, small with black body and blue tip to tail)
	Darter (unidentified, small with ruddy colour)
Other aquatic taxa	
	Frog (unidentified)
	Cyprinid fish (abundant but not identified)

Source: Everard, 2014.

(personal communication) report that the fish and other aquatic organisms were not stocked, but naturally colonised the pools. Lack of local knowledge about their taxonomy is in part due to local people being overwhelmingly vegetarian.

At a wider landscape scale, Rathore (2003) reports an increasing area under forest in Thanagazi Tahsil from 8.4 per cent in 1989/90 to 14.37 per cent by 1998/99, with the area under agriculture rising from 42 per cent to 54.9 per cent. The convergence of interests between TBS and the parallel process of emerging collective management of forest resources has been noted by Subramaniam (2014).

Environmental outcomes beyond the rewetted catchment area comprise a balance of positive and negative observations. Jayanti (2009) observed enhancement in grazing in the buffer zone adjacent to Sariska National Park (a tiger reserve), reducing grazing pressure in the core zone of the Park. Kanhaiya (personal communication) reports that leopard regularly come down from the Park's mountainous perimeter to drink at *johads*, with other wildlife also exploiting the open water. However, Kandal (personal communication) argues that reinstating moisture in a naturally dry location may work against tiger conservation through stock encroachment into the Reserve.

5 Economic outcomes

Significant economic benefits are observed in terms of food sufficiency and overall wealth as villagers are able to engage in profitable farming, with economic uplift following improved soil moisture and catchment hydrology. The area under single and double cropping increased from 11 to 70 per cent and from 3 to 50 per cent respectively, improving significantly the livelihoods of farmers. Forest was also reported as increasing from 7 to 40 per cent through agro-forestry and social forestry, providing sufficient fuel wood and sequestering atmospheric carbon.

As well as brokering relationships between communities, TBS has also been successful in drawing upon wider funding sources although TBS insists upon a minimum of 30 per cent funding from local communities as an assurance of communal ownership and continued maintenance. 'Sweat equity', in the form of volunteer labour, or *shramdan*, a form of collective labour for local good closely linked to Gandhian ideals of self-sufficiency and mutual aid, is generally at the centre of water structure construction and maintenance.

Agrawal (1996) also compared the costs of water conservation work with their benefits. Community-based collaboration in WHSs design and construction was cheap, assessed in over half of the villages in the Arvari catchment as 0.5–2 rupees (0.01–0.04 US dollars) per m^3 storage area. Agrawal (1996) found a strong correlation between per capita increase in the value of the gross village product and investment by villages in water conservation work, with a ratio around 4:1, and also between recharge capacity and groundwater rise. Agrawal (1996) concluded that the *johads* stood the test of time and 'are, by and large, engineering-wise sound and appropriate', concluding that '[t]here can be no better rural investment than on *Johads*'.

At village scale, distribution of benefits and shares of costs of WHS construction and management are key issues. Whereas common lands are grazed, croplands are privately-owned and Tiwari (2008) Subramaniam (2014) and others found some indications of economic marginalisation based on access to land.

5.1 Political (governance) outcomes

Kumar and Kandpal (2003) observed that successes achieved by TBS had impacts on state and national level water policies including: formation of a national water network addressing issues of community ownership of water; influencing a refocussing of state drought relief works on water harvesting structures; contributing to the Sariska Tiger Reserve's soil conservation programme; spreading learning to other states; and educating officers within government. Kumar and Kandpal (2003) also observed that TBS was active in policy advocacy for water management at Rajasthan state level, attempting to steer state water policy in a more equitable direction particularly through *Jal Biradari* networks at nested scales linking regional, state and national levels of government.

However, the political news is not all sunny and harmonious. The hydrological recovery of the Arvari catchment has triggered conflict between local communities and the state. Singh (2009 and personal communication) reports that as fisheries fall under central government control, once fish had colonised the newly restored perennial open waters of the Arvari River, the government of Rajasthan issued a license permitting fishing rights to contractors from outside the region. Subramaniam (2014) reports that, in 1996, the government issued fishing rights to a private contractor for Rs18,700 (approximately US$526 at the time), although TBS believed the market value to be over Rs100,000. Hameerpur residents resisted this take-over of their resource, in part recognising that central control of fisheries could be followed by central control of surface waters, also technically subject to central government management. Thus, they regarded the fisheries question as the thin edge of a wedge that could reignite the cycle of disempowerment, ecosystem degradation and socio-economic decline. Village residents protested and kept vigil so that the contractor would not have access to the river. This led to a conflict between the villagers, the government department of fisheries, and the fishing contractor. The contractor then reportedly put the pesticide Aldrin into the river to kill the fish, creating a dangerous situation as the presence of fish may be partly responsible for the lack of incidences of malaria despite creation of new, substantial surface water resources (Singh 2009). Pressure on the government led to the annulment of the contract, and no licence for fishing has since been granted in the Arvari.

Moreover, in strictly legal terms most of the WHSs constructed by the TBS are illegal because under the Rajasthan Drainage Act of 1956: 'Water resources standing collected either on private or public land (including groundwater) belong to the Government of Rajasthan.' However, recognition of successes at both village and higher political levels up to the Indian presidency has ensured that notices issued by the irrigation department have not resulted in follow-up activities. Singh (personal communication) has also campaigned with success using India's public-interest litigation (PIL) process against water-intensive industries, such as distilleries and mines, moving into water-sparse parts of Rajasthan. Narain and Agarwal (2007, p.100) describe current relations between the official state water management bodies and the unofficial (and technically illegal) local *Gram Sabhas* as underpinned by an 'unwritten understanding'.

The processes of commodification and privatization by the state as a form of economic neoliberalism have sparked a wide array of popular counter-movements often targeting corporate and state power, seeking to return power to local levels (Haugerud 2010), in which citizens resist accumulation by dispossession by framing their struggles as efforts to 'reclaim the commons' from perceived constraints on livelihoods by the state or private agencies (Subramaniam 2014). Indigenous people across India have been found commonly to oppose privatization and to support collective ownership of natural resources such as forests and water (Fenelon 2012). NGOs such as TBS have an important and influential role to play in mobilizing citizens and contributing to accumulative practices seeking to achieve local neoliberalism (Subramaniam 2014).

6 A new epistemology of water in the Arvari catchment?

This review of community-based water management in the Arvari catchment highlights close links between productive ecosystems, social and economic outcomes, and local and participatory governance in ensuing technology choice and management. This strong interdependence between livelihood choices, locally integrated governance structures and the capacity of ecosystems to provide the necessary suite of supportive services is, on our analysis, more in line with Gandhian than Nehruvian principles.

Of course TBS is not the only organisation to have recognised the value of traditional, local-scale water management systems in semi-arid and arid regions of India (see for example, Boyce *et al.* 2007). However, empirical evidence reviewed in this paper suggests that TBS has been extraordinarily successful in generating demonstrable long-term benefit to ecosystems and community structure, and the social and economic benefits that flow from them, sometimes in the face of central government direction. Successes result from the connected reconstruction of social and technological infrastructure appropriate to geography and socio-economic needs, creating a self-reinforcing cycle of ecosystem restoration and enhancement of socio-economic wellbeing. This approach identified the value of traditional technologies and devolved forms of consensual governance, including land-use zoning, use of pastureland to avoid ecological degradation, and modification of crop production and other uses to better integrate with environmental 'carrying capacity' and the livelihoods of others in the community.

TBS has achieved this by addressing governance and the connection of local solutions at nested scales, from village level (*Gram Sabha*) to catchment scale (*Arvari Sansad*), and also through influencing both state and national policy by positive example and more directly via *Jal Biradari* and – sometimes – direct legal challenge. This nesting is essential to ensure that the functioning of whole river ecosystems remains central, whilst empowering local communities and making use of their context-specific knowledge about environmental conditions and needs. Since the 1990s, the decline of traditional water harvesting systems based on indigenous knowledge and technology has been prominent in the development discourse (Agarwal and Narain 1997), with technological changes, such as the introduction of electric well pumping leading to individualism and the breakdown of community stewardship (Appadurai 1990).

TBS adopted a learning approach based on practical outcomes, developing from this a set of guiding principles. In Table 6.3 we show how, in effect, TBS has brokered development of an Ostrom-type CPR management regime. Table 6.3 highlights a high degree of congruence between CPR principles and observed factors behind the success of community-based groundwater recharge in the Arvari basin. There is close concordance between practices

Table 6.3 Congruence between Ostrom's CPR principles and observations in the Arvari catchment

Key CPR principles (summarised) from Ostrom (1990, 1997)	Experience in the Arvari catchment
Clearly defined boundaries to the common	The focus is on linked surface and groundwater and associated terrestrial systems bounded at village scale, but also at catchment scale
Consistent appropriation and provision rules	*Gram Sabha* set rules allocating shares of water and also designs of WHSs for water capture through a process of village consensus
Participatory/collective choice arrangements	*Gram Sabha* and Water Councils make decisions on the basis of participation and consensus, guided by elders
Effective monitoring by accountable parties	*Gram Sabha* and Water Councils also monitor practices and outcomes through consensus
Graduated sanctions	There are no formal sanctions other than the potential opprobrium of the village community; evidence in the Arvari is that this is effective Shared investment in WHSs also leads to potential exclusion from benefits, including failing to build structures near the land holding of defaulters However, the absence of formal sanctions may mean that this CPR principle is less strongly observed than other principles
Accessible conflict resolution mechanisms	The *Gram Sabha* and Water Councils serve as a forum for discussion and resolution of disagreements
Minimal recognition by the state of rights to organise	The social structures have not received the recognition of the state, and the physical structures are also technically illegal given centralisation of legislation
Nested governance with local CPRs as their base	*Gram Sabha* and Water Councils operate respectively at nested village and catchment scales, with TBS also actively influencing State and National scales through mechanisms such as Jal Biradari

found to be successful in the Arvari catchment and principles advanced by Ostrom, the absence of strong graduated sanctions perhaps representing the weakest area though conflicts are acknowledged and resolved by consensus in village and catchment-scale parliaments. Lopez-Gunn (2012) also found strong connections between successful community-based groundwater management initiatives and Ostrom's CPR principles, emphasising that positive social capital underpins the key factors identified by Ostrom (1990) in self-governance systems and the factors that bond and bridge social capital in two Spanish case studies (see also, Ostrom and Hess 2007).

Lessons learned from the Arvari catchment may be adapted to other semi-arid and arid contexts around the world. Without diminishing the many accomplishments of TBS on the ground in Arvari, we want to emphasise that nothing would have been possible without an epistemological revolution. The Nehruvian water epistemology saw water in terms of

abstract national interests, and therefore quite reasonably vested control over water and water infrastructure at the state and national levels. In the terms of this hydro-epistemology, spontaneous local involvement in water engineering and management is both perverse and backward-looking. The Gandhian water epistemology espoused by TBS and other like-minded civic organisations sees progress in terms of local rather than national priorities and interests.

Acknowledgements

The authors acknowledge the support of the Lloyd's Register Foundation, a charitable foundation helping to protect life and property by supporting engineering-related education, public engagement and the application of research.

Notes

1   In our reading, Nehru sought to use the power of a strongly centralised and hierarchically organised Indian state to rapidly achieve development aims, albeit at the cost of bureaucratisation and effacement of traditional administrative systems. Conversely, Gandhi sought to inject already existing local management systems with a stronger sense of equality (especially between castes and between men and women) and an ethic of self-help.
2   Not related to the author G. D. Agrawal 2009 cited elsewhere in this chapter.

References

Agrawal, G. D. (1996) *An Engineer's Evaluation of Water Conservation Efforts of Tarun Bharat Sangh in 36 Villages of Alwar District*, Tarun Bharat Sangh, Alwar, India.

Agarwal, A. and Narain, S. (eds) (1997) *Dying Wisdom: Rise, Fall and Potential of India's Traditional Water Harvesting Systems*, Centre for Science and Environment, New Delhi, India.

Appadurai, A. (1990) 'Technology and the reproduction of values in rural western India', in F. Apffell-Marglin and S. A. Marglin (eds) *Dominating Knowledge: Development, Culture and Resistance*, Oxford University Press, Oxford, pp. 185–216.

Birkenholtz, T. (2009) 'Groundwater Governmentality: Hegemony and technologies of resistance in Rajasthan's (India) groundwater governance', *The Geographical Journal*, vol. 175, no. 3, pp. 208–220.

Boyce, J. K., Narain, S. and Stanton, E.A. (2007) *Reclaiming Nature: Environmental Justice and Ecological Restoration*, Anthem Press, London.

British Geological Survey (2004) *Community Management of Groundwater Resources: An Appropriate Response to Groundwater Overdraft in India?* British Geological Survey, London.

Budds J and McGranahan G. (2003) 'Are the debates on water privatization missing the point? Experiences from Africa, Asia and Latin America', *Environment and Urbanization*, vol. 15, no 2, pp. 87–113.

Census2011.co.in, 2011, 2011 Census of India: Alwar District, http://www.census2011. co.in/census/district/429-alwar.html

Central Ground Water Board (2006) Dynamic Ground Water Resources of India, Ministry of Water Resources, Government of India, New Delhi, India.

Cochran, J. and Ray, I. (2009) 'Equity re-examined: A study of community-based rainwater harvesting in Rajasthan, India', *World Development*, vol. 37, no. 2, pp. 435–444.

Downing R. A. (2002) *Groundwater: Our Hidden Asset*, UK Groundwater Forum and British Geological Survey, Nottingham, UK.

Down to Earth (1999a) 'Coming back to life', *Down to Earth*, 15 March 1999, http://www.downtoearth.org.in/node/19493, accessed 3 September 2014.

Down to Earth (1999b) 'To hell and back', *Down to Earth*, 15 March 1999, http://www.downtoearth.org.in/node/19488, accessed 3 September 2014.

Forsyth T. (2002) *Critical Political Ecology*, Routledge, London.

Glendenning, C. J. and Vervoort, R. W. (2010) 'Hydrological impacts of rainwater harvesting (RWH) in a case study catchment: The Arvari River, Rajasthan, India. Part 1: Field-scale impacts', *Agricultural Water Management*, vol. 98, no. 2, pp. 331–342.

Government of India (2007) 'Report of the expert group on Groundwater management and ownership submitted to Planning Commission, September 2007', Government of India, Planning Commission, New Delhi, 61 pages.

Gupta, S. (2011) 'Demystifying "tradition": The politics of rainwater harvesting in rural Rajasthan, India', *Water Alternatives*, vol. 4, no. 3, pp. 347–364.

Harvey, D. (2003) *The New Imperialism*, Oxford University Press, Oxford, UK.

Haugerud, A. (2010) 'Neoliberalism, satirical protest, and the 2004 US presidential campaign', in C. Greenhouse (ed.) *Ethnographies of Neoliberalism*, University of Pennsylvania Press, Philadelphia, PA, pp. 112–127.

Jha, B. M. and Sinha, S. K. (2009) 'Towards better management of ground water resources in India', *Bhu-Jal News Quarterly Journal*, vol. 24, no. 4, pp. 1–20.

Jayanti, G. (2009) *25 Years of Evolution: Restoring Life and Hope to a Barren Land*, Tarun Bharat Sangh, Alwar, India.

Kumar, P. and Kandpal, B. M. (2003) *Project on Reviving and Constructing Small Water Harvesting Systems in Rajasthan*, Sida Evaluation 03/40, Swedish International Development Cooperation Agency, Stockholm, Sweden.

Kumar, V. (2011) 'Innovative loan products and agricultural credit: A case study of KCC scheme with special reference to Alwar District of Rajasthan', *Indian Journal of Agricultural Economics*, vol. 66, no. 3, p. 477.

Lopez-Gunn, E. (2009) 'Governing shared groundwater: The controversy over private regulation', *The Geographical Journal*, vol. 175, no. 1, pp. 39–51.

Lopez-Gunn, E. (2012) 'Groundwater governance and social capital', *Geoforum*, vol. 43, no. 6, pp. 1140–1151.

Moench, M. (1994) 'Approaches to groundwater management: To control or enable?' *Economic and Political Weekly*, vol. 29, no. 39, pp. A135–A146.

Ostrom, E. (1997) Self-governance of Common-pool Resources W97-2, Workshop in Political Theory and Policy Analysis, Indiana University, Bloomington, IN.

Ostrom, E. (1990) *Governing the Commons: The Evolution of Institutions for Collective Action*, Cambridge University Press, Cambridge, UK.

Ostrom, E. and Hess, C. (2007) *Understanding Knowledge as a Commons: From Theory to Practice*, MIT Press, Cambridge, MA.

Postel, S. (1999) *Pillar of Sand: Can the Irrigation Miracle Last?* W. W. Norton, New York.

Ramusack, B. N. (2004) *The Indian Princes and their States*, Cambridge University Press, Cambridge, UK.

Rathore, M. S. (2003) *Community Based Management of Ground Water Resources: A Case Study of Arvari River Basin*, Institute of Development Studies, Jaipur, India.

Seckler, D., Barker, R. and Amarasinghe, U. (1999) 'Water Scarcity in the Twenty-first Century', *Water Resources Development*, vol. 15, no. 1–2, pp. 29–42.

Singh, R. (2009) 'Community Driven Approach for Artificial Recharge –TBS Experience', *Bhu-Jal News Quarterly Journal*, vol. 24, no. 4, pp. 53–56.

Staddon, C., Langberg, S. and Sarkozi, R. (2015) *Water Governance*, Water Research Commission, South Africa.

Staddon, C. (2010) *Managing Europe's Water Resources: 20th Century Challenges*, Ashgate, Farnham, UK.

Staddon, C. (2009) 'Towards a critical political ecology of human-forest interactions in Bulgaria', *Transactions of the Institute of British Geographers*, vol. 34, no. 2, pp. 161–176.

Subramaniam, M. (2014) 'Neoliberalism and water rights: The case of India', *Current Sociology*, vol. 62, no. 3, pp. 393–411.

Tarun Bharat Sangh (2013) TBS Website, www.tarunbharatsangh.in, accessed 16 September 2013.

Terry, A., Mclaughlin, O. and Kazooba, F. (2015) 'Improving the effectiveness of Ugandan water user committees', *Development in Practice*, vol. 25, no. 5, pp. 715–727.

Tiwari, M. (2008) 'Water management through people's participation in Rajasthan: A sociological study', PhD thesis, Jawaharlal Nehru University, New Delhi, India.

Vaidyanathan. A. (1996) 'Depletion of groundwater: Some issues', *Indian Journal of Agricultural Economics*, vol. 51, no. 1–2, pp. 184–92.

Wani S. P. and Ramakrishna Y. S. (2005) 'Sustainable management of rainwater through integrated watershed approach for improved livelihoods', in B. R. Sharma, J. S. Samra, C.A. Scott and S. P. Wani (eds) *Watershed Management Challenges: Improved Productivity, Resources and Livelihoods*, IWMI, Colombo, Sri Lanka, pp. 39–60.

Wani S. P., Ramakrishna Y. S., Sreedevi T. K., Long T. D., Wangkahart, T., Shiferaw B., Pathak P. and Keshava Rao, A. V. R. (2006) 'Issues, concept, approaches: Practices in the integrated watershed management: Experience and lessons from Asia', in *Integrated Management of Watershed for Agricultural Diversification and Sustainable Livelihoods in Eastern and Central Africa: Lessons and Experiences from Semi-Arid South Asia*, Proceedings of the International Workshop held during 6-7 December 2004 at Nairobi, Kenya, pp. 17–36.

Wani, S. P., Sudi, R. and Pathak, P. (2009) 'Sustainable groundwater development through integrated watershed management for food security', *Bhu-Jal News Quarterly Journal*, vol. 24, no. 4, pp. 38–52.

7 Traditional knowledge and modernization of water

The story of a desert town Jaisalmer

*Chandrima Mukhopadhyay[1] and
Devika Hemalatha Devi[2]*

1 Introduction

Fresh water is identified as the most crucial and quickly depleting resource on the globe today. Failure to incorporate traditional knowledge about water into the modernization process has been cited as fundamental to this situation (Goodall 2008; Weir *et al.* 2013). Weir (2009) points this out as a methodological question when water is investigated as an object and an economic resource, as against its cultural interpretive meaning. Weir also considers the ecology/economy dichotomy destructive as these are perceived to be opposites.

With this background, the chapter discusses the case of Jaisalmer. Jaisalmer, a desert town, located in the western Thar region, close to the India-Pakistan border, is the only living fort in India and a popular tourist destination. It has had limited access to water; moreover, due to its location at India-Pakistan border, as part of defensive strategy, Indian government kept this region as a physical boundary between India and Pakistan, and made little investment for development until the late twentieth century. There are other regions on the north-eastern part of India where such similar strategy was followed. The case significantly contributes towards the debate on positive versus interpretive epistemologies of water, and the ecology/economy dichotomy.

Section 2 briefly introduces the wider debate in the specific area of research. Section 3 introduces Jaisalmer, both in terms of its geological context and socio-economic profile. Section 4 discusses the methodology and presents three main research questions. Section 5, 6 and 7 are based on the three main research questions of the study. Section 8 concludes and summarizes the contribution of the study.

2 Modernization of infrastructure versus complexity of understanding water

The modernization of infrastructure, which has been largely facilitated by technological advancement, has, to a great extent been ignoring traditional knowledge and methods, and denying traditional populations' cultural understanding (Niebuhr, this volume). Weir (2009) provides an excellent account of the foundation to understanding 'water as resource', also discussed by Niebuhr (this volume). 'Resource', she argues, involves an understanding from only an economic perspective and she questions this way of thinking, arguing that water is central to people's way of life. She argues that water is more than a resource as recognized in modern planning and engineering/technology. In particular, she raises a voice against the dualism of ecology/economy and their destructive contradictory positions. Strang (2004) and Weir *et al.* (2013) take the same position. Weir (2009, p. 5) claims 'water planning and management uses a language that carries cultural assumption'.

In relation to indigenous people's right to water, the phrase 'cultural flow' has been used in the Australian context to acknowledge the use of traditional contextual knowledge around water in modern planning. Their traditional knowledge of water and land must be considered in the modernization process, which was not the case when the regulatory framework was formulated (Weir *et al.* 2013). Strang (2004) shows that cultural values and meanings associated with water are constituted by day-to-day activities. She focuses on the relationship between people and water and the way they value water, and how such values are passed down through generations; these questions are examined below in the study. Strang's (2008) research compares two groups who are involved in a scientific management of a water catchment area and alternative ways of managing it. Baghel's (2014) study criticizes irrational planning of large dams by the Indian government that considered engineering as the starting point without taking the ecology of a river into consideration.

Water is investigated in its different forms (ice, snow, flowing, static) (Strang 2005; Bell 2015) and usability, with regard to the diverse actors and their different stakes at hand (Star 1999); urban water is investigated in different forms, mainly in the sense of its usability (drinking water, storm water etc. through an engineering perspective – Bell 2015). Bell's (2015) study investigates water in an even more complex manner showing that the coding of water provides a better tool to understand how urban water can be understood in a technological perspective. She speaks about changing social relationships and bodily functions related to water. Moreover, restricted access to water is often associated with one's social status and physical cleanliness (Brown *et al.* 2008).

Hurst (2013) argues that the cultural aspect of water management is crucial to finding sustainable solutions for the global water crisis. She recommends considering the local planning scale since information about the informal and formal use, categories and qualities of water are more accessible through the knowledge of local communities.

3 Introduction to Jaisalmer

The Thar is the ninth largest subtropical desert in the world, situated across the borders of India and Pakistan, with an area of almost 320,000 km² area falling within India. In terms of population, it is the densest and most heavily used desert in the world, with a mean density of 151 persons per km², and a human footprint of 33 (Global Desert Outlook 2006). It has approximately 14 urban areas with numerous rural settlements interspersed between them.

Jaisalmer presents a different physiography from the greater sand dunes of the Thar. Jaisalmer town is located on a non-sandy plain that contains tertiary and pre-tertiary beds of sandstone, limestone and shale. The region suffers from little ground water availability, has one of the lowest numbers of rainfall days in India and lacks natural springs or rivers. There is little natural drainage and no permanent surface water because of limited rainfall. Jaisalmer town is the capital of the Jaisalmer district and is located approximately at its centre.

According to the 2011 Census of India, the municipal area covers 126.27 km^2 in total with a population of 66,919 and an annual foreign tourist load of 105,204. However, excluding the walled city, most of the area consists of rocky hillsides and uninhabited sandy areas. Gross population density in the fort is as low as 457 persons per km^2, but there are locations of very high density, particularly within the fort and walled city.

Jaisalmer is the only living fort in Asia. Its economic base has flourished because of the tourism industry. In the late 1990s, following interest from major conservation bodies such as United Nations Educational, Scientific and Cultural Organization (UNESCO), Indian National Trust for Art and Cultural Heritage (INTACH) and World Monument Fund (WMF), the city was included in national tourism strategies (Jain 2011). The cultural landscape is a major attraction, making Jaisalmer popular amongst both domestic and international tourists. However, the continuous influx of tourists also contributes towards both domestic and international migration, and the process of urbanisation.[3]

3.1 Gadsisar Lake

Access to water is critical in Jaisalmer and thus, the water supply infrastructure has played a significant role in its history and development. The Gadsisar Lake located to the south-east of the Jaisalmer pediplain[4] is an artificial water body, constructed around 1400 AD. The lake was constructed using the catchment of ephemeral streams and has a 4.8 km long drainage basin. The lake was a water reservoir for the town for a long period. A dike was constructed to direct water from miles around into the lake, with 75 per cent of the water collected by the dike used by farmers upstream for agriculture and the remaining 25 per cent was directed into the lake. The force of the incoming torrent of water was broken by a stone curtain before entering the lake. Even as late as the 1950s, prior to the Indira Gandhi Canal project period, the water thus collected was said to last the population of Jaisalmer for up to three years. This was important as the region used to have rainfall only once in two to three years. Water exiting from the lake was led into a series of nine interconnected lakes dispersed in and around the Jaisalmer area. Concerns over construction and governance of the lake included the maintenance of the lake. Jaisalmer's king formulated a series of rules regarding the cleanliness of the lake's basin, the usage of the water itself and the defence of the water bodies in the area. As can be seen, the

Gadsisar Lake was planned and constructed by local rulers, aided by the benefit of in-depth traditional water knowledge.

3.2 Indira Gandhi Canal Project

Jaisalmer's case demonstrates how political, development and technical decisions are intertwined. Due to its location at the India-Pakistan border, the town did not receive any support for infrastructure from the Indian government as the government marked this region as a buffer region between India and Pakistan – especially in terms of water – for a long time. This constituted a strategic defence policy by the central government. In 1961, the government of India intervened against prolonged drought in the region; the Indira Gandhi Canal Project (IGC) was part of a desert greening scheme, which aimed to introduce agriculture in the desert as an economic base in the region. This intervention – an engineering marvel – was made possible by technological advancement and initiatives taken at the national level. Partially operating since 1994, the canal carries water from Harike Barrage, located at the confluence of Sutlej and Beas rivers of the Indus river system. Water is thus sourced from approximately 650 km away. The IGC Project is the largest of its kind in India and has a feeder network of 9,060 km in length. On completion, it will have a command area of 19,630 km^2, and besides Jaisalmer, it covers the supply for regions such as Ganganagar, Hanumangarh, Churu, Bikaner, Jodhpur and Barmer (Indira Gandhi Nahar Pariyojana 2015). The IGC is mainly identified as an irrigation project, but the water from the canal is also used for drinking and domestic purposes. Besides economic development, which includes changes in socio-economic condition and rise in household income, other associated benefits are control over drought, introducing an influx of new population (that isn't familiar with lifestyle in desert) due to the softening of the climate, improvement in the micro-climate, the controlling of sand storms, and a rise in the ground water table (0.8 m per year).[5] However, as claimed by the local communities, Gadsisar Lake became non-functional over time due to aspects of the IGC project which were insensitive to the local conditions.

4 Applied interpretive approach as a methodology

Positivism as an epistemological approach investigates water as an object and as a resource. Positivists believe natural (pure scientific) truth exists irrespective of our understanding of the environment or the phenomena, and is to be investigated through observation. On the contrary, constructivists and interpretivists believe that one should interpret the world of meanings to understand them. They would investigate the subjective and culturally constituted understanding of water through an examination of rituals, celebrations and cultural practices. As discussed in Section 2, and following Weir's (2009) position, we investigate epistemologies of water through an interpretive approach. However, we argue that, as evident in Jaisalmer, the interpretive understanding of water does not exist in isolation, but is intertwined with its positivist understanding (which is about its pure scientific, and hence, objective knowledge); therefore, positivist and interpretivist knowledge are inseparable. We adopt an applied interpretive approach as the methodology (Thorne 2014). This approach allows one to interpret socially constructed meanings while having one's foot set firmly on the ground in the sense that it acknowledges natural/environmental/climatological realities.

There are three main research questions. First, which traditional knowledge fragments about water are useful for decision making on development? Second, how did traditional water knowledge produce hard and soft spaces? (Hard spaces are built up areas and soft spaces are spaces created through the relation amongst actors and use of hard space). Third, to what extent was such traditional knowledge considered to define Jaisalmer's water problems and included in the planning process of the IGC project? The unit of analysis here are community and interest groups.

Considering the significance of interdisciplinarity in landscape research, this study draws upon data from natural sciences, and qualitative data based on interpretive methodology of the humanities. Multiple methods were used to collect primary data from two field trips conducted in early 2015. Data for the study was collected through semi-structured interviews, focus-group interviews, non-participatory observation and document analysis. We conducted a total of 34 interviews and a list of interviewees who are quoted in the paper is enclosed in the appendix. Other sources such as images and blogs were also used as supplements, as the study draws on data from historical practices that are no longer prevalent. Addressing water as a women–dominated domain (Shiva and Jalees 2005), focus group interviews were conducted with older women to understand their role and experience with water. This was done with the help of a key informant who helped translate Rajasthani to Hindi, as most of the older women in Jaisalmer predominantly speak in the local Rajasthani dialect.

5   A collection of traditional knowledge around water

Traditional knowledge around water was useful for decision-making on development in the region before the IGC project. Section 5.1 enlists the traditional knowledge about typologies of water, depending on its sources, usability, and community perception about its holiness. Section 5.2 discusses practices of the local community that were informed by a lack of access to water and reflects their perceived value(s) of water. Findings show that such knowledge is constituted by local communities, conforming to their natural scientific understanding. Bayly (1999) provides similar evidences in case of Australian indigenous population.

5.1   Typologies of water influencing peoples' livelihood

Knowledge around water and its various categories impact people's choices of development strategies. It also shows how the form of water, the type of water, its usability, its source and the religious and cultural values around it are constituted.

5.1.1 Three types of harvested water

Traditionally, water harvesting was extensively promoted, not only in Jaisalmer but in all western Rajasthan. Due to variations in the soil character of the region, typologies of harvested water varied, leading to different implications for development. Harvested water was classified into three types: *palar pani*, *patal pani* and *rejani pani*. Palar pani is surface water and water harvested directly from rainfall. *Patal pani* is subterranean water; it is drawn from the section of the ground water table accessible through wells. *Rejani pani*, the third type, is water that accumulates due to higher levels of gypsum in the soil. In certain areas of the Thar, rainwater percolates through the earth but does not reach the groundwater table due to a semi-permeable gypsum layer. This percolated water forms a secondary subterranean layer from which water is then harvested through carefully constructed well structures.[6] These three terms are derived from local language to indicate varying types of water depending on its source, quality, and usability. *Paar* is a practice of constructing structures for improving water harvesting by reducing evaporation, which is popular amongst locals.[7]

5.1.2 Salty water and fresh water

Local peoples' knowledge further classifies water in terms of fresh water and salty water. Those categorizations of water are closely related to its particular purpose. Interestingly, the idea of cleanliness, and hence, drinkability, is related to people's spiritual belief. For instance, the categorization of fresh water stems from the belief that the ancient leeway of *Saraswati*, a mythical river venerated by Hindus, existed in areas around Jaisalmer. In addition, fresh water in this region was used sparingly. As mentioned by a key informant, the quality of water in the Gadsisar Lake was believed to cure tuberculosis, which may be due to the perceived cleanliness of water sources maintained by communities in Jaisalmer (Interview 04).

Salty water was found in areas in the immediate vicinity of Jaisalmer Fort, where ground water is infused by salt deposits occurring in the Thar. The water here was used for other domestic purposes including bathing, washing and feeding animals, and household chores, except for cooking. Apparently, the categorization as clean water used to determine whether water is holy or not (Interview 01). However, it also depends on the source of water and has a temporal dimension and is subject to transformation. The quality of water has changed over time, due to development and other reasons, and the community also acknowledges this. The following quote shows how clean water used to be considered holy.

Actually, here, all clean water itself is holy. All the wells are considered to hold holy water; even now they are used in the temple. Before we used to consider Gadsisar's water also holy, but now it has become dirty. Fishes have been introduced there; once fishes came, people started dropping grain into the lake.

(Interview 01)

5.1.3 Well and lake

Besides the classifications of water in terms of ecology and usability, the distinction between well and lake as the source of water is important as it has significance in terms of reliability of water supply, especially in the desert region, where lake water evaporates easily due to high temperature.

Interestingly, the categorization of well and lake water intersects with the requirement of fresh water and salty water. Wells were typically recognized as having salty water, appropriate only for animal consumption and household chores. Lakes were considered to be filled with fresh water and hence several regulations were related to their use. The use of lake water was preferred over wells, also because they were available only for a limited period, and regulations prohibited the hoarding of fresh water. In addition, the lake was a common pool resource open to all members of society, regardless of caste. Wells, on the other hand, were allotted community-wise and, hence, were more zealously guarded. The use of a well by a member of a different community was only allowed after the allotted community had satisfied their daily requirement. The wells still exist in Jaisalmer – with saline water 0.3048 to 0.6096 km deep – but were blocked by the government with the advent of water supply from the Dabla tube-well field (Interview 06). However, considering the quality of water, lake water was preferred for drinking. As the lake used to be dry in summer due to dry periods with no rainfall for two to three years, this posed one of the major challenges with regard to drinking water.

5.2 Traditional practices reflecting value of water

The scarcity of water was deeply reflected in people's way of life in the region. Starting from the construction methods used for their buildings; their day-to-day routine; and their rituals and spiritual beliefs, it was reflected in all aspects of peoples' lives.

5.2.1 Innovative dry construction method

Interestingly, both interviews with the key informants from the local community and observations reveal that the unavailability of water was reflected in the construction methods in the region. As one of the interviewees pointed out:

Yes, before there used to be only dry construction here; it used to be constructed using an interlocking system – the entire fort and the *havelis*[8] outside have been built using dry construction techniques. Nowadays, recently, people have begun building using cement. In the older buildings, if you look through, at night you could see to the outside landscape. Lights from outside used to filter into the houses. There is no cement or mud mortar used in between old houses. But the fort's four walls are still like this.'

(Interview 01)

On the one hand, such practices were influenced by the unavailability of water, while on the other hand, such kind of construction techniques imparted a particular kind of architecture to their fort and the *havelis*. As the interviewee mentioned, during the night, one could see the glimpses of light inside the fort from outside due to the slit gaps between its masonry.

5.2.2 Reuse and recycle

Even though the settlement is in a desert region that faces extreme scarcity of water, Jaisalmer was also known to have very rich merchant families living in luxurious *havelis*. An outsider would wonder how a rich family survived with such restricted water supply. It is not uncommon to connect access to water with socio-economic status (Brown *et al.* 2008). However, evidence shows that this was not the case in traditional Jaisalmer, but it varied depending on caste (Interview 01). Water sources used to be located in public spaces to assure that everyone had equal access to water irrespective of their socio-economic status, as mentioned by an interviewee:

Q: The richer people here, how did they use water?
A: It was the same source, there used to be servants to lift the water – that was the only difference. In Patwa Haveli, there are these kundis.[9] There is a long broad kundi there. The kundi system used to be that every house had a kundi kept in front that used to be filled.

(Interview 01)

However, one of the interviewees mentions that Jains (followers of the ancient religion of Jainism) were privileged due to their social status for a certain period (Interview 09). Residents were guaranteed a pot of water, and it was up to them to make the most efficient use of this with as much reuse as possible (Interview 33). There was a sequence of activities to be carried out using such water, and the building structures were integrated with elements that helped them recycle the water. The water would first be used for bathing, which was carried out in a shallow *kundi* located on a plinth using minimal slope. The shallow *kundi* was drained via an outlet connected to a pot kept at a lower level. Thus, all water that fell in the *kundi* was collected again.

As explained by a tourist guide (Interview 21), the bath water that was collected in the second pot was then used to wash clothes in the *kundi*. After the clothes were washed and the water recollected in a third pot, it was used to wet pats of dried manure. These were then stuck to the courtyard and walls of the house to cool the microclimate. Any water left over after this activity was minimal and sprinkled over the ground to regulate the ambient temperatures. Water for cooking was similarly recycled to water plants or feed animals. However, even though such knowledge is preserved in *havelis*, and presented to the tourists by the tourist guides as history of the region, the local younger generation neither practices nor acquires such kind of knowledge. Taking into consideration that there are recent engineering innovations in water recycling that struggle to gain public acceptance (Dolnicar and Scafer 2009; Hartley 2006), the findings illustrate how water recycling was carried out historically in Jaisalmer.

6 How did the traditional knowledge of water create unique hard and soft spaces around water?

The built environment of Jaisalmer, especially the fort and the *havelis*, shows a combination of built environment elements integrated with specific cultural practices and rituals reflecting the regions' experience with water. This section presents findings that show how hard spaces (i.e. built-environment) and soft spaces (i.e. rituals, celebrations and relations amongst people) were created surrounding water.

6.1 Hard spaces/built environment created around water knowledge

The following is a compilation of elements integrated in the built environment, starting from individual private properties to public spaces at varying scales, that were especially used for water collection and harvesting. Such built environment components are referred to as hard spaces created around water knowledge in Jaisalmer.

A *kund* is the smallest of all structures; they were used to collect spilled water from wells and excess water from houses. They used to be manually filled, and also provided water to birds and animals (and in some cases, travellers) in the arid climate.

The *angan* was a sloping courtyard from which *palar pani* (surface water and water harvested directly from rainfall) was directed into a corner or central *kund*. The *angan* was rigorously attended to and cleaned daily to prevent pollutants from entering the water thus collected.

The *tanka* collected roof-top water in a similar fashion to the *angan*. Water was strained before it reached the *tanka*. Even *churros* (interstices between sand dunes) were tapped to feed underground *tankas*.

Kuin were narrow wells, delicately constructed on collective property to tap *rejani pani* (the type of water that accumulates due to higher levels of gypsum in the soil). Yet, knowledge about *kuin* construction is limited since *rejani pani* is available only in certain areas.

Beri were well-like structures constructed around larger water bodies. They acted as ground water reservoirs – collecting rainwater, tapping ground water and recharging the water table too.

The *talab* is the largest of all water harvesting structures; it could also be termed as open water tank, harvesting rainwater. The construction depends on the type of water harvested. The *talab* is a community effort in the deepest sense. In erstwhile kingdoms, like Jaisalmer, the rulers themselves participated in the creation of these *talabs*.

The *nadi*, a smaller version of a *talab*, is a structure meant for water harvesting. It taps water from *churros* (interstices) and *thalis* (small dunes) and may be mud-lined or unlined. They are also called ephemeral streams and are active during some seasons in the year.

6.2 Soft spaces, cultural practices and relations created around traditional practices

The local communities integrated traditional practices and rituals with their built environment elements. The following discussion unfolds how novel spaces were created surrounding the scarcity, use, accessibility and celebration of water. The region was characterized by the experience of extreme hardship in livelihood due to long-term unavailability of rain over three to four years at a stretch. Thus, people celebrated their access to water during the monsoon or when it rained. Such creation of built-up space and soft spaces, through practices of water use has also been discussed elsewhere. Strang (2008), for example, speaks about social groups involved in creative activities that express ideas about community and attachment to place, celebrating spiritual and aesthetic interaction with the environment.

6.2.1 Medbandhi

Medbandhi was the process of building an 8 km-long dike, to bring water to Gadsisar lake, which was the biggest *talab* constructed. Popularly known as *Kesri Singh Baandh* (the word '*baandh*' means dam), it collected rain and water from the surrounding catchment areas to the north of Gadsisar.

6.2.2 Gorbandh

Gorbandh is a stone column consecrated near any water source that indicates the presence of water (see Figure 7.1). Once a well was dug, it was blessed with prayers asking for the well to be filled with water throughout the year. Then the *gorbandh* was installed with the symbol of a deity invoking protection; it also signified the sanctity of the water body. To outsiders, *gorbandhs* were also indicators of the nearby temple and served as a warning that the water body was being watched (Interview 02).

Figure 7.1 A view of a Gorbandh.

6.2.3 Raasleela

All water harvesting structures constructed within the town limits were considered common-pool resources. This stipulation was necessary for the development of customary codes, which regulated schedules of irrigation, the division of water resources and their equitable distribution (Rudolff and al Zekri 2014). *Raasleela* was an occasion celebrated around the common pool wells amongst the male and female members of a community centred around the use of water. Such celebrations in the designated spaces also helped constitute a sense of public space.

6.2.4 Panghat

The *panghat* was a network of routes through the city, travelled by women twice a day to collect water. While the network of routes is a physical element, the practice of water collection by groups of women and the associated activities contributed towards the creation of a space with special character. The *panghat* was decked every morning with stalls and the route covered with extended shades; the entire process took on the aspect of a festival. It remained closed in the afternoons when the women could not go out due to the extreme heat and re-opened once again in the early evening so that the women could, simultaneously, complete their (daily) tasks (Interview 01).

6.2.5 Pakhal

The *pakhal* was another practice related to water collection. It involved a bullock or camel cart slung with skein bags and a couple of water bearers who brought water to the lower town. Being predominantly livestock rearers, almost every house in Jaisalmer possessed cattle. The need to feed these animals was recognized as a necessity and incorporated as a built-in feature of wells and the water distribution system. In *pakhal*, water bearers were required to fill the *kundis* at every house, after the household requirement was met. There were specific points of distribution also outside the fort from where water could be collected. Often, households without women managed *pakhal* or were given priority over other residents to collect water (Interview 21).

Besides outlining how public spaces were created, this section also shows how women had a specific role to play in relation to water. Feminist scholars have discussed the comfort brought to women due to easy availability of water for private use (Wajcman 1991; Bell 2015). Development planning literature also identifies water as a gender-specific sector (Rathore 2002), recognizing women's greater role in managing household work, especially for low income communities in developing nations; hence, their better knowledge about water requirement and management. O'Reiley's (2006) study in particular speaks about how modernization of water contributes towards women's mobility, and hence contributes towards modernization of women's way of life. The discussion on spaces created by water also reflects a dominance of women in the water sector, and a sense of a gendered job split in relation to collection, transport, usage and management of water.

7 Planning of the Indira Gandhi Canal project – Who defines Jaisalmer's water problem?

The IGC project was an initiative taken by the central government with the aim of using technological advancement, to benefit regions that have suffered economically due to the lack of access to sweet water. However, this study raises some questions about the planning process of the canal project that may have failed to take both the local scientific knowledge and the rich culture of water conservation in Jaisalmer into account.

7.1 Ownership in source, supply and management of water

Depending on one's interest, one can understand water in terms of its source, management (both transfer and budgeting) and supply. In the case of the Jaisalmer Fort, the source of water supply has chronologically shifted from well, to lake and now to a distant source of water (Harike Barrage). One criticism of the IGC project is that because of the modernization of the water supply through the canal, local water sources have been completely abandoned and ignored. This is especially peculiar considering the water history of the region. Moreover, even though tangible structures like the fort and the *havelis* are preserved for their architecture, hard spaces like *kundis*, and soft spaces like *panghat*, created around water are not preserved or maintained. The change in water source is also of concern in terms of its reliability over a long period of time, and in relation to the climate pattern, since – in Jaisalmer's dry and hot climate – water evaporates easily (Interview 01). This was also the reason that Gadsisar Lake could not serve the town during summer.

Water management is also intertwined with water source, transfer and supply; previously it used to be local sources supplied and managed by local rulers. On the one hand, communities were aware of their sources of water, and understood the limitation and budget for the same. On the other, they had a sense of gratitude to the local rulers who looked after their interest. Because of the IGC project, local communities assume that they now have an unlimited water supply. Although they don't know where the water is coming from, they know very clearly that this has been possible due to a marvel of engineering. In terms of management, this leads not only to excessive use of water, but also ignores the fact that they actually do not have an uninterrupted water supply throughout the day, or the year, to afford such wastage (Interview 17). While wastage due to mismanagement is one of the concerns, echoing one interviewee's words, one would not care about the wastage of water if the water supply were perceived to be unlimited (Interview 01).

It cannot be ignored that ownership and control of water have always been political (Coelho 2004; Strang 2008; Kratz 1994; Parkin and Kaplan 1996). In the present case, changes in the source, management and supply of water can be read in many ways: local versus distant, well versus piped, local ruler versus central government (as opposed to local or regional government), and at a local scale, in public space versus private space (as the wells are now included into private properties).

7.2 Intersecting meanings of water

The tourism industry is known to have increasing demands for water supply. However, local communities, including the older generation do not complain about this, as tourism is the main base of the economy with all other sectors of the local economy grounded in tourism.

The older generation is much more critical towards the younger generation who does not, according to them, consider the traditional value of water in a desert town like Jaisalmer. This frustration has been reflected in many interviews, where concern is expressed more about a shift in the value of water, and not just the loss of knowledge of water, both being related; especially as the region has a history of a lack of access to water, which lies in a not very distant past (Interview 01 and 17). However, easy access to water allows the population to live a different life, and to invest their time in economic activities and activities related to improving their quality of life. Hence, the younger generation may value water in a different way. Van Vliet, Chappells and Shove (2005) discuss the role of infrastructure in determining the pattern of consumption. Supporting this study, interviews with the older generation in Jaisalmer confirm that with an abundant water supply, the traditional way to value water in its restricted availability disappears (Interview 01). The authors of the study suggest that a combination of technology and institutions should be developed to remove the gap between centralized and decentralized sources of water.

8 Conclusion

The study examined how traditional knowledge around water in Jaisalmer influenced decision making for development, as well as how people's behaviour in their daily lives adapted due to extremely restricted access to water. The hard and soft spaces created around water in traditional Jaisalmer were co-constructed, and discussed as tangible and intangible heritage elsewhere (Mukhopadhyay and Devi, under review). Only a few of the numerous hard spaces are preserved as heritage for tourists' experience. This is because built structures are considered architectural heritage, while the special spaces created surrounding water knowledge are not given a similar status as heritage. Whereas the analysis showed that even rich merchant families used to recycle water, it also shows that such knowledge is not being transferred, but being transformed over time, shifting the 'value' of water across the generations.

The last question about whether the traditional knowledge about water has been considered while framing the water problem of Jaisalmer and planning of the IGC project is an important one and would benefit from further research. Clearly, evidence from Jaisalmer shows that water is not a mere natural resource, and traditional water knowledge produced specific built environments, cultural practices and social structures that would be transformed, if such knowledge is not considered in framing water problems. As discussed elsewhere, besides destroying heritage, there are serious implications in terms of sustainability of cultural landscape (Mukhopadhyay and Devi, under review). This would have an impact on the local economy of Jaisalmer, which is mainly based on tourism. Moreover, both insensitive desert-greening planning and climate change may have an adverse impact on the natural environment of the settlement, bringing into question its very existence.

Acknowledgements

The chapter emerges from research on Jaisalmer that was carried out for Devika Hemalatha Devi's Master of Urban and Regional Planning Dissertation under the supervision of Chandrima Mukhopadhyay at CEPT University Ahmedabad, India.

Notes

1   Faculty of Planning, CEPT University, India.
2   Programme officer at Indian Heritage Cities Network (IHCN-F), India.
3   Bhatia, V. 2015. Pakistani Hindus flock to Jaisalmer http://timesofindia.indiatimes. com/india/Pakistani-Hindus-flock-to-Jaisalmer/articleshow/49477600.cms, accessed 4 January 2016.
4   A pediplain is an extensive plain formed in a desert by the coalescence of neighbouring gently sloping bedrock surfaces extending outward from foot of mountain slope.
5   Raja, K. (2012) 'Short notes on Indira Gandhi Canal Project' http://www.preserveart-icles.com/2012020122326/short-notes-on-indira-gandhi-canal-project.html, accessed 4 January 2016.
6   Such varied forms of water depending on diversity of environment have been discussed by Strang (2005). She mentions them in the form of frequent precipitation, immobilized in ice, monsoonal flood and hidden spring.
7   CSE, no date, http://www.rainwaterharvesting.org/Rural/Traditional1.htm, accessed on 4 January 2016.
8   Havelis are traditional townhouses and mansions of rich merchant classes in Rajasthan.
9   Kund is the smallest of all structures; they were used to collect spilled water from wells and excess water from houses. They used to be manually filled, and also provided water to birds and animals (and in some cases, travelers) in the arid climate.

References

Achyuthan H. (2003) *Bedrock Channel Incision in the Arid Tracts of Jaisalmer, Rajasthan*, Proc Indian National Science Academy, vol. 69A, no. 2, pp. 241–250.
Baghel, R. (2014) *River Control in India: Spatial, Governmental and Subjective Dimensions*, Springer, Dordrecht, Heidelberg, New York and London.
Bayly, I. A. E. (1999) 'Review of how indigenous people managed for water in desert regions of Australia', *Journal of the Royal Society of Western Australia*, vol. 82, no. 1, pp. 17–25.
Bell, S. (2015) 'Renegotiating urban water', *Progress in Planning*, vol. 96, pp.1–28.
Brown, R., Keath, N. and Wong, T. (2008) 'Transitioning to water sensitive cities: historical, current and future transition states', in *11th International Conference on Urban Drainage*, vol. 10.
Coelho, K. (2004) 'Of engineers, rationalities, and rule: An ethnography of neoliberal reform in an urban water utility in South India', PhD thesis, University of Arizona, AZ.
Dolnicar, S. and Scafer, A. (2009) 'Public perception of desalinated versus recycled water in Australia', 2006 AWWA Desalination Symposium, 7–9 May, Honolulu, HI.
Goodall, H. (2008) 'Riding the tide: Indigenous knowledge, history and water in a changing Australia', *Environment and History*, vol. 14, no. 3, pp. 355–384.
Hartley, T. W. (2006) 'Public perception and participation in water reuse', *Desalination*, vol. 187, no. 1, pp. 115–126.

Hurst, K. F. (2013) 'Cultural waters values of water resources in Hidalgo', Mexico, PhD thesis, Texas A&M University, College Station, TX.

Jain, S. (2011) *Indian Heritage Cities Network: Walking into the microcosm of Jaipur*, UNESCO, New Delhi, India.

Kratz, C. (1994) *Affecting Performance: Meaning Movement and Experience in Okiek Women's Initiation*, Smithsonian Institute Press, Washington DC.

Mukhopadhyay, C. and Devi, D. H. (under review) *A conceptual framework of Cultural landscape: A case study on Jaisalmer*.

O'Reilly, K. (2006) '"Traditional" women, "modern" water: Linking gender and commodification in Rajasthan, India', *Geoforum*, vol. 37, no. 6, pp. 958–972.

Parkin, D. and Caplan, L. (eds) (1996) *The Politics of Cultural Performance*, Berghahn Books, Providence.

Rathore, M. S. (2002) *Pro-Poor and Gender Sensitive State Water Policy–NGO's Perception on Water Resource Management in Rajasthan*, Institute of Development Studies, Jaipur, India.

Rudolff, B. and Al Zekri, M. (2014) 'A network of traditional knowledge: the intangible heritage of water distribution in Bahrain', https://opus4.kobv.de/opus4-UBICO/frontdoor/index/index/docId/12889, accessed 4 January 2016.

Shiva, V. and Jalees, K. (2005) *Water & women: A Report by Research Foundation for Science, Technology, and Ecology for National Commission for Women*, Navdanya/RFSTE, New Delhi, India.

Star, S. L. (1999) 'The ethnography of infrastructure', *American Behavioural Scientist*, vol. 43, no. 3, 377–391.

Strang, V. (2004) *The Meaning of Water*, Oxford, Berg, New York.

Strang, V. (2005) 'Common senses: Water, sensory experience and the generation of meaning', *Journal of Material Culture*, vol. 10, no. 1, pp. 92–120.

Strang, V. (2008) 'Wellsprings of belonging: Water and community regeneration in Queensland', *Oceania*, vol. 78, no. 1, pp. 30–45.

Thorne, S. E. (2014) 'Applied interpretive approach', in P. Leavy (ed.) *The Oxford Handbook of Qualitative Research*, pp. 99–115.

United Nations Environment Program (2006) *Global Desert Outlook*, UNEP, Nairobi, Kenya.

van Vliet, B., Chappells, H. and Shove, E. (2005) *Infrastructures of Consumption*, Earthscan, London.

Wajcman, J. (1991) *Feminism Confronts Technology*, Allen and Unwin, Sydney, Australia.

Weir, J. (2009) 'Our understandings of water and how they translate into our decision making', In ACT Government Workshop 'Planning for our Future – Securing Canberra's Water', Australian National University, Canberra, vol. 10, p. 1A6.

Weir, J., Ross, S. L., Crew, D. R. and Crew, J. L. (2013) 'Cultural water and the Edward/ Kolety and Wakool river system', Research Report, AIATSIS Centre for Land and Water Research, Australian Institute of Aboriginal and Torres Strait Islander Studies, Canberra, Australia.

Appendix: List of interviewees

Interviewee No.	Gender	Age	Location	Remarks
Interview 1	Male	55	Jaisalmer	Tourist guide and shop owner within Jaisalmer Fort.
Interview 2	Male	40	Burna	Restaurant owner and tour operator within the fort; resident within the fort upwards of 15 years.
Interview 4	Female	70	Amar Sagar Pol	Resident of the Jaisalmer fort since the age of 20 (married).
Interview 6	Male	70	Amar Sagar Pol	Shop owner outside of Jaisalmer Fort; born and brought up in Jaisalmer; at the lower end of the caste spectrum.
Interview 9	Male	45	Amar Sagar Pol	Shopkeeper within the fort; migrant from the nearby Barmer
Interview 17	Male	50	Burna	Uncle of Interviewee 2; farmer at Burna and trader at Jaisalmer over the past 40 years.
Interview 21	Males	(63, 81)	Patwon ki Haveli	Father-son; residents of Jaisalmer town and the lower end of the caste spectrum.
Interview 33	Male	70	Amar Singh Pol	Ex-teacher and renowned educator within Jaisalmer town.
Interview 34	Males	(62, 63)	Rangrej Pol	Neighbours born and brought up within Jaisalmer; former and ex-government servant, and the latter a doctor.

8 The hydro-ecological self and the community of water

Anupam Mishra[1] and the epistemological foundation of water traditions in Rajasthan

Daniel Mishori and Ricki Levi

1 Introduction

Sustainability is a challenge the complexity of which seems to overwhelm modern science and current dominant social, economic and ideological structures. Policy makers usually expect that solutions be found at our culture's dominant institutions: science, and the free market economy. However, sceptics argue that these very institutions, science and the free market, are the ones to blame for many of the ecological hardships facing our planet and that, therefore, new ideas and practices should be sought elsewhere – for instance, in non-Western ideas, ways of life, and traditional ecological knowledge (TEK) that may prove superior to current dominant structures of science and technology. Anupam Mishra (born 1948), an Indian author and a Gandhian activist, could serve as such an inspiration for an alternative paradigm.

According to the Indian eco-feminist scientist and anti-globalization activist, Vandana Shiva:

> Over the past three decades, Anupam Mishra has created a silent but permanent revolution. He has changed the dominant paradigm of water and shown that water security and insecurity is a product of nature plus culture, not just a given of nature. There can be water scarcity in high rainfall region and adequate water in low rainfall regions like the Rajasthan desert.
>
> (2001, p. 1)

Shiva justly describes Mishra's achievements in rediscovering the indigenous water systems of Rajasthan as 'a work of poetry as well as a work of science' (ibid.), since Mishra creates a unique combination of archeology and engineering mixed with social activism, ethics and deep-rooted spirituality. His work is an homage to the achievements of prior generations, and an opportunity for humility and learning in the face of past accomplishments. In this, Mishra presents an intriguing challenge to current dominant paradigms and conventional ways of thinking, and show the likelihood of non-Western routes of sustainability.

Mishra questions accepted conventions – that the modern is better than the past, and that science and technology are best represented by experts whose qualifications rely on university certifications – and thus enters a discourse on epistemology: on ways of knowing and on types of knowledge. Therefore, this paper begins with a discussion of water as an epistemological challenge. In Section two we briefly describe the work of Anupam Mishra. We do not elaborate on the varieties of Rajasthani water systems and infrastructures, or on their cultural, spiritual and religious background, which are discussed in depth elsewhere (Levi and Mishori, 2015). Instead we focus on the epistemological aspects of Mishra's work, in Section three, and elaborate on their significance.

2 Water as an epistemological challenge

2.1 The challenges of dominant paradigms of thought and praxis

Faced with detrimental consequences of human misuse and mistreatment of nature, the Brundtland Report (the United Nations World Commission on Environment and Development, 1987) came up with a novel notion, which was supposed to become the conceptual foundation of any responsible environmental policy: sustainable development. This concept was never clearly defined; the report *Our Common Future* (Brundtland *et al.* 1987), described it as 'development that meets the needs of the present without compromising the ability of future generations to meet their own needs'. However, the exact nature of such development, except the normative principle of intergenerational responsibilities and ethics, was never defined.

Part of the concept of environmentally sustainable policy has always been education. Already in 1972, the United Nations Conference on the Human Environment, held in Stockholm, affirmed in article 19 that

> Education in environmental matters, for the younger generation as well as adults, giving due consideration to the underprivileged, is essential in order to broaden the basis for an enlightened opinion and responsible conduct by individuals, enterprises and communities in protecting and improving the environment in its full human dimension.

Likewise, the Brundtland Report is loaded with references to 'education'. These references mostly rely on the same intuitions which were manifested already in Principle 20 of the Stockholm Declaration which stated that:

> Scientific research and development in the context of environmental problems, both national and multinational, must be promoted in all countries, especially the developing countries. In this connection, the free flow of up-to-date scientific information and transfer of experience must be supported and assisted, to facilitate the solution of environmental problems; environmental technologies should be made available to developing

countries on terms which would encourage their wide dissemination without constituting an economic burden on the developing countries.

In other words, 'education' is the easy-to-digest dissemination of environmental knowledge and technologies, which were commonly presumed to be mostly modern and Western. This conception was contested by David Orr (1991), who criticized the goals and presumptions of modern education. Orr argued that traditional educational methodologies reproduce old conceptual and mental models and practices, in which the greater whole, connections, interrelatedness and affinities are being lost in a system which in principle is based on a myriad divisions and fractions of knowledge and expertise. The fragmentation of knowledge in disciplines and sub-disciplines results in students without any integrated sense of the unity of things. This lack of integrated understanding is presented as inherent characteristics of modern higher education, and is manifested in modern science and technology.

The alternative model which Orr advances is rooted in virtue ethics.[2] Orr claims that 'education is no guarantee of decency, prudence, or wisdom. Our education up till now has in some ways created a monster. More of the same kind of education will only compound our problems. It is not education that will save us, but education of a certain kind'.[3] In other words, real sustainability is based on wisdom and ethics, which does not necessarily follow the Western dominant models of science, technology or ethics. Hence, sustainability is a challenge also on the educational and epistemological levels, and calls for fresh rethinking of technology, praxis and ethics. In this, Anupam Mishra's work on water in India makes significant and intriguing contributions to the study of the epistemological foundations of sustainability.

2.2 Water and epistemology

There is hardly any literature on the epistemology of ecology in general, and of water or resources in particular. Infatuation with epistemology was the hallmark of modern philosophy, and its keen interest in metaphysics.[4] In order to create a modern science, which will enable humans to dominate nature, ways of knowing (i.e. epistemology and methodology) were presumed to be of major importance. Ontologically, modern philosophy hailed atomism, a world of separate particles, governed by the laws of mechanics, which could be analyzed into their primitive (simple) constituents. It was assumed that wholes were nothing but the aggregation of their parts, and could be fully understood by analysis and study of their minute constituents.

With the ecological turn, atomism gave way to wholism, to the idea that reality is made up of organic or unified wholes, and that if we aspire to understand the world, we should study also relations and interdependencies. This view emphasized ecosystems, systems and processes (e.g. Capra 1982, 1987). Capra (2007) characterized ecological thinking in terms which are epistemological in essence:

From the parts to the whole. Living systems are integrated wholes whose properties cannot be reduced to those of their smaller parts. Their "systemic" properties are properties of the whole that none of the parts has …

From objects to relationships. An ecosystem is not just a collection of species, but is a community. Communities, whether ecosystems or human systems, are characterized by sets, or networks, of relationships … The properties of the parts are not intrinsic, but can be understood only within the context of the whole. Since explaining things in terms of their contexts means explaining them in terms of their environments, all systems thinking is environmental thinking.

From objective knowledge to contextual knowledge … The shift of focus from the parts to the whole implies a shift from analytical thinking to contextual thinking … *From structure to process.* Systems develop and evolve. Thus, the understanding of living structures is inextricably linked to understanding renewal, change, and transformation.[5]

(p. 12, emphasis in the original)

Besides emphasizing relational and process epistemology – systems, context, subjectivity, and interconnectedness – deep ecologists discussed two more relevant conceptions: 'sense of place', and the 'Ecological Self'.

'Sense of place' (SoP) represents the emotional and spiritual bonds which people form with certain locations, landscapes and spaces (Williams and Stewart 1998). It is a centre of meaning, and reflects the complex web of lifestyles, meanings, and social relations endemic to different places: 'What begins as undifferentiated space becomes place when we endow it with value (Tuan 1977, p. 6). Williams and Stewart (1998) suggest thinking of SoP as

the collection of meanings, beliefs, symbols, values, and feelings that individuals or groups associate with a particular locality. Familiarity with places intensify sense of identification and vice versa, if one is an "outsider" who is unfamiliar with the place. The set of meanings associated with places are "actively and continuously constructed and reconstructed within individual minds, shared cultures, and social practices.

(ibid.)

Rodaway (2002) emphasizes the sensuous experiences which construct our geographies, based on immediate experience of the world through touch, smell, hearing and sight, while Heise (2008) discusses the role of 'sense of place' in forming ethical commitment to the local and the global environment. Finally, Stedman (2003) argues that former definitions overemphasized the social construction of sense of place and neglected important contributions of the physical environment to place meanings and attachment. He suggests that this concept include:

1   characteristics of the environment,
2   human uses of the environment,

3   constructed meanings, and
4   place attachment and satisfaction.

In Section 2.3, we discuss how such insights are manifested in, and contribute to, the understanding of Mishra's work.

Another key epistemologically-related concept is the 'ecological self'. It first appeared in a book which discussed the new ecological metaphysics, with an emphasis on ontology (Matthews 1991). It soon became, however, loaded with epistemological connotations. Wilson (1996) presented this concept as 'based on the premise that our individual ecological identities are determined by how we extend our sense of self in relationship to the world of nature'. Besthorn (2002) related it to eco-centric and biocentric intuitions, arguing that 'the new ecological self does not ignore identification with nonhuman beings or with nature' while Bill Devall simply stated: 'I am a forest being' (1995, p. 101). Such insights recur in Mishra's discussion of the intimate relations between the Rajasthani people and water.

2.3 Epistemology and traditional ecological knowledge

Traditional ecological knowledge (TEK) is a term that refers to knowledge derived by indigenous communities and cultures from their intimate relation to nature and through accumulated ecological experiences over many generations. TEK is a source of 'knowledge about species, ecosystems, or practices held by people whose lives are closely linked to their natural environment' (Rist *et al.* 2010, p. 1). Nature is being perceived in TEK from a holistic point of view, as an integrated whole, comprised of all natural elements (and sometimes super-natural entities) such as animals, plants, landscapes, and humans. Occasionally, this 'whole' is being referred to as a 'community', or a 'family', which, according to Pierotti and Wildcat (2000, p. 1333), implies a perception which attributes intrinsic value to nature. According to this definition, two central notions emerge: First, everything is connected to everything else, and second, non-anthropocentrism, in which 'nonhuman organisms are recognized as relatives whom the humans are obliged to treat with respect and honor' (ibid.). As we will show, Mishra argues for the superiority of the water TEK in Rajasthan over its Western modern science. This Rajasthani TEK is rooted in epistemological foundations.

3 The work of Anupam Mishra

3.1 The mission of Mishra

Anupam Mishra is known for his work on water and environmental conservation in Rajasthan. In two books (1993, 2001), Mishra describes a water ethics and world-view, which supports highly successful traditional water practices. These practices enabled Rajasthani society and culture to thrive in the arid

Thar desert, in which some parts receive average of only 16 to 25 mm of rainfall annually. The Thar desert is located in North–West India, bordering Pakistan, and is part of the Afro–Asian desert belt that stretches from the Sahara to the Gobi desert. Despite its warm climate and water scarcity, it is and historically has been the most populated desert in the world.

Decades of research into time-tested systems of water conservation and harvesting in several Indian states, principally in Rajasthan, persuaded Mishra that truth and knowledge, including ecological wisdom, practices and technologies, emerge from local and traditional experiences. However, this rich TEK is being undermined and suppressed by the gigantic forces of the market and the state, processes exacerbated by globalization and the general belief in the superiority of modernism and progress over so-called 'primitive' traditional technologies and practices. In the case of India, these foundations of sustainability were being suppressed by the British rule since the nineteenth and the beginning of the twentieth centuries, and later by the new state of India.

Mishra's philosophy is rooted in Gandhian philosophy and values (Mishra 1995), including the pursuit of truth as both a personal and collective endeavour, perceived in spiritual and religious conceptions. Following Gandhi's praises of simple living (which were manifested in his personal dressing code: the Khadi [homespun cloth]), Mishra endorses modesty and small-scale traditional practices, manifested by the water ventures he promotes, which also encourage self-reliance and empowerment of both individuals and local communities. In this he is a disciple of Gandhi, and a living proof of the relevance of Gandhi's teaching in the twenty-first century.

The water ethics Mishra describe have different aspects, but a dominant and a recurring theme is a kind of virtue ethics, which according to Mishra enabled the Rajasthani people to develop their water traditions (Levi and Mishori, forthcoming).[6] In this Mishra is in perfect agreement with David Orr's views on education. Moreover, this ethics is a part of an integrated whole, which enables the Rajasthani people to acquire an in-depth understanding of water in the desert.

3.2 Varieties of water infrastructures

Mishra describes in detail the varieties of water infrastructures in Rajasthan, of which the most common and important we will shortly describe below: *kund*, ponds; *kundis*, small ponds; and *tanka*, huge reservoirs. Two unique water bodies are the *talais* or *johar-joharis* (special pond) and the *kuin* (special well). It should be noted that similar infrastructures are also found in other places in India (e.g. in Banaras, see Lazzaretti, this volume), although *Kuins* are to be found mainly in the Thar desert.

The *kund* and the *kundis*, as well as every other types of pond (e.g. *tanka*), are built with slopes to create a drainage space, an *agor* (*agoma* means collecting), designed to catch every drop of rain or moisture. This area is kept spotlessly clean on a regular basis, especially before the rainy season starts, and there is a strict prohibition on entering the *agor* with shoes on.

The *tankas* were huge artificial lakes usually built far from human settlement, and their purpose was to serve as emergency reservoirs, in case of war or drought. An average *tanka* contained between 200,000 and 300,000 litres of water, and enabled shepherds to water cattle in areas remote from human settlements (ibid., p. 62–63). The largest *tanka*, in Jaigarth Fort (near Jaipur) was built in the fifteenth century. It is an enormous reservoir, containing up to 3 million litres of water, and is supported by a complex construction of pillars and a highly developed ventilation system.

According to Mishra, contrary to the famous proverb 'running water, pure water', in Rajasthan, water in closed water storages remains pure and fresh. Most water bodies were kept sealed to prevent evaporation, and to keep them clean and pure (ibid., p. 53).

The Rajasthanis differentiated between three types of water: *palar pani*, rainfall water, absorbs directly into various water bodies; *patal pani*, groundwater, accessible from wells and closed-deep ponds; and *rejani pani*, capillary water, trapped in the soil, below the surface, before it reaches the salty underground water table (Mishra 2001, p. 45). To use these different types of water, proper infrastructures were needed, which required sophisticated craftsmanship and joint community efforts.

A unique practice of the Rajasthani water tradition, not found anywhere else in the world, is the ability to extract drinking water from the *rejani pani*, the layer of moisture (usually from rain water), which is trapped in the soil and cannot percolate into the groundwater, which in many cases in Rajasthan is very salty (Mishra 2001, p. 45). *Rejani pani* water can be found in certain areas, at depths of 20 to 30 meters below the surface, where there is a gypsum layer; the water is captured between the surface and the gypsum, and forms a layer of sand absorbed with moisture. It cannot be easily spotted from the surface, but miners know how to locate it, according to special characteristics, for example, the presence of stagnated rain water, or the presence of certain trees (ibid., pp. 41–44).

Chelvanjis, well-diggers, dug very narrow and deep wells known as *kuin*. The excavation process was risky and slow. Once completed, such wells collected water even when there was no rain. Each *kuin* can convert moisture into water at a rate of about 2 to 3 buckets a day. Usually, in areas where the gypsum layer is located, up to 30 to 40 *kuins* can be dug. The gypsum layers are very common in the heart of the Thar desert; Jaisalmer, Bikaner, Churu and Badmer (ibid., pp. 45–51).

Other unique ponds are the *talais* or *johar-joharis*. Their uniqueness is manifested in the fact that they are found in the most extreme geographical conditions, in salty soil and lake areas. There, every drop of water touching the ground immediately turns salty; the subterranean water and even the *rejani pani* is saline. Therefore, the *agor* of the *talais* was built above the ground, two to four hands in height, catching the rain water and directing it into the pond, where the water remains clear and sweet (ibid., p. 68).

4 Water praxis and epistemology

Mishra does not only describe the different types of water infrastructures. He focuses on the cultural and spiritual foundations of water knowledge, skills and practices. In this he makes a series of presumptions, assertions and inferences regarding the epistemological foundations of water conservation and harvesting, which we will critically examine below. These epistemological foundations enable Mishra to argue for the superiority of Rajasthani water TEK over modern methods and technologies of water.

The epistemological assertions he makes are intimately linked to the unity which seemingly exists between community, culture and the environment, which is presented by Mishra as a precondition for a sustainable management of water resources. He explores traditional practices and technologies of water, emphasizing their dependence on social-cultural infrastructures, and depicts them as rooted in spiritual-religious conceptions, including personal and social virtues (Levi and Mishori, forthcoming) which Mishra sees as necessary for surviving and thriving in the arid desert.

Throughout his work, there are recurring epistemological discussions and concerns which underlie his analysis of water craftsmanship, management and conservation. This paper surveys these issues, and regards Mishra's philosophy of water as preoccupied and rooted in questions of knowing, know-how and knowledge.

4.1 Knowledge as inherently collective, and indistinct from its cultural, religious, local or environmental circumstances

Mishra sees knowledge and skills as inherently collective (social), rooted and cultivated in the community and its culture. He argues that

> It is thanks to the local society and indigenous culture that lack of rain was not translated into scarcity. The people of Rajasthan did not mourn the lack of rain Nature bestowed upon them. Instead they took it up as a challenge and decided to face it in such a way that from top to toe the people internalized the nature of water in its simplicity and its fluidity.
>
> (2001, pp. 23–24)

According to Mishra, water has to be understood in the broadest terms, as a whole 'philosophy' of life. For this reason, Mishra explores in depth these various contexts. He claims that 'The people of Rajasthan scaled the peaks of trade, culture, art and standard of living because of the depth of their philosophy of life. This philosophy gave a special space to water' (ibid., p. 24).

Knowledge of water exists in the society as a whole. Therefore, he argues that 'To access this ambrosia [water] the people of the desert ... elaborated a whole science to translate their experience into practicality' (ibid., p. 44). This 'experience' is fundamentally collective. The 'whole science' they elaborated was formed in the Rajasthani culture as a whole: 'This culture which knows how to propitiate earth, water and heat through its asceticism, also knows how to keep flowing water as well as still water pure' (ibid., 104). Mishra speaks of water TEK as a 'science', and thus put it on par with Western modern science, only to claim later that the Rajasthani water TEK is (in the particular case of Rajasthan) better than modern technologies; it is better technically (more adapted to the Thar desert) and it is better since it is in accord with and an expression of a holistic philosophy:

> In Rajasthan, more particularly in the desert, water work was never considered as work but came to be viewed as a moral duty; that is why it was able to rise far above what is today called community project, to take the beautiful shape of a *samagra jal darshan*, a perfect water philosophy.
>
> (ibid., p. 105)

4.2 *The superiority of local, traditional knowledge over modern foreign knowledge*

From the outset, Mishra is preoccupied with the contrast between local, traditional and indigenous knowledge and technologies, and their modern 'foreign' alternatives. He gives several examples to prove this claim, that the Rajasthani TEK is better than modern water practices. For instance, according to Mishra,

> Some time back, a certain department launched a new project according to which it was decided to innovate upon the *kundis* by replacing *phog* (desert shrub) with cement. Those who were experimenting must have thought that this modern *kundi* would be stronger. However it did not turn out to be so. The cement domes of these ideal *kundis* could not withstand such strong heat and caved in. Even the inner walls of those *kundis* instead of being coated as usual with sand and lime were coated with cement. Numerous cracks appeared on them too. To rectify them, tar was used to fill up the cracks; however under the blaze of the desert, the tar melted away. All the water collected during the rainy season evaporated.
>
> (Mishra 2001, p. 59)

Thus, local materials, which were known only to local experts, are better than modern foreign techniques and materials. According to another example:

> Various governments of the 20th century, including the government staking a claim to take the country into the 21st century, have not been able to provide sweet water to the villages of these [Rajasthani] saline areas.
> ...Some years ago new and old governments tried to make a similar new *talai* near the *talais* of the *Banjaras*: however they were not successful in segregating the 'fusing' properties of salt and water.
>
> (ibid., p. 69)

Hence, the ability to separate water from salt, known to the traditional craftsman of Rajasthan, is unknown to modern so-called experts, whose skills were acquired in colleges and universities, based on Western practices.

Mishra applies the same criticism to a water project in Botswana, where a Canadian organization helped building *kundi*-like structures (1975–1981): 'some high government officials, foreign engineers, water experts went around some villages and then they made a slope to the courtyard used for drying grains in the fields and dug a ditch in one corner so that the rainwater would be collected in it. About ten such were made with 100% foreign aid and using material coming from very far' (ibid., p. 109).

Built in a square shape (with foreign aid and materials), these *kundis* suffered from pressures and their structure tended to break. Mishra claims that experts agree now that Indian-style round *kundis* are better. For Mishra, it is a paradigmatic example of how international aid with expensive 'ready-made' foreign solutions cannot improve local conditions, which require local solutions and expertise, adapted to local conditions

All these testify, according to Mishra, that regarding water structures and technologies, traditional Rajasthani knowledge is superior to modern engineering and expertise:

> My travels made me aware that our forefathers were far more aware and educated than we are … Just imagine; the enormous water tank of Jaigarh fort in Jaipur can contain several hundred million litres of water! Look at Gadasisar in the heart of the Jaisalmer desert or the three-century old Toda Rai Singh Tank. Now, students from IIT are studying the engineering of these old water tanks and ponds.
>
> (Bajpayee, n.d.)

4.3 Neglect and forgetting

Mishra suggests that indigenous knowledge is an edifice which accumulates over generations, and seems to resent modern knowledge constructions since they allegedly abolish completely old forms of expertise. The rejection of new Western knowledge and expertise is based on two reasons: first, such knowledge is not necessarily superior to traditional practices; second, the adoption of modern practices often causes 'forgetting', a situation which is often aggravated by what Mishra sees as the bias of the state and the market for modern methods, materials and conceptions. In this, Mishra's criticism strongly resembles Vandana Shiva's arguments against modern industrial corporate-based agriculture (1993, 1996, 2015). For example, regarding Gharsisar Lake, Mishra notes that:

> The 668 years old Gharsisar has still not died. Those who had built it made it strong enough to withstand the blows of Time. Those who created the fine tradition of maintaining their lakes through sand storms

did not probably realise that one day these would have to face storms of neglect.

(ibid., p. 80)

The same 'neglect' also causes the forgetting of the Rajasthani nomenclature, created to sustain water: 'This rich diversity of names and tasks is slowly fading out as on some wells electric and diesel pumps are being fixed' (ibid., p. 101).

The same criticism is directed towards current water management agencies for their inability to comprehend the value of traditional water reservoirs and to preserve them:

> [T]hose who govern Jaisalmer today have forgotten the very meaning of Gharsisar [lake] so how can they ever remember the *neshta* [spillway] linked to those lakes? In the *agor* of Gharsisar an air pad has been made for the Air Force, which is why the water of that part of the *agor* no longer goes to the lake but flows elsewhere. Around the *neshta* and the nine lakes which fall on its course, have sprung widely growing housing colonies, new housing societies and to top it all the office and the housing colony of the employees of the modern water department, the Indira Nahar Pradhikaran. Slowly, the *ghat* schools, kitchens, *baradari* and temples have fallen apart just like the upkeep of this place. Today, the town does not even play the game of *lhas*, which king and subjects would play together while cleaning Gharsisar and removing the silt. Even the *jalstambh* (water tower), erected by the lake, is leaning on one side, having been slightly dislodged. The stones of the *burj* meant for the battalion protecting the lake have also fallen down.

(ibid., p. 79)

Criticisms of the inability of the Indian and Rajasthani governments to appreciate and to sustain the ingenuity of water TEK recurs in Mishra's books. Mishra seems to be astonished by the total absence of TEK in the school system and in scholarship in general:

> Present throughout Rajasthan, this water culture has been quasi absent in the modern education system, books and bookshops. All the various governments of Rajasthan, and to top it all even new social action groups have almost erased from memory such a sophisticated work of their society … I have only been able to understand slowly, drop by drop this *smriti* (memory), *shruti* (revelation), and *kriti* (creation).

(ibid., p. 106)

Such forgetting and neglect were caused, according to Mishra, first by the long colonial rule of India, and secondly by the inability of the modern state of India to depart from habits of subservience to the intellectual primacy of the West. When Mishra realized that no parallel water traditions are found in other parts of desert states in the worlds, he presumes that 'If it ever existed, it must have been dismantled and fragmented during the long colonial period of subjugation' (ibid., p. 107).

4.4 *The ability to perceive water in the desert*

Mishra repeatedly discusses the epistemological achievement of the ability to perceive water in the desert, which he attributes to skills, described in terms of individual spiritual achievements and collective cultural heritage. The ability to perceive water is repeatedly narrated in spiritual terms, as asceticism or as *sadhana*:

> The sage asked: "What is the greatest form of asceticism?" The simple cowherd replied: "The asceticism of the eyes". The asceticism of the eyes is indeed the greatest. The experience of looking upon our surrounding world in a proper way and the collective point of view which emanates from that experience down the generations, such asceticism facilitates the passage of life from this world to the other. In the desert, the asceticism of the eyes is behind the unusual (devotion) of collecting food grains together with water.
>
> (ibid., p. 88)

The 'asceticism of the eyes' is not an individual achievement, but a 'collective point of view which emanates from that experience down the generations', and 'this is the eternal script of water and grains, which the asceticism of the eyes has written' (ibid., p. 91).

As we have seen, the Rajasthani people differentiated between different types of water, and while the *palar* water, which falls on the surface, is visible, *patal* (underground) water is not (ibid., p. 93). The ability to perceive unperceivable water requires the cultivation of unique skills:

> To be able to perceive this unperceivable water, to be able to perceive the ground water table, one needs an exceptional vision indeed. One of the names of the water which flow deep down in the *patal* is *sir* [vision], and the one who can perceive it is called *sirvi* [one who can see, vision]. It would not suffice to have the vision capable of seeing underground, *patal* water; the whole society, too, had to have a special point of view, a point of view which combines the capacity of seeing, searching, drawing and obtaining underground water with that of being able to guard oneself from the frightful mistake of losing it forever, once it has been found.
>
> (ibid.)

And, just like the one who can perceive unperceivable water is called *sirvi*, one who has vision, the experts which can dig and reach the underground water also have special name, *kiniyas,* who also possess a special type of knowledge:

The action of digging is known as *kinna* and those who dig are called *kiniyas*; *kiniyas* are people who know each particle of soil. The *sirvis* with empowered vision (*siddha drishti*) first 'see' the underground water and then the sure handed (*siddhahasta*) *kiniyas* start the digging. *Kinniyas* do not belong to a specific caste; any person skilled in this art can be a *kinniya*.

(ibid., p. 94)

The same kind of expertise is attributed to understanding clouds, and Mishra asks with admiration: 'How could the people who have such perfect knowledge of clouds, who are aware of each of their activities, including their rest, a people who have so much love for them, not count their rain drops as auspicious?' (ibid., p. 36).

4.5 Words and nomenclature as capturing primordial knowledge-culture

An intriguing part of Mishra's effort to rediscover water TEK in Rajasthan is his preoccupation with words and nomenclature. It is as if language creates and captures cultural reality and, therefore, the first thing a person should do in trying to penetrate the primordial world of Rajasthani water traditions is to become completely conversant with its language and conceptions.

There are numerous examples for this in Mishra's writings. We therefore give only few such instances. For instance, Mishra comments on the sheer number of words and names of clouds:

> It is not surprising that in this region where there are the least clouds one finds the most names for them. Any number of words have been created from the word *badal*, where the phonetic variant of by and the gender opposition come into play: *vadal, vadli, badh, badli* - a plethora of words have also poured in Sanskrit: *jalhar, jimut, jaldhar, jalwah, jaldharan, jalad, ghata, viksar, sarang, vyom, vyomchar, megh, eghadambar, meghmala, mudir, mahimandal* However, in the dialect, the names of clouds literally seem to overcast the sky: *bharabad, pathod, dharmandal, dadar, ambar, dalvadal, dhan, ghanmand, jaljai, kalikanthal, kalahan, karayan, kand habra, mainmat, mehajal meghan, mahghan, ratnyo, seher.* The nomenclature for clouds is so vast that clouds might fall short of them. Moreover any cowherd can add a name or two to this meticulously prepared list.

(ibid., pp. 34–35)

And Mishra adds that 'The empirical and linguistic richness of this rain specialist society does not stop at these forty-plus nomenclatures. The clouds are also classified according to their form, style, behaviour, character' (ibid.).

According to Mishra, it is not just the linguistic richness which is impressive, but the meanings which convey a totally different cultural outlook on the environment than the conventional technical terminology of water and infrastructure, which modern science uses. For instance, Mishra says that

In the language of the desert people the terms *mati* soil, *varsha*, rain and *taap*, temperature have a totally different meaning from the new technical definitions of soil, rain and temperature. In their culture, the asceticism of the soil, rain and heat is reflected and in this asceticism we can find both the radiance of life and its coolness.

(2001, p. 33)

Likewise, regarding the name of the ground water (*patal*), Mishra says that 'water is but one of its various names, there are also *geuo, sejo, sota, vakal pant, valiyo, bhuinjal*, add to these *talsil* and *sir*. Besides ground water, *sir* has two other meanings, one of them being sweet and the other source of regular income' (ibid., p. 97).

4.6 Symbol of the hydro-ecological self: The Talaab[7]

An expression of the place of water in Rajasthani culture is the *talaab*, a 2000-year old tattoo design, usually tattooed in the inner part of calves (the lower part of the leg). The meaning of the word *talaab* in Hindi is a pond or a pool:

> The *talaab* (or *Sita Baawadi*) is mainly rectangular in shape. There are waves in it. In its center is a point which is symbolic of life. Outside the rectangle are the steps and on all the four corners are flowers. The flowers signify the fragrance of life. It is very difficult to depict so many things in a single but simple sketch. But the mind of the engravers and the engraved have so much been imbued with love for the *talaab*, that eight ten lines, eight ten points portray the whole scene effortlessly on the body.
>
> (Mishra 1993, pp. 69–70)

The idea that people tattoo symbols of water and ponds on their bodies is an indication for the importance of water and ponds in their culture: 'One who is heart and soul with the pond, does not view it as a pit of water only. For him it is a throbbing tradition [sic] a family with a number of kins. He is well aware of who is to be remembered at what time so that the pond lives on' (ibid.). The *Talaab* signifies a close relationship between humans and nature. From the drop in the centre, waves reach the edges of the pond, enabling the growth of flowers and trees, the beauty and fragrance of life. There is a continuous movement between survival and the aesthetic and spiritual aspects of life, which give life meaning, beauty and pleasure.

5 Discussion and conclusions

What is water? Unquestionably it is a basic resource. However, by characterizing something as a resource we understand very little of its nature. This concept merely refers to our needs. A basic distinction regarding resources is between renewable and non-renewable. Water is supposed to be a renewable resource, and shortage of water often implies mismanagement. This consideration makes water a product of culture just as it is a product of nature. Perhaps, therefore, the best way of looking at resources is to observe the intricate relationships and affinities which exist between resources and the cultures which depend on them, to view them not as merely the necessities of life but also in a way the creation of cultures and societies. In other words, water may well be the creation of culture just as it is part of nature. Then we suddenly find ourselves in need of ideas discussed above, of 'sense of place' and of the 'ecological self'.

The whole of Mishra's work is nothing but a celebration of a unique 'sense of place': the sense of existing in the desert, or becoming intimate with water, and of a culture that breeds a special type of 'hydro-ecological selves'. These selves, the inhabitants of the Rajasthani Thar desert, were fully aware of their debt to the ancient traditions that enable them to sustain themselves in the desert, and regarded these traditions as sacred and their practitioners as accomplishing ethical and spiritual achievements, which was made possible by their ability to learn from nature:

> There can be water scarcity in a high rainfall region like Cherrapunji and hilly regions like Uttarakhand and there can be sufficient water available in low rainfall regions like Rajasthan - *it all depends on how we follow the directions provided by nature...* Mother Nature has always provided us enough water to survive.
>
> (Bajpayee, n.d., emphasis added)

According to Mishra, water requires 'The Commitment of Body, Soul and Wealth' (ibid., p. 134), and he adds that 'In Rajasthan all the water work has been the result of the spiritual commitment of the whole society which also enjoyed its fruits' (p. 135). Water harvesting and conservation is portrayed as requiring "ascetic" attitude and *sadhana* (devotion). These give rise to a particular form of intuition which enables one 'to perceive this unperceivable water' (p. 93). Thus, according to Mishra, water practices in Rajasthan are based on a distinct and unique epistemology, which he contrasts with Modern instrumentalist and intrusive knowledge and techniques.

The context of Mishra's work is the Gandhian intuition which emphasizes the collective, localism and time-tested indigenous traditions that enable sustainable commons to thrive in the desert. For this reason, Mishra sympathizes with criticisms of Modern technologies and methods, even when such criticisms are debated, especially regarding the assessment of the British water policies and heritage (D'Souza 2006; Levi and Mishori 2015, pp. 17–18). Nevertheless, Mishra's ideas are still of great importance in terms of narrating traditional knowledge as sustainable, and in questioning the common belief in technological progress and in the alleged superiority of the new over the old, or of Western science and know-how over the South's (Third World) traditions and ecological practices.

In this, Mishra's ideas contribute to the discussion of sustainability and resources governance, including the role of localism (I limes 2000) in our global era, on the background of other ecological philosophies, especially in India. This work, described as *sadhana*, a spiritual commitment and effort, is regarded by Mishra as bearing global importance, whose relevance stretches beyond the borders of India:

In matters of water, the depth and height which the people of Rajasthan have been able to achieve through several years of *sadhana* and through their own means, is a knowledge which must be given to regions of the country where even after good rainfall people remain thirsty. This work also seems relevant for the various deserts of the world.

(Hines 2000, pp. 106–107)

We strongly believe that the *sadhana* and knowledge which Mishra explores, and such collective achievements of TEK, represents epistemological disciplines which could serve as valuable inspirations for sustainability and for meeting pressing ecological challenges of the coming decades.

Notes

1   We wish to dedicate this chapter to the late Shri Anupam Mishra. A seeker of profound wisdom, diverse knowledge and multitude truths.
2   Virtue ethics is a moral approach that emphasizes the building of moral character and personal virtues (and vices), in the context of a concrete community, practices and traditions. Adherents of such ethics emphasize the role of model figures which are supposed to exemplify moral character and personal virtues. Discussions in this tradition tend to focus on issues such as moral constraints, the good life and the practicality of ethics. For a discussion of virtue ethics and environmental (water) ethics, see Levi and Mishori (forthcoming).
3   Quoted from an online version of Orr (1991). http://www.context.org/iclib/ic27/orr/
4   Metaphysics involves two key topics: the nature of things (ontology), and how do we know or can study them (epistemology).
5   Capra (2007) also gave other characteristics of the new ecological thinking: From quantity to quality; from structure to process; and from contents to patterns (pp. 12–13).
6   We consider Mishra as a virtue ethicist, who contributes to environmental virtue ethics and particularly to water ethics (Levi and Mishori, forthcoming).
7   This Hindi word for pond is alternately written in English as *talab*, *talaab* or *taalab*. We will use here the form *talaab*, as in Mishra (2001) and in Nawre (2013).

References

Bajpayee, Nitika (n.d.) '"Mother Nature has always provided us enough water": An interview with Anupam Mishra', *Harmony Magazine*, http://www.harmonyindia.org/hportal/VirtualPageView.jsp?page_id=11291&index1=7, accessed 15 May 2016.
Besthorn, F. H. (2002) 'Radical environmentalism and the ecological self: rethinking the concept of self-identity or social work practice', *Journal of Progessive Human Services*, vol. 13, no. 1, pp. 53–72.
Brundtland, G., Khalid, M., Agnelli, S., Al-Athel, S., Chidzero, B., Fadika, L., Hauff, V., Lang, I., Shijun, M., de Botero, M. M. and Singh, M. (1987) *Our Common Future* (\'Brundtland report\'), Report of the World Commission on Environment and Development, http://www.un-documents.net/our-common-future.pdf.
Capra, F. (1982) *The Turning Point: Science, Society, and the Rising Culture*, Simon & Schuster, New York.
Capra, F. (1995) 'Deep Ecology: A new paradigm', in G. Sessions (ed) *Deep Ecology for the 21st Century: Readings on the Philosophy and Practice of the New Environmentalism*, Shambhala, Boston & London, pp. 20–25.

Capra, F. (2007) 'Sustainable living, ecological literacy, and the breath of life', *Canadian Journal of Environmental Education*, vol. 12, no. 1, pp. 9–18.

Devall, B. (1995) 'The ecological self', in A. R. Drengson and Y. Inoue (eds) *The Deep Ecology Movement: An Introductory Anthology*, North Atlantic Books, Berkeley, CA, pp. 101–123.

D'Souza, R. (2006) 'Water in British India: The making of a "Colonial Hydrology"' *History Compass*, vol. 4, no. 4, pp. 621–628.

Heise, U. K. (2008) *Sense of Place and Sense of Planet: The Environmental Imagination of the Global*, Oxford, UK.

Hines, C. (2000) *Localization – A Global Manifesto*, Earthscan, London.

Levi, R. (3 October 2014) Personal interview with Anupam Mishra, Delhi, India.

Levi, R. and Mishori, D. (2015) 'Water, the sacred and the commons of Rajasthan: A review of Anupam Mishra's philosophy of water', *Transcience: A Journal of Global Studies*, vol. 6, no. 2, pp. 1–25.

Levi R. and Mishori D. (forthcoming) 'Water, virtue ethics and TEK in Rajasthan: Anupam Mishra and the rediscovery of water traditions', in R. Ziegler and D. Groenfeldt (eds) *Global Water Ethics: Towards a Water Ethics Charter*, Routledge, London.

Matthews, F. (1991) *The Ecological Self*, Routledge, London.

Mishra, A. (2001) *The Radiant Raindrops of Rajasthan*, Research Foundation for Science, Technology and Ecology, New Delhi, India.

Mishra, A. (1993) *Aaj Bhi Khare Hain Talab* [The Lakes Are Still Alive] (in Hindi), Gandhi Shanti Pratishthan, New Delhi, India.

Mishra, A. L. (1995) *Fundamentals of Gandhism*, Mittal Publications, New Delhi, India.

Nawre, A. (2013) 'Talaab in India multifunctional landscapes as laminates', *Landscape Journal*, vol. 32, no. 2, pp. 137–150.

Orr, D. (1991) 'What is education for? Six myths about the foundations of modern education and six new principles to replace them', *In Context: A Quarterly of Humane Sustainable Culture*, vol. 27, Winter, pp. 59–64.

Pierotti R. and Wildcat D. (2000) 'Traditional Ecological Knowledge: The third alternative (Commentary)', *Ecological Applications*, vol. 10, no. 5, pp. 1333–1340.

Rist, L., Uma Shaanker, R., Milner-Gulland, E. J. and Ghazoul, J. (2010) 'The use of traditional ecological knowledge in forest management: an example from India', *Ecology and Society*, vol. 15, no. 1, p. 3.

Rodaway, P. (2002) *Sensuous Geographies: Body, Sense and Place*, Routledge,, London.

Shiva, V. (2001) Foreword, in Mishra, A., *The Radiant Raindrops of Rajasthan*, Research Foundation for Science, Technology and Ecology, New Delhi, pp. 1–2.

Shiva, V. (1996) *Biopiracy: The Plunder of Nature and Knowledge*, South End Press, Cambridge, UK.

Shiva, V. (1993) *Monocultures of the Mind: Perspectives on Biodiversity and Biotechnology*, Zed Books, London and Third World Network, Penang, Malaysia.

Stedman, R. C. (2003) 'Is it really just a social construction? The contribution of the physical environment to sense of place', *Society &Natural Resources*, vol. 16, no. 8, pp. 671–685.

Tuan, Y.F. (1977) *Space and Place: The Perspective of Experience*, University of Minnesota Press, Minneapolis, MN.

Wilson, R. A. (1996) 'The development of the ecological self', *Early Childhood Education Journal*, vol. 24, no. 2, pp. 121–123.

Williams, D. R. and Stewart, S. I. (1998) 'Sense of place. An elusive concept that is finding a home in ecosystem management', *Journal of Forestry*, vol. 96, no. 5, 18–23.

9 Epistemological undercurrents

Delhi's water crisis and the role of the urban water poor

Heather O'Leary

The multi-dimensional importance of water

Unprecedented rates of urban growth are challenging traditional models of development and natural resource management worldwide. Successful urban growth is largely measured through the macro lens of political economy, particularly those in fast-paced developing economies like Asia. These macro-level shifts of urbanization are built upon micro-level changes to cultures and environments which are often more profoundly woven into the mundane experiences of everyday life. By understanding the growth of Asian cities through the flows of people, water and hydrosocial rhetoric, water becomes more than just a resource, but rather, its control becomes a socio-cultural marker of legitimacy in participation in development. For the millions that live in Delhi's underserved water communities, water allocation can be traced alongside claims to legitimacy of lifestyle and legitimacy of life. As Delhi becomes increasingly committed to its world–class urban identity, alternate forms of living within the city are discouraged and enforced through water: stagnating water flows that directly impact lives (O'Leary 2016) and silencing the epistemologies which interpret and valuate water. These hydrological epistemologies, or ways of knowing water, translate onto the aspirational futures of millions of water users, impacting waterscapes across multiple spatial and temporal scales.

This paper explores epistemic legitimacy at the intersection of justice and water access in Delhi, India through the perspective of the marginalized urban water poor. It draws on 18 months of ethnographic fieldwork among Delhi's water poor, tracing the water narratives that mark people's deeper relationships to their homes, their city and the larger circuits of their urban ecosystems. This ethnographic research substantiates the importance of discredited 'flows' in the city. These marginalized flows are literal and metaphoric, and are systematically excluded from water planning and discourse. As such, these legitimate flows can be considered 'undercurrents', reflecting their exclusion from 'top-down' planning mechanisms and recognizes their enduring impact on formal models. Epistemic undercurrents emerge around the 'flows' of water, water discourse and resources that map onto water, uniquely unmooring the problematic categories underlying larger narratives regulating unjust global flows.

In contemporary Delhi, urban residents have been exposed to and trained in evaluating water through its functions in the world-class city, which presupposes middle-class levels of water access. This is even more relevant in the lives of domestic workers, who cross literal and symbolic thresholds during their lives as employees in middle-class domestic waterscapes. Water produces these new spaces of flow. In parallel, just as the urban water poor work to produce and reproduce world-class water relationships, so too do they produce rituals of resistance and alternate water philosophies, including the re-incorporation and adaptation of traditional water practices and beliefs. The production of these 'undercurrents' of water allocation is imbued with valuable water epistemologies, the rich implications of which transcend the category of mundane water and spillover into the emerging realms of urban cultures and ecologies.

Epistemologies of urban development and governance in Asia

By 2050, 70 per cent of the world's population is predicted to be living in cities, with most of the absolute growth occurring in Asia, where the urban population is already facing difficulties due to inadequate infrastructures (OECD 2014). Political economists fear this rapid growth might result in an urban crisis if infrastructure and civic services cannot keep up. To ensure the security of these important nodes of global politics and trade, there is increased focus on developing systems of risk evaluation at international, federal and municipal levels, all interpreting adequate markers towards 'development', many of which prioritize advanced infrastructures, including water and sanitation. These indicators are rooted in understandings of urban development and governance which are driven by external, globally-informed models (e.g. United Nations Sustainable Development Goals, India's Millennium Development Goals and Five-Year Plans) that universalize urban development.

Universal urban development epistemologies also encompass international finance narratives, which focus on investment pathways in infrastructures, institutions and information systems to meet standards to propel 'developing' cities forward into the category of 'developed' or world-class cities. International finance institutions, namely the World Bank and Asian Development Bank, have 'built the ideological foundation ... [using] decentralization as a narrative for capital' bringing international investment and public private partnerships to the forefront of development ideologies (Baindur and Kamath 2009, p. 6). The centrality of capital to this narrative has a profound impact on the ways development ideologies affect urban governance and justice – namely refashioning governance mechanisms to respond to citizens based on their capacity to become consumers, rendering others invisible (Mankekar 1999). This 'politics of forgetting' underscores how the state exacerbates exclusionary, middle-class models of cultural citizenship (Fernandes 2003). This is a crisis of governance – it presumes that the role of the urban poor is to either become middle-class consumers or to disappear either politically or physically. The urban poor, unable to fit the consumer model, are systematically left out of development discourse.

Externalization limits the understanding of development to not only a neo-liberal but also a Westernized conception of the term. Understanding development in the Asian context is a critical step to the framing of global social realities. Reimagining development as multi-contextual can be transformative to understanding increasingly complex social and political realities. Ho (2006) argues for a different narrative of interconnected urban development in Asia – stemming not from globalization and Western development models, but rather from historical Asia. This is a late after-effect of decolonization, according to Ho: the turn away from the Euro-centric model of imperial metropoles is a 're-turn' to pre-imperial social organization, the diaspora of which was rife with deep interlinkages through dispersed kinship and commonplace religious and economic travel that spanned thousands of miles. Interconnected urban Asia – viewed as neither modelled on nor deviant from universal urban development – has the capacity to enrich development epistemologies.

Asia has not always succumbed to universal development practices – this is particularly exemplified through water services where physically different hydrology (variance in waterscapes and dynamics) helped make a new space for understanding the relevance of local context even to positivist ideology. Gandy observes, 'the episodic nature of the hydrological cycle in India—epitomized by the monsoon—also thwarted the imposition of an engineering model derived from the European experience in which a universalist impulse occluded any close engagement with the cultural, political, and also biophysical complexities of other urban contexts' (Gandy 2008, p. 126). The universal models of water services, like development models, assume homogeneity rather than complicated realities like fragmentation or cyclicality. These contextual complications mark water as inherently local, but also development, as both are reciprocally affected by the politics of place (ibid). It is not only the monsoon, but differing cultural contexts which impact development. The contemporary rapid rate and non-uniform development of India's urban spaces does not fit typical development models, and this can have major impact on building infrastructures such as water and sanitation whilst also increasing socio-political fragmentation and undermining cooperation (Ranganathan *et al.* 2009). The political and economic fragmentation in India's urban areas directly contributes to water and sanitation crises, which again affect fragmentation, as disparities further discriminative development.

Many urban poor have been evicted and denied their citizenship through many governance institutions, notably in post-millennial Delhi, which hinged on the systematic incitement of arguments of 'encroachment' and hygienic risk (Ramanathan 2004; Ghertner 2008; Bhan 2009; Dupont 2011). This strategy of externalization labelled the urban poor as polluted, unproductive, illegal sub-citizens instead of critical producers of world-class city development. The crisis of urban governance was made worse by emergent middle-class citizen activism. Middle class Resident Welfare Associations became recognized as representative of Delhi's citizens' voices, despite only existing for planned, legal colonies (Fernandes 2006; Chatterjee 2004). The exclusionary practices of middle-class citizen groups are not only physical and infrastructural, pushing the urban poor out to areas without civic services, but are also political insofar as they encroach on subaltern platforms of political representation – in a process which 'mimics' the same externalization practiced by elites in other global cities (Chakrabarti 2007). As the urban poor are rendered physically invisible through processes like slum clearing, an act in itself which rapidly increases disparities of income and development, uneven representation of discourse in governance is also masked, further impacting the agency of the urban poor (Benjamin and Bhuvaneswari 2001). As Benjamin and Bhuvaneswari suggest, successful pro-poor politics recognize the poor as longstanding agents in actively developing urban economies. This is to say that successful governance relies on recognizing the role of the urban poor as autonomous agents.

The urban poor actively produce the world-class city and are critical to its function. In the externalization of the poor, an important opportunity to improve urban development and governance epistemologies is missed. The urban poor, though externalized in epistemology, are significant agents in the development of cities. For example, the urban poor are not completely without agency upon government mechanisms (Benjamin 2004; Kaviraj 1991) because they frequently rely on informal governance epistemologies which 'lie below the radar of formal planning' (Ghertner 2011). Access to civic services in these interstitial, developing contexts is often worked out illegally or informally through 'porous bureaucracy' processes, which structure and govern 'socio-politically fractured' cities like Delhi (Benjamin 2004). In this way, informality is produced by the state (Roy 2005, 2009). As informal systems operate in tandem with the formal system, understanding their interdependency is critical to forming accurate conceptions of development.

Urban development can be enriched by recognizing informality as 'an important epistemology of planning' (Roy 2005). This provides a productive model of restructuring development to better account for distributive justice beyond Asia. For these reasons, it is paramount that discourse and epistemologies of development are understood in local context – particularly through the informal systems that are typically externalized by models. This means using new tools to deconstruct classic tropes of development categorization and classification. In Delhi, informality is a critical economic and governance epistemology that is used to structure and enrich access to essential urban civic services. While informality impacts many services, none are more relevant than water because of its role in redefining the urban poor, who, through the incongruous logic of urban development, must contend with externalization due to pollution and hygienic risk in systems which actively deny their basic right to water.

Water and fluid epistemologies

The development of urban Asia can be read through the politics of water; to know water is to know the development trajectory of the region through shifting historic flows at the macro- and micro-level. Imagining urban water use as coterminous with middle-class water use is grounded in a trajectory of water development that has a deep history in cities like Delhi. Colonial structures of social and infrastructural water disparities carry through to contemporary hegemonic systems (Guha 2006; Gupta 1986) that conceive of development in Asia as a straightforward matter of purification. World-class cities were imagined through their provision of 'world-class 24/7 European-style [water] service' which, between 1990 and 2015, transformed 1.5 billion urban city-dwellers in the global South into world-class water consumers (Goldman 2011). Through the lens of water, the intentional and hegemonic cultivation of developed water infrastructures, institutions and information systems became more than a resource management system; the control of water at municipal levels became a marker of the city's legitimacy as a world city and its status as a node in global flows of resources. Urban development was also marked by infrastructural violence, where certain types of citizens were defined by their lack of developed water by a system which, paradoxically, both created and prolonged the water poverty to which it condemned people.

Flows of water are an indicator of ongoing disparities of social justice – they cannot be understood outside of this context (Gandy 2006). The epistemologies of urban development and governance directly impact urban planning, physically building water infrastructure in its own image and defining the ways its citizens are expected to use water. Social injustices are not always concrete, but in the case of water, cultural materialism can use water as an object which reifies complex, abstract issues of disparity at the core of development. Water becomes a tangible substance through which intangible cultural values can be synthesized into a nuanced picture of the way social structures impact individual lives and broader cultures through processes of resource valuation and allocation. Water allocation operates through culturally constructed systems of social stratification, and as such, water is a traceable resource through which these structures can be examined.

Yet beyond this, water can also be used as an ontological tool to unmoor epistemologies from the rigidity which undermines accurate ways of knowing. Metaphors of water, specifically of flows, inform distinct epistemologies by foregrounding the fluidity of systems. As such, the water metaphor of 'flow' has become a common mechanism to understand the complexities of contemporary global development. Yet, this comes with challenges. Rockefeller argues, as 'flow' was appropriated into globalized development, it no longer explicitly articulated *what* was flowing (Rockefeller 2011). Because of this, Rockefeller ontologically differentiates flows of water from flows that are made of multiple objects or abstractions, highlighting the lack of specificity in the contents of flows such as 'cultural flows' or 'transnational flows'.

I argue that while these flows cannot be entirely conflated, water does provide a unique ability as a fluid object proven capable of flow through social systems and infrastructures which shape it. Water, through its flow, is both imbued with the values of these systems and also allows for epistemic growth of these systems by applying the metaphor of flow. The complexity of structures enabling both flows of water and development require epistemic fluidity. Tsing likens other, abstract flows to water flows precisely because of parallel complexity, highlighting the shared difficulty in accurately capturing the cyclical, subjectively defined, multi-stated contexts which shape them (Tsing 2002). In this conception, complex developmental flows, such as internet connectivity, human migration, international capital and global cultural ideologies, move like water in a creek bed, suggesting that 'to tell the story of the landscape requires an appreciation not only of the changing landscape elements but also of the partial, tentative and shifting ability of the storyteller to identify elements at all' (ibid.). Taking these perspectives together, the story of Delhi's waterscape requires an epistemology which integrates long-term, multi-vocal inter-subjectivity – an epistemology which cannot be achieved but for fluidity in both appraising the object and reflecting on the subjectivity of knowing.

This also falls prey to Rockefeller's second critique of flow: flow is often characterized by smooth, 'pure', agentless movements which minimize the importance of local, spatially grounded cultures while emphasizing instead broad, sweeping movements as the engine behind increased interconnectivity and social change (Rockefeller 2011). This undermines the acknowledgement of the significance of local instances where globalization is carried out by people who are the agents who drive the small encounters which become cumulative, broad flows (ibid; Escobar 2001). Stretching this further, this could amount to epistemic violence – erasing the agency of local actors through the use of broad and sweeping metaphors. However, this can be alleviated when rooting these metaphors in the physical flows of water which are shaped by embodied and epistemic acts of agency by people who direct the flow of water and the flows of conversations about it.

I argue that water is a rich object for understanding the critical relevance of subaltern epistemologies of development. In Delhi, the stark unevenness of global economic and political 'flows' can be framed in relation to water pipes; it is exemplary of the exclusive connections and widespread disconnections which establish places with uneven political, economic, social and infrastructural relations (Cooper 2005; Ferguson 2006). The uncertainty, which regulates the rate and direction of flows, can be explained at the macro-level as 'awkward connections' of development (Inda and Rosaldo 2002) and also at the micro-level, like the myriad ones that interconnect urban waterscapes as systems are repaired, merged, built over, atrophied and rendered defunct. Drawing parallel between the systems of water infrastructure and social structures which shape vital urban flows can reveal how the urban poor are agents in producing and directing flows to tangible sites. The billions of small encounters directing water make up a substantial 'undercurrent' which, I argue, shapes the waterscape of Delhi.

The physical 'undercurrents' of water in Delhi – the water which is not accounted for in government planning and formal systems – are often discussed at odds with formal models. However, they run in parallel, merging, dividing and rendering separation invisible. Undercurrents, enacted on the individual, cultural or spatial level, have dynamic power in transforming the fluidity which is used to understand key, hyper-connected nodes in global flows and the 'black holes' of polarized exclusion described by Trouillot *et al.* (2001). The role of people as agents in creating undercurrents is crucial to capturing the effects and dynamism of flows. The undercurrents, rendered invisible by development epistemologies that seek to mediate and control the outer surface, instead directly impact modelled flows. The undercurrents are grounded in specific, local conditions that regulate who receives flows of water, how the flow is accessed and controlled, to what purpose, and at what rate. The metaphor of undercurrents therefore recognizes the potential for agency and spatial embeddedness in the control over the rate and composition of flows.

As global water standards flow into Indian homes as cultural practices, they are modified by the specific cultural context of Delhi and by the systems and individuals who direct the flow of water. In the historical political-economic context of Delhi, ways of understanding water are affected by class, which impacts flows of people, water and epistemologies through highly stratified spaces. In the same way the global standards for water allocation are reinterpreted, or absorbed, into the Indian context, so too are these standards reproduced in Indian homes – the middle-class homes with world-class access to water and the informal settlements of the water poor – often by hands which are neither international nor world-class.

Epistemologies of world-class urban water

The knowledges and epistemologies of diverse planners and global development consultants define the trajectory of urban water, prioritizing 'world-class' water services as markers of development and insurance against crisis, however, these epistemologies are taken up by citizens, who incorporate these values into their everyday practices. In contemporary Delhi, urban residents have been exposed to and trained in the water of the world-class city, which often prescribes values and practices that require middle-class levels of water quantity, which meet global standards of quality. This means that global epistemologies of urban water are transmitted into some of the most private spaces in cities, mediating personal and family practices behind closed doors. However, these practices are not uniformly adopted. There is negotiation in ideologies as local people actively weigh competing values for water. The emerging, resistant practices are critiques of a form of hydrological colonialism; the new water practices are not passively accepted, rather put into local context, and contested internally. Critiques of global hydrological colonialism happen at every level of citizenship, however, groups who are marginalized from discourse, but through whose hands the water flows, are of most importance.

The most compelling negotiations of hydrological colonialism come from those who are in constant negotiation between the water worlds of different classes. Domestic workers in particular are professionals in class negotiation (Dickey 2000; Ray and Qayum 2009). Inter-class relationships involve the spillover of physical bodies and materials, and also intangible things like ideas. In a domestic worker's transition between their employers' culture and their own, the culture of the employers' homes and neighbourhood is rarely able to be simply left at the gate. The lessons and the chiding that come with the inculcation of modern, bourgeois aesthetics and ethics follow domestic workers home and occupy their mental spaces, challenging and creating a mélange of cultural markers for how to, as a middle-class wife and mother, manage a home properly, and hence, how to manage water properly. Simply put, domestic workers must grapple with middle-class employers who, at times, ask them to use more water to wash their floors than the domestic workers' families have access to in a week. This exposure affects their identity as urban citizens and as stakeholders of the city's water.

Urban water is refined both through a process of material purification and also of semantically decontaminating the concept of 'urban water'. The refinement of the term is further reproduced via mundane water rituals like those in the private, domestic spaces of urban homes. This space is an important urban node, especially with the attainment of development measured by units of middle-class families as Mankekar (1999) suggests, but more critically, as a space which is co-produced by two different types of water stewards, stratified by their relationship with urban water: middle-class employers and their (nearly always water poor) domestic workers.

The multiple spatial and temporal scales of urban water are felt by domestic workers every day and are a unique epistemological challenge to those who work so closely with disparate water systems. Domestic workers cross literal and symbolic thresholds into middle-class domestic waterscapes every day and the onus is on them to master the transition into a different way of knowing water: its properties, its applications and its limits. These mixed spaces become rich nodes of trans-cultural interaction and entanglement. Here, Harvey's (1989) insights can be applied to the urban waterscape: The reproduction of middle-class water epistemologies is based in the understanding of water through the annihilation of space and time through water access technologies.

In contrast to the refinement of urban water experienced in middle-class homes, for the larger majority of Delhi's inhabitants, urban water is inflected with epistemic value rooted in their participation in water collection not only as consumers, but foremost as part of the infrastructure itself. Water becomes a substance which people experience through temporal and physical parameters by waiting at taps and roadside pickup points and by pouring, lifting, carrying it to their homes, storing, scooping and filtering it by hand. In these neighbourhoods, the shortcomings of the government and city planners in constructing adequate piped water input and sewage output, force residents to instead pick up where their infrastructure leaves off, resulting in a different way of knowing water: through its wait and weight. People become stationary pipes, waiting to become temporary conduits, and they become the motors, as they power and push the water to the place where it is needed, and sweep it into the open gutters and guide it out once more. This resonates with Massey's (1994) power geometries. In the underserved areas, absence remains core to the conceptualization of water, but instead of an absence that marks purity, the absence is marked by empty pipes, late trucks and missed opportunities in broader integration in to the city.

Epistemologies of water are paralleled by systems of water infrastructure, which dually reflect the limits and validity of knowledge production. In the case of Delhi, ways of knowing water – through different modalities of space and time – are streamlined, just as are the idealized technologies which deliver it. This silences the potential for knowing alternate kinds of water. As the urban water poor work to produce and reproduce the technologically-mediated world-class water relationships, so too do they produce rituals of resistance and alternate water philosophies, including the re-incorporation and adaptation of traditional water practices and beliefs, like water offerings in reverence for trees. Domestic workers' transmittance and negotiation of world-class water over urban cultural boundaries occurs in three important sites: middle-class homes that are produced by domestic workers, domestic workers' own homes and the shared public spaces of the city.

In their employers' homes, domestic workers often practice clandestine water conservation. In extensive ethnographic interviews, many explain that using water to clean does not happen in gradations; if water is used, things became clean. The function of water in cleanliness is more of a binary. Urmi sums this up, 'more water is just more water, not more clean', which is contrary to the daily chiding of her many employers. In an employer's stairwell, scrubbing marble steps with a one-inch triangle of sandpaper and splashes from a one-litre bottle, Rajesh explains that he is asked to clean the floor of his employer's four room house with 30 fresh litres of potable water for each room. Like many supervisors, his employer is a vigilant housewife who frequently corrects his techniques even though they have been working together for over 20 years. Rajesh chortles and explains, 'On the days when she does not watch, I use only ten liters total—she cannot tell, only thinks she can. The floors are still clean.' The resistance to use more water is explained through phrases which ultimately come back to some of the core discrepancies of development. Domestic workers say that water over-allocation takes the water out of a neighbour's bucket. Through extended conversations about resource allocation in the city, domestic workers come back to this metaphor about water as a finite communal resource to underscore their ideology of healthy urban development coming from interdependent conservation.

These small-scale, clandestine conservation acts are demonstrative of resistance to hydrological colonialism. In this case the colonialism of 'automatic' water is tempered by the physicality of manual collection. Although many workers behaviourally comply, their internal contestation and periodic acts of clandestine resistance are part of the undercurrents of water epistemologies disregarded in traditional models. Through empowering these stories as legitimate discourse about the development of water resources in the city, domestic workers are recognized as reflexive agents who control their absorption of these values. This reflexivity both produces and is produced by differences in flow, founded on the primacy of time and space as a way of knowing. These changes in relationship to water persist with the attempt to recreate them in spaces of autonomy, such as in their own homes. Water is reinterpreted in domestic workers' homes as they replicate technologies, but not the quantities expended, of world-class water. To many, the transition to being middle-class is, like macro-development discourse, marked outwardly by water infrastructure: a sink, a water-filter, a washing machine, an air conditioner or even a toilet – despite still living in a slum settlement that would be labelled as poverty by others. In this transition, what is salient from middle-class water epistemology is the growing concern over distance from polluted waters, though the increase in water use to eradicate pollution is not glorified.

To know water through interstitial development is to document ways of conspicuously combating contagion without increased water use. In one informal settlement, washing machines, though not ubiquitous, were one of the first major water-distancing appliances that were adopted, since they are conspicuous in their prominent location outside homes' front doors and in the form of the new variety of clothes they facilitate families to wear. Though the women must manually pour in and bail out the loads of water because there is no piped water attachment, many prefer bailing to touching the dirty water much more in traditional handwashing. To treat water before drinking, some domestic workers, concerned about the black flecks that sometimes speckle their municipal water, modelled their employers' water filters by using old sari material – allowing water to pass through a few gauzy membranes to remove the visible contamination. Though few, some homes adopted water coolers for their drinking water, buying one-by-one the 20 litre Bisleri-brand pre-packaged jugs that their employers' have a subscription to have delivered. Others simply bring home drinking water in salvaged soda-pop bottles, both the ubiquitous world-class vessel and its contents brought back from their employers'. These practices ensure that their drinking water meets a middle-class standard of purity and that it is kept separately from water used for domestic chores. Two homes built separate rooms and installed toilets to create further distance from polluted water. The toilets were always western-style (seated) and the inconvenience of a lack of piped water was solved by a mug and bucket of water to rinse the toilet bowl. The toilet bowl's contents flow out into a pipe, through the wall, and into the informal, open sewer outside.

Water was understood as an object whose quality and quantity were secondary to middle-class material culture of purification. These fixtures, to many, represented a distinct transition to *pukka* life, living 'properly' in the city.

Although they do not mark the transition of families from the informal water poor into developed, world-class citizens, they represent the growing number of interstitial identities and spaces in Delhi where water epistemologies are intermingling to form into local transitional negotiations. The slower adoption of drinking water purification pills, despite their low investment cost and high efficacy, do not provide the same outcome as a carbon filter, even if the carbon filters had long since been removed. In a way, this parallels the global predilection towards visible, world-class water infrastructure, but it is reinterpreted in the context of water scarcity rooted in physical collection and financial hardship, which has limited buyout due to their limited ability to invest and their inurement to many issues of contamination. The incorporation of these technologies created an interstitial space, neither devoid nor representative of water development, though very representative of many areas of the city and the changing epistemologies of water under development.

The translation and negotiation of urban water allocation is imbued with valuable water epistemologies the rich implications of which also transcend the category of mundane domestic water and spill over into the most pressing realms of sustainability in urban cultures and ecologies. For example, in the current discourse on the Green Delhi movement and environmental consciousness in the city at large, the urban water poor are largely left out. Experts and politicians, particularly the high-level representatives interviewed at the municipal Delhi Jal Board and the national Ministry of Water Resources, assume either wanton wastage in the informal slum communities where water is delivered by tanker for free or, on the opposite end of the continuum, they assume incapacity to conserve due to destitute water poverty. Both of these ascribed categories deny the role of the urban water poor as conscious conservers of water, and in turn, this denies their participatory agency in creating the urban waterscape through directing water according to any systematic logic – whether learned or internally generated. This silences a great number of hydrological epistemologies and undermines the profound impact that the urban poor have as stewards of the city's urban landscapes and waterscapes.

Seemingly in contrast to their water conservation, many water poor of Delhi champion trees and parks in Delhi because they serve critical functions in their lives which sets the foundation for an environmental ethos that drives their everyday interactions and water allocation. As a municipal gardener described to my colleagues and I at the municipal corporation's annual garden show in the lush park abutting their slum, that one park received one and a half times the water allocation when scaled by size, meaning that any eight-by-eight foot area of grassy, topiary-studded park received one third more water than a family of seven living in a *jhuggi* (a hand-built informal hut) only a few meters away. This highlights the priorities of water allocation and its logic in the urban waterscape, however, it says much more if we listen to the interpretation of this value in the perspective of the water poor. When prompted to critically think as though they were the water 'experts' in whose hands allocation decisions rest, their proposals for re-allocation were nearly uniform.

Although many 'know' water through their painful and sacrificial relationship of scarcity, they also deeply know water through the life of plants and see the life of the city as dependent on water in this form. In this view, water becomes the green spaces of Delhi that enhance urban vitality in its multiple forms. These living, green vessels of water are spaces that are part of the extended function of the home. This form of water is the lifeblood for the livelihood of the poor as sites of networking and transportation hubs, and as a source of water in an emergency. Relationships with trees represent relationships with other people, and, sometimes, with the divine. This is corroborated by active practices of conservation, reallocation and sharing of water by the urban water poor with the city's public green spaces and through linguistic markers that accept the plants as relatives or divine incarnates. A domestic worker, Varsha, explains her water allocation, 'this tree is my mother, this tree is my goddess … I offer her the water she needs; her body is the same as my own'. In over half of the homes in a quantitative survey parallel to my ethnographic research, people gave water to plants everyday – even on days when no water came; even after four-day stretches without water. Families still reported watering the plants and trees in their neighbourhood on water-use surveys and could be seen with full mugs pouring water onto the roots of trees and plants in their neighbourhoods.

For domestic workers, the water poor and in development discourse, the idea of value is present in water allocations. What lives and lifestyles matter are those that receive water access. In the middle-class conception of water, water can be used on a continuum to assist with cleanliness, a value which takes form in these other two spheres, too. Yet it is not universal; the way value is constructed is what differentiates the stakeholders. As invisible allocators of this world-class city's water, domestic workers and the larger group of the urban water poor experience the value of water through their difficulties in acquiring it. This translates into understanding water through its ability to clean, through its ability to confer status and citizenship, and its value to unify urban fragmentation as multi-fold vitality.

Epistemological undercurrents produce new spaces where water allocation and justice are reconfigured into dynamic ideas about the role of everyday people in development. Although these perspectives are not used to inform larger political-economic discourse, macro-level systemic models of development could benefit from incorporating new perspectives of spatial and temporal scale. Delhi's water poor have the power to substantially shape the city over time, even though these 30 per cent of the city's population living in informal communities must wait for 3 per cent of the city's water allocation (CSE 2012). However, the water poor have the capacity to control greater flows of water as they are increasingly entering middle-class waterscapes in their employment, with the aim of transitioning into this category through the city's rapid economic growth in the longer-term. This heightens the pressure to understand water and urban development through systemic transition, which starts by understanding undercurrents.

Epistemological undercurrents as systemic necessity

In Delhi, systemic partiality manifests when gaps in knowledge production are ignored to build resilience universals, which further marginalize people and the realities of their unique social situation. Systemic partiality – incomplete models legitimized through traditions of epistemological disembodiment which veil the relevance of situated, multi-vocal perspectives – is a form of epistemic violence. The dominant decision-making discourse and consumer stakeholders reduce authority and legitimacy of participation in water decisions to only those who are paying consumers. Ultimately, formal water discourse rests on the global narratives that champion world-class development for middle-class consumer-citizens. However, as documented in Delhi, taking in the perspective of domestic workers – those who actively make the city world-class on a micro-level – control over flows is, in reality, multi-modal. As such, the exclusion of these perspectives is a failure of current developmental epistemes, which by limiting discourse also limit systemic resilience and security.

Opening knowledge production to include the diverse stakeholders of Delhi's waterscapes not only maintains and deepens agency among those most vulnerable through system externalization, but also creates new potential for addressing systemic gaps. The telescoping, multi-spatial and multi-temporal scales of urban water produced by fluid citizens and their fluid epistemologies can produce new solutions to securing water systems. Undercurrents produce spaces of negotiation and tension which, if given authority within system planning, can broaden both systemic waterscape discourse and increase urban rights and epistemological justice. The interdependency of formal systems and informal undercurrents does not need to be denied in emergent urban planning. Rather, the future of successful water governance recognizes the poor as longstanding, autonomous agents in developing urban economies, and that their externalization undermines urban development, creating unbalanced cities without security for all. Partial systemic security both enhances the reliance on extra-governmental flows and also puts formal systems at risk when these informal methods fail.

The incorporation of informality, including state-produced informality, in the epistemologies of planners and global development consultants could instead redefine the trajectory of urban water and development. These ways forward would prioritize accurate conceptions of development intertwining flows of resources, flows of resistance discourse and externalized flow patterns. This involves re-imagining development as multi-contextual – neither modelled on nor deviant from universal urban development – which has the capacity to enrich development epistemologies.

Development epistemologies have a great opportunity to strengthen through understanding the undercurrents of epistemologies of water in Delhi, in Asia and beyond. In the contemporary era of unsustainable growth, cities can no longer afford to discourage alternate forms of living by enforcing the hydrological infrastructural violence and epistemic violence that undermine the productivity and agency of its most vulnerable citizens. As diverse perspectives and knowledge-systems contribute to revising methodological practices and the core assumptions about research objects, categories and systems, social inquiry is strengthened. Epistemological empowerment increases the multi-vocal reframing of practical issues like water justice, whose effects ripple outward into the equity necessary to sustain our collective socio-environmental future.

References

Baindur, V. and Kamath, L. (2009) *Reengineering Urban Infrastructure: How the World Bank and Asian Development Bank Shape Urban Infrastructure Finance and Governance in India*, Bank Information Centre, New Delhi, India.

Benjamin, S. (2004) 'Urban land transformation for pro-poor economies', *Geoforum*, vol. 35, no. 2, pp. 177–187.

Benjamin, S. A. and Bhuvaneswari, R. (2001) *Democracy, Inclusive Governance and Poverty in Bangalore*, University of Birmingham, Birmingham, UK.

Bhan, G. (2009) '"This is no longer the city I once knew": Evictions, the urban poor and the right to the city in millennial Delhi', *Environment and Urbanization*, vol. 21, no. 1, pp. 127–142.

Chakrabarti, P. (2007) 'Inclusion or exclusion? Emerging effects of middle class citizen participation on Delhi's urban poor', *IDS Bulletin*, vol. 38, no. 6, pp. 96–104.

Chatterjee, P. (2004) *The Politics of the Governed: Reflections on Popular Politics in Most of the World*, Columbia University Press, New York.

Cooper, F. (2005) *Colonialism in Question: Theory, Knowledge, History*, University of California Press, Berkeley, CA.

CSE – Centre for Science and Environment. (2012) *Excreta Matters* vol. 1, New Delhi: Centre for Science and Environment.

Dickey, S. (2000) 'Permeable homes: Domestic service, household space, and the vulnerability of class boundaries in urban India', *American Ethnologist*, vol. 27, no. 2, pp. 462–489.

Dupont, V. D. (2011) 'The dream of Delhi as a global city', *International Journal of Urban and Regional Research*, vol. 35, no. 3, pp. 533–554.

Escobar, A. (2001) 'Culture sits in places: Reflections on globalism and subaltern strategies of localization', *Political Geography*, vol. 20, no. 2, pp. 139–74.

Ferguson, J. (2006) *Global Shadows: Africa in the Neoliberal World Order*, Duke University Press, Durham, NC.

Fernandes, L. (2004) 'The politics of forgetting: Class politics, state power and the restructuring of urban space in India', *Urban Studies*, vol. 41, no. 12, pp. 2415–2430.

Fernandes, L. (2006) *India's New Middle Class: Democratic Politics in an Era of Economic Reform*, University of Minnesota Press, Minneapolis, MN.

Gandy, M. (2006). 'Water, sanitation and the modern city: Colonial and post-colonial experiences' in Lagos and Mumbai (No. HDOCPA-2006-06). Human Development Report Office (HDRO), United Nations Development Programme (UNDP).

Gandy, M. (2008) 'Landscapes of disaster: water, modernity, and urban fragmentation in Mumbai', *Environment and Planning A*, vol. 40, no. 1, pp. 108–130.

Ghertner, D. A. (2008) 'Analysis of new legal discourse behind Delhi's slum demolitions', *Economic and Political Weekly*, 17 May 2008, pp. 57–66.

Ghertner, D. A. (2011) 'Gentrifying the state, gentrifying participation: elite governance programs in Delhi', *International Journal of Urban and Regional Research*, vol. 35, no. 3, pp. 504–532.

Goldman, M. (2011) 'Speculative urbanism and the making of the next world city', *International Journal of Urban and Regional Research*, vol. 35, no. 3, pp. 555–581.

Guha, R. (2006) *How Much Should a Person Consume? Environmentalism in India and the United States*, University of California Press, Berkeley, CA.

Gupta, N. (1986) 'Delhi and its hinterland: The nineteenth and early twentieth centuries', in R. E. Frykenberg (ed.) *Delhi Through the Ages: Essays in Urban History, Culture and Society*, Oxford University Press, New Delhi, India, pp. 250–269.

Harvey, D. (1989) *The Conditions of Postmodernity: An Enquiry into the Origins of Cultural Change*, Blackwell, New York.

Ho, E. (2006) *The Graves of Tarim: Genealogy and Mobility Across the Indian Ocean* (vol. 3) University of California Press, Berkeley, CA.

Inda, J. X. and Rosaldo, R. (2002) 'Tracking global flows', in J. X. Inda and R. Rosaldo (eds) *The Anthropology of Globalization: A Reader*, Blackwell, London, pp. 3–46.

Kaviraj, S. (1991) 'On state, society and discourse in India' in J. Manor (ed.) *Rethinking Third World Politics*, Longman, London, pp. 72–117.

Mankekar, P. (1999) *Screening Culture, Viewing Politics: An Ethnography of Television, Womanhood, and Nation in Postcolonial India*, Duke University Press, Durham, NC and London.

Massey, D. (1994) *Space, Place and Gender*, University of Minnesota Press, Minneapolis, MN.

OECD (2014) *Cities and Climate Change: National Governments Enabling Local Action*, jointly prepared by OECD and Bloomberg Philanthropies.

O'Leary, H. (2016) 'Between stagnancy and affluence: Reinterpreting water poverty and domestic flows in Delhi, India', *Society & Natural Resources*, vol. 29, no. 6, pp. 639–653.

Ranganathan, M., Kamath, L. and Baindur, V. (2009) 'Piped water supply to Greater Bangalore: putting the cart before the horse?', *Economic and Political Weekly*, 15 August 2009, pp. 53–62.

Ramanathan, U. (2004) *Illegality and Exclusion: Law in the Lives of Slum Dwellers, Working Paper*, International Environmental Law Resource Centre, Geneva, Switzerland.

Ray, R. and Qayum, S. (2009) *Cultures of Servitude: Modernity, Domesticity, and Class in India*, Stanford University Press, Stanford, CA.

Rockefeller, S. (2011) 'Flow', *Current Anthropology*, vol. 52, no. 4, pp. 557–578.

Roy, A. (2009) 'Why India cannot plan its cities: Informality, insurgence and the idiom of urbanization', *Planning Theory*, vol. 8, no. 1, pp. 76–87.

Roy, A. (2005) 'Urban informality: Toward an epistemology of planning', *Journal of the American Planning Association*, vol. 71, no. 2, pp. 147–158.

Trouillot, M., Hann, C., Krti, L. and Trouillot, M. (2001) 'The anthropology of the state in the age of globalization 1: Close encounters of the deceptive kind.' *Current Anthropology*, vol. 42, no. 1, pp. 125–138.

Tsing, A. (2002) 'The global situation', in Jonathan Xavier Inda and Renato Rosaldo (eds) *The Anthropology of Globalization: A Reader*, London: Blackwell

10 'Being-in-the-water' or socialisation through interactions with water in the thermal baths of Taipei

Nathalie Boucher

Introduction

Solitude, by Chou Meng-tieh	寂寞-周夢蝶
Tailing the dusk	寂寞躡手躡腳的地
Quietly creeping up from behind to enfold me, wrap around me	尾著黃昏
An incomplete moon hangs lonely in the sky	悄悄打我背後裡來裏來, 裏來
Reflected in the reed-tangled bed of the stream	缺月孤懸天中
The surface of the stream as clear as the glass in a mirror	又返照於荇藻交橫的溪底
With now and then a floating wisp of white cloud	溪面如鏡晶澈
And the shadows of birds crying softly as they skim the water ….	紙偶爾有幾瓣白雲冉冉
	幾點飛鳥輕噪著渡影掠水過……
Sitting cross-legged	我趺坐著
I see myself on the bank	看了看岸上的我自己
See my shadow in the water	在看看投影在水裏的
Roused from contemplation, I smile	醒然一笑
And with a twig snapped from a willow	把一根斷秸的柳枝
On the unblemished surface of the water	在沒一絲破綻的水面上
I carefully trace the character for "man"	著意點畫著 「人」字
Once, twice, three times…	一個,兩個, 三個……

(Leroux 2006)

The Pacific Rim is a rich and fertile area for studying the socialisation virtues of water, in that many of its cultural groups integrate aquatic public spaces into their daily lives. This has inspired many researchers from the Asia-Pacific area (e.g. Lahiri-Dutt 2006; Leybourne and Gaynor 2006; Strang 2004, 2005, 2006, 2010; Toussaint 2008; Wagner 2013) to work on the social dimensions of water, with a focus on topics such as power, management, supply, history, religion, gender and global discourse. And yet, despite the fact that in some regions, aquatic public pools are some of the most commonly frequented public spaces (Mitchell and Haddrill 2004), and despite an emergent interest in Asia for urban green spaces (Shan 2014; Tan *et al.* 2013; Zhang *et al.* 2013), we do not know much about the specific contribution of water in public space when it comes to the social fabric of the contemporary city. In Taiwan only, studies focus on tourism management and environmental issues (Chang and Holt 1991; Deng 2007; Erfurt-Cooper and Cooper 2009; Lai and Nepal 2006; Lee *et al.* 2009). We know that ever since the thermal baths of antiquity, Western aquatic public spaces have been sites where social values are expressed, where social roles are negotiated (Eliav 2000, 2010; Lehmann *et al.* 2003; Scott 2009; Watson 2006), and where ideologies and doctrines are displayed, particularly through their built form (Mumford 1964; Zitzmann 2007). However, the pertinence of using this natural resource for leisure activities in a context of exhaustion of fresh water supplies and/or privatisation (Norcliffe *et al.* 1996) (as it is the case in Taiwan) is rarely discussed, and even less studied, perhaps because many aquatic leisure activities (such as hot springs resorts and spas) have been a practice of privileged classes and countries in the Western world (Tabb and Anderson 2002).

This chapter aims to explore the social dimensions of Taiwanese hot springs, the social interactions to which they give rise and the cultural knowledge of water. The site discussed here is contemporary the Běitóu (北投) area, where locals (mostly retired – on a daily basis, and young adults – on special occasions) and tourists can choose from more than 40 hot springs facilities. The study of Běitóu's hot springs sheds light on social behaviours in water and hence remedies the lack of scientific knowledge on intimate sociality around water and its relevance in the production and reproduction of hydrosocial knowledge.

In the following pages, the study of water offers a unique way to apprehend local and global water epistemologies. In aquatic public spaces, the body becomes the main tool for communication as well as the main channel for acculturation through the sensorial experience of water (Strang 2004). Water appears to be the main interlocutor with which users interact when using a thermal bath; indeed, people precisely use aquatic public spaces to be with (around and in) water. Building on the 'definition of the situation' as applied by Richardson (2003) and drawing on the interactionist perspective and Ingold's ideas on engagement with the environment (2000), I suggest that in aquatic public spaces, water is a part of the environment with which users interact, very much as described in Meng-tieh Chou's poem. The setting, the interactions (between users and between water and bathers) and the image of the baths are three steps in the definition of the situation process through which people build knowledge about their environment and their culture. In other words, water is part of the material environment with which users interact, and these interactions contribute to the process of incorporating material and non-material knowledge.

Situated consciousness of the body

Reflections on the production of knowledge suggest that we learn what we know because we are conscious beings. Thinking in terms of consciousness (as we have in Western terms since the ancient Greeks) also supposes the existence of unconscious beings and things, associated with all that is not human, all that belongs to nature (Coleridge 2000, p. 21). There are more recent attempts to overcome this dichotomy. Bateson has suggested that things and events are unconsciously perceived by the senses, then processed and finally scanned and brought to the attention of the conscious (Bateson 1999, p. 438). Ingold goes a step further by suggesting that we know what we know because we are able to situate the information that we accumulate, and 'understand its meaning, within the context of a direct perceptual engagement with our environments' (Ingold 2000, p. 21). I would like to explore the path of our perceptual engagement with the environment. On this subject, Ingold writes: 'Placed in specific situations, novices are instructed to feel this, taste that, or watch out for the other thing. Through this fine-tuning of perceptual skills, meanings immanent in the environment – that is in the relational contexts of the perceiver's involvement in the world – are not so much constructed as discovered'(Ingold 2000, p. 22). This interactional engagement with our environments (we shape them and they shape us – Ingold 2000, p. 20) is also explored in a study of being in South America by Richardson (2003), who is himself inspired by Heidegger's phenomenology of being-in-the-world (1958). From a phenomenological perspective, to know is to become aware of one's bodily presence in the world.

Making a break from a being-focused ontological phenomenology, in which the body, as criticised by Ingold (2000, p. 168) seems to be too passive, I suggest we follow Richardson's proposition, which involves a more interactional and situated perspective. Richardson exposes the process by which human beings engage with their environments – a process that does not necessarily play out in a successive order. He has named this process the 'situation', a term coined by the symbolic interactionist Thomas (1990), for whom a situation is defined as the examination and deliberation phase that precedes any self determined behaviour. Richardson (2003) details the process as follows: first, the environment, through its material culture, presents a first definition. Richardson defines the environment as 'not so much in the isolated setting but in the manner in which the physical and thematic features distinguish, in a quasiphonemic fashion, that setting from others in the community' (Richardson 2003, p. 76). Second, an interaction takes place in a specific environment. When approaching a place such as a hot spring facility, first impressions become imbedded into behaviours. As interactions with other users and the environment take place, these behaviours make sense and interactions carry symbolic meanings. Richardson qualifies this moment as 'being-in-the-world', as opposed to being-out-of-place, that is, the initial moment when interactions clash with the first impressions (Richardson 2003, p. 80). Finally, an image emerges from the interaction that completes the definition of the situation. As Richardson (2003, p. 85) explains, situations are established through a phase of objectifying the components of a social experience into the environment. It is through interactions, the development of meaning, and the acquisition of knowledge that the environment becomes a complete and explicit image.

This way of engaging involves a dynamic process through which one feels and discovers a world *shown* by others in a specific situation. Such a perspective echoes the concept of interaction as defined by the symbolic interactionists to describe engagement with other conscious (human) beings. Goffman (1973, p. 23), for instance, writes that face-to-face interaction is the reciprocal influence that actors have on their respective actions when they are physically co-present. Interactionists (e.g. Eliasoph and Lichterman 2003; Goffman 1973; Winkin 2001) emphasize the dynamic impact of interactions by stating that the behaviours and actions of all parties involved will change over the course of interactions in order for the parties to reach common ground. My aim is to unveil how the interactions may also involve water as a dynamic actor.

I argue that sociability in water implies an interactional engagement between users (as some individuals direct others in their experience of the premises and of water) and also between water and users (as water and bathers are involved in continuous interaction). This happens in such a way that the overall state of users and water will be impacted by their interactions, and this experience yields new knowledge. Just as people incorporate their environment into the situation in order to build unity between the situation and the environment (Strang 2004, p. 50), they will also incorporate other people's knowledge and uses of water. A soaking session in the hot springs of Běitóu is a unique context that involves water in a physical environment, and, most likely, other users. The presence of other bathers transforms this 'holistic engagement of the simultaneously biophysical and social person with a particular environment' (Becker 1983, p. 50) into a very social experience. In this context, water is a material artefact, and so is the building in which the bathing takes place. But water is not only the link between the body and the material environment of the facility (Strang 2004, p. 62); nor is it only 'the physical expression of the world in which we are' (Richardson 2003, p. 76). Water is the flow and it is the content of flows (O'Leary, this volume). Water is something with which bathers interact, through their bodily perception and senses, and through which they interact with others. It is through these interactions that meaning is presented and knowledge is acquired. In this view, 'being-in-the-water' is a mode of engaging in the world that goes beyond cultural boundaries. It is the knowledge that is created out of these interactions that is culturally meaningful.

The waters of Běitóu

Because it is located at the crossing of two faults, Taiwan is home to one of the world's highest concentrations of hot springs (541 across the country – Jacobs 2010). On the Northern area of Taipei, at the foot of the mountain ranges called the Tatun Volcano Group, is the Běitóu district. The name 'Běitóu' comes from the Plains Aborigines, the Ketagalan, who call the area *Patauw*, or 'home of the witches'. Indeed, the area's atmosphere is known for being somewhat supernatural; Běitóu itself is home to the highest density of hot springs in Taiwan (Coulson 2013). The first written mention of Taiwan's hot springs dates to the seventeenth century, and describes the hot spring waters as being toxic (Erfurt-Cooper and Cooper 2009, pp. 214, 269). At the beginning of the twentieth century, James W. Davidson, a US diplomat in Taiwan, used powerful words to describe the countless number of extinct volcanoes, fumaroles and crater lakes, their smoky jade and milky waters that give off sulphurous smells (Davidson 1903, pp. 496–498).

The area's first spa was built in 1894 by a German businessman, to whom the millennial practice of bathing in European spa towns was probably very familiar. But the development of hot springs facilities grew principally under Japanese rule (1895–1945) (Lee *et al.* 2009). The need to feel at home may be one of the motives behind the Japanese investments in bathhouses in Taiwan (Jennings 2006). The Japanese practices of bathing, and even the built form of bathing buildings, have been greatly influenced by Buddhist rituals that travelled from India through Tibet and Turkestan, mingling with local customs on their way to Japan (Schafer 1956). But the Japanese practice of bathing as a mundane activity especially developed during the Edo era (1603–1868) – notably due to the need for hygiene facilities in a context of growing urban population – and in various permutations: steam baths, tubs, private baths, public baths, mixed and non-mixed baths, and baths with entertainment, baths with prostitution, and so on.

Today, the bathing practices and bath houses of Japan are largely patterned after practices dating to this period (Clark 1994). The common practice is to start with a rinse outside the bath at a station provided for this purpose (on a small bench, using a tap installed low on a wall, and a bowl), followed by a soaking period in hot water that can reach 42 degrees Celsius (109.4 Fahrenheit) (Erfurt-Cooper and Cooper 2009, p. 135; Clark 1994, p. 4). This is also the most common way to use Taiwanese baths today. In both countries, some variations have always been found, depending on the region and the facility, whether private, public, mixed or gender-divided.

During the colonial era in Taipei, the Japanese built an efficient road system, erected barracks for military staff, opened plants that extracted up to 200 tons of sulphur monthly, and set up multiple hot springs hotels in what would become a well-known sanitary resort (Davidson 1903). The first years of Japanese rule were times of segregation between Japanese citizens and Taiwanese natives (Morris 2004, p. 15), and the baths were only welcoming to the former and to successful Taiwanese businessmen. In 1905, the Bath Improvement Association started advocating for cleaner sanitary bathhouses and well-kept nearby parks, an initiative that continued on with varying degrees of success over the ensuing years (Taipei Hot Spring Association 2006).

Although the Japanese set the bar in terms of quality hosting and tourism, the hot springs gradually lost their attractiveness after Taiwan's unification with China in 1945. The reasons range from the lack of regulation over the facilities' quality, to the discomfort about naked bathing and to the irrelevance of this practice in Taiwanese culture (Lee *et al.* 2009). There are very few accounts on how locals used and represented bathing then. But for the Chinese, who were present in Taiwan before and after the Japanese colony, mundane activities (and eventually legislation) included daily washing of the hands and face, and weekly washing of the body. Special focus was put on the periodic washing of hair. Although there are some exceptions, and geographic variations can be noted throughout China, water was rarely used for relaxation (Schafer 1956). Schafer summed up the utilitarian function of Chinese bathing mainly to remove dirt (Schafer 1956, p. 60).

In Europe and North America, in recent decades, the health and wellness spa industry has become very active in marketing the benefits of 'taking the waters'. Similarly, hot springs resorts have been touted by the Taiwanese government as a significant touristic industry (Erfurt-Cooper and Cooper 2009, p. 29). This initiative stemming from Taipei, which is also an aspirant to the title of 'global city' (Deng 2007; Erfurt-Cooper and Cooper 2009; Huang and Kwok 2011; Kwok 2005; Kwok and Hsu 2005; Lai and Nepal 2006; Thorpe 2014), raises an issue that concerns all neoliberal cities (Lee *et al.* 2009), but especially Taiwan, as its recent touristic niche of hot springs and hotels brings into play Japanese tradition (as per collective memory), domestic Taiwanese practices, and foreign users' expectations (Chang & Holt 1991). The City of Taipei has created a Hot Spring Certification Mark (currently granted to more than 40 hot springs facilities) to ensure compliance with the current regulations under the Hot Spring Act, with regard to ownership, water quality, sanitation and safety. More recently, the government has considered legislating on the discharge of sewage into local streams (Mo 2013). The consequences of neoliberalism (Douglass 2000; Low and Smith 2006; Marcotullio 2003; Sennett 1976) on aquatic public spaces such as hot springs speaks to the elite standardization and commodification observed in other public spaces (Allen 2006; Cybriwsky 1999; Iwata and del Rio 2004; Loughran 2014; Low 2000; Zukin 1998, 2006). These examples contradict the received wisdom of the democratic definition of public spaces, which perpetuates a normative utopia in which people of all classes rub shoulders (Boucher 2013).

It is especially important to situate the uses of Běitóu's hot springs in the context of an urban Taipei district. For Bateson (1999, p. 451), cities are places where individuals have to seek an equilibrium, but such dense environments are new and unique in the sense that humans are the dominant species. Yet, the city surrounding the baths defines the environment that people interact with. The Běitóu area, part of which is a Taipei City heritage zone, is presented in the advertising of the Department of Information and Tourism of Taipei as a place of 'nature's bounty on exhibit, of cultured living, of hot spring refinement'. Local advertisements promote the area's gardens, parks, museums, historic neighbourhoods, hotels, restaurants and new eco-friendly library.

Methodology

This chapter focuses on a two-month pre-fieldwork period (mid-May to mid-July 2014) and a three-month fieldwork period (September to November 2015) completed in Taipei[1] as part of a broader postdoctoral project on sociability in urban aquatic public spaces. The aim of this is to shed light on the sociability of such public facilities in the context of privatization, merchandizing and homogenization of users and design. In the case of Taiwan, the data collection was based on a methodology that allowed me to grasp the social interactions and spatial patterns in play in such closed environments. I performed participant observation in five public spaces (indoor/outdoor, mixed/gender divided) with entrance fees ranging from NT\$40 to NT\$800, particularly noting the users background, the frequency peaks, the spaces appropriated and the interactions between users. My experience of bathing is tinted by the fact that I am a woman and that I was, on some occasions, accompanied by my daughter (as a baby and later as a toddler). Even in the case of mixed bathing, I observed that men and women tended to organize themselves separately in space, and even adopt different practices around the tubs. Additional fieldwork in the near future will allow me to focus on such particularities. In this chapter, I submit to the reader an initial exploration of this rich and complex terrain.

In his work dealing with South American markets, Richardson (2003) details 'situations' according to the three steps of a given occurrence: the setting, its attendant interactions and its image. I will use the same steps to present the Taiwanese baths, with a strong focus on interactions. Richardson underlines the challenge of unpacking the differences between being-in-the-two-markets without exaggerating their differences; my aim, following on Strang (2004) in her cross-cultural comparison of interactions with water, is rather to unveil the unique water-related interactions that are common to human beings who are experiencing the hot springs of Běitóu. I thus do not detail each bath's distinct features, but tease out common elements in the interactions between the baths' users, and between the physical body and the water.

The settings – the multiple forms of, and interactions with, water

Because of the design of the fieldwork and for the purposes of this chapter, I will mainly describe the water itself, as it is the feature that differentiates certain material settings from other spaces in the city. I draw on Strang's thick description of the multiple characteristics of water, which are also noted by Laozi, the philosopher and founder of Taoism, who describes water as being invincible because of its ability to change forms, its softness and strength, that shape surfaces and even solid rock (Salpeter 2013).

The visitor's first contact with water when entering a bath is with the adjacent shower or cleaning station, usually in a separate but open-air section; in the case of the cleaning station, the water comes out of a tap on the lower part of the wall. The sulphurous waters of Běitóu come in different compositions (green acidic sulphate-chloride water, white sulphurous water, and iron-neutral carbonate water), but the water in the hot springs I visited is close to a clear white, with a very light sulphurous smell. In the most luxurious places, fragrances or natural flowers can be added to the water for further relaxation. The water is held in man-made baths or ponds of different sizes (but never deeper than a man's height) and materials (ranging from wood to concrete). Waters of different temperatures are isolated in various tubs; the hottest waters can usually be found in the smaller baths. Only in one place was the hottest water to be found in a bath on an upper level, from which water poured down into other baths below, thus cooling the water in a natural process. In that case, paradoxically, the thermal water, sometimes called the 'hot tears of the Earth' (Erfurt-Cooper and Cooper 2009, p. 269), is heated underground but comes from higher levels. In many facilities, manually activated running water pours down from a top level into specific sections, for relaxing and massaging purposes. All this rushing water, in conjunction with people's constant wading into and out of the water, creates repetitive noise that, as Strang underlines, tends to prompt a trancelike state of mind (Strang 2004, pp. 52–53). Yet, this noise does not mute the sounds of ambient voices, as Běitóu's baths do not enforce a no-talking policy, a topic I will return to in the section on interaction.

Having gone through an initial bodily rinse, when preparing to enter a bath, visitors must pour a small bucket of water (from the basin) onto their feet in order to acclimate part of the body to hot water. I argue that starting from this first major contact between the body and very hot water, each individual develops a sharpened consciousness of the body, both in terms of its parts and as a whole. Strang has noted the sense of disembodiment that some experience when the body is submerged: water gives a feeling of isolation that is close to the feeling of being in the womb, in addition to bringing about a sense of unity with other elements (Strang 2004, pp. 54–57). Instead of disembodiment – and especially because of this feeling of unity with other elements – I prefer to think in terms of the emergence of a situated bodily consciousness. I can think of three ways in which this consciousness arises in the context of hot spring baths. First, water challenges gravity. It is a physical fact. But this is uniquely palpable in the case of Běitóu, whose baths are located at high elevations; in some cases, with an astonishing view of the Taipei Basin that can be enjoyed while floating in the water. Other facilities are located in valleys, which gives the impression of being at the centre of the world, directly connected with the earth through geothermal-heated fluids. Second, because water blurs visitors' vision, modifies ambient noise and alters the texture of the ground beneath one's feet, and because it provides a new context of depth, temperature, floating matter and speed of movement, water requires heightened vigilance. In the case of public bathing, water gives rise to a new perception of one's limits. The water surrounding one's body seems to be an extension of the self; the individual's bodily fluids mix with the surrounding water, body temperature might affect the water temperature, and any bodily movements are reproduced by the water until their kinetic energy dies away in a final wave. Chou Meng-tieh's poem, *Solitude*, aptly expresses how the body and shadow seem to merge with water, which also acts as a mirror of one's movements and temperature. In a social context, such as that of public baths, moving, splashing or sweating will affect other users, even in the absence of direct contact. The round-shaped bath also allows the creation of an ephemeral community of bathers experiencing together high temperature of waters, sweating or freezing.

In most places, drinking water from the baths is prohibited and visitors are urged to shower in cold waters after bathing to cleanse their skin, to rehydrate and to rebalance the body temperature.

Interactions – towards an engaged presence

The social dimension of bathing in hot springs in the nude or semi-nude makes it a highly complex coded place fraught with vulnerability, which is not comparable to what takes place at any other aquatic public space (Davidson and Entrikin 2005; Laurier and Philo 2005). For example, in her observation of the social code in swimming pools, Scott (2009) refers to the negotiated order theory and to Foucault's disciplinary power to describe how the social order is implemented through micro-actions and mundane negotiations. The social factor weighs heavily in the matter, since, as Goffman puts it, actors are self-conscious of their self-presentation (Scott 2009, p. 128, quoting Goffman 1959 pp. 108–140). Given that the image of the body is less preoccupying, at least for the Taiwanese male (Yang *et al.* 2005), and that Scott's outlook on British swimmers does not reflect the social interactions I observed in the hot springs of Běitóu, I instead adopt the perspective of Richardson in the way he qualifies interactions, namely as engaged participation and intense action vs disengaged observation and serene action (Richardson 2003, p. 84).

The hot springs of Běitóu do not strictly enforce silence as in other thermal facilities in the Western world. Their bathers interact without discretion, sometimes loudly, from one bath to another, often commenting on or addressing themselves to other users. One object of intervention is undesired contact between water and specific body parts. During one observation period, while I carried my then eight-month-old daughter, I noted the following:

> Everyone comes over to talk to us, women come to take A [my daughter] in their arms and go introduce her to other users. She's crying. An elderly woman close to us begins discussing our case with a man on the other side of the bath. We are warned not to drink the water, and to rinse A's hands before she puts them in her mouth. We are advised to rinse A's skin often. Everybody takes care of us.

I observed a similar level of engagement on the part of bathers, both towards myself, and towards others users, whether local or not. This engagement was never discreet, and showed a different level of 'respect for personal space' (Scott 2009, p. 128) than what can generally be expected in Western public spaces. One element that is given special attention during almost every bathing experience I observed is hair; bathers frequently warn others to tighten their hair, or to use the hairnets available for this purpose.

New bath users, then, are monitored and shown the dos and don'ts and how to behave by other users. I observed a very meaningful event in this regard. In a very small (nude, gender-divided) bath facility, three women were bathing and chatting in a familiar and friendly fashion. Another bather came in, almost immediately demonstrating signs that she was there for the first time. After much hesitation, and having read the outdated instructions on the wall, she headed towards the tap for a rinse. She was obviously confused as to which tools (bucket, brush, bench or towel) were provided for her use (and did not belong to other users). At one point, she discreetly asked another bather for shampoo. She was loudly admonished and told in no uncertain terms by the three women not to wash her hair, even though one of the women was actually doing so. After sitting by the bath in a state of humiliation and utter confusion for a few minutes, she promptly exited the bath section. In the dressing room, she said to me that as a first-time visitor from Korea, she did not know what procedure to follow at this facility. She was troubled by the hair issue, and subsequently shared her experience with acquaintances outside the bath. A few months later, a Taiwanese woman related to me a very similar experience at this same hot springs that she used to visit with her mother. She expressed a strong dislike for this place where regular bathers played with the rules in such a way that younger and irregular bathers feel unwelcome.

This echoes episodes related by Lin (2004) involving the subordination of daughters-in-law and maidservants in the traditional Han family, or the subordination of 'lower classes' and 'races' – an important criterion in establishing close relationships with relatives and friends in contemporary Taiwanese culture (Tsai 2006). The issue at stake here is not the degradation through domestic chores, but of bodily treatments and of water uses and access. There can be little doubt that the mixed signs and instructions given to the new visitor were meant to confuse them, to make the interactions clash with their first impressions of the facility and hence to elicit a sense of being-out-of-place and act upon this feeling by leaving the premises.

Bathers' engaged presence can be overwhelming, and one may not wish to fall under the heightened scrutiny of other users. In the case of both dressed (gender-mixed) bathing and nude (gender-divided) bathing, the visitor's body and physical behaviour will not be overlooked – although such scrutiny might be specific to women (Yang *et al.* 2005), despite the rule of etiquette that 'it is impolite to stare at other people's bodies' as set out explicitly in tourist brochures. Women who may be pregnant will be looked after, while individuals with scars or other marks will be questioned. Entering a public hot spring pushes the visitor to become attentive to his or her body and its sensitive issues, which will be put on display for everyone's gaze and comments. Some facilities do, however, offer private bathing where one can avoid public examination.

Despite a refined décor devoid of heavy decorations and ritual elements, the engaged presence of bathers corresponds to the Chinese term *re'nao* (热闹) (or Taiwanese *lau-jiat*) which literally mean 'hot and noisy', commonly used in the sense of 'lively'. It usually qualifies the human, spiritual and divine structure of the most successful religious procession. Causing heat and noise, convoy members and the audience repel negative energies and attract positive forces (Allio 2000; Yuet Chau and Moore 2013). As such, it also describes the personal experience associated with the event. It is 'the emotion that transforms formal occasions into warm and interactive events. *Re'nao* is also considered to be a manifestation of the "human flavor" (*renqing wei*) that is generated from enthusiastic human interactions' (Yu 2004, p. 138). With various spiritual connotations, we can find the same principle of 'hot and noisy' applied to sporting events (Soldani 2012), shopping (Warden and Chen, 2009), at weddings and night markets (Yu 2004).

Beyond the obvious hot and noisy characteristics of (the water in) hot springs in general, *re'nao* applies acutely to Běitóu's public baths in particular. In Běitóu's hot springs, even when the facilities are not really crowded, bodies are found in and around various water facilities (showers, hot, warm or cold baths), users are found in close proximity to one another, and ambient noises include water splashing and bathers talking. Again, even if this space of communal life is sought after in order to balance out the strong individuality characterizing the modern world (Tuan 1982, cf. Yu 2004, p. 139) – especially in a dense urban context – public bathing in hot springs does require a sharp awareness of one's physical body, personal space, and limits. Bathers have to know their tolerance to high water temperature, find their spot in a busy bath, avoid taking up too much space (some baths can hold a maximum of two persons), and walk around from one tub to another, without touching or splashing – or feeling – other's people bodies or other people's fluids.

At the same time, the rumble of the surrounding activity, compounded with the repetitive swishing sounds of the water, can be mesmerizing and call forth contemplation, as in Chou's poem *Solitude*. The bath's *re'nao* atmosphere, the warmth of the water, the closeness of other people's skin and the presence of surrounding bodies stripped of any social representations may actually prompt a feeling of isolation that encourages personal confidences and warm, attentive interactions. Some may go to the baths to find comfort and intimacy in a busy atmosphere. This has been eloquently attested to by the Korean novelist Shin Kyung-sook, who wrote about such a personal moment between two characters in a way that recalls what I have occasionally seen in the baths of Běitóu. In the novel, a woman invites her younger sister to a bathhouse, washes and combs her hair, strokes the back of her neck, puts lotion on her back, as they used to do when they were kids (Shin 2010, pp. 190–191). Given that filial piety is taken quite seriously by the Taiwanese, as an enduring Chinese tradition (Jordan 2004), hot spring baths can be expected to be a part of the close relationships maintained between relatives. The baths offer an opportunity for warm contact between close family members, for showing devotion and caring through touch, bodily pampering, and physical presence.

Despite the chaotic first impression associated with these baths, some users may also wish to have intimate interactions in these settings (although not sexual intercourse, which is another topic (Simon 2004)). Because of the unique features of water and of the social context of bathing, they may be able to do so. Interacting with and around water in the hot springs allows for meaning to emerge, for cultural knowledge to be transmitted and for socialization to take place.

The image – from Taiwanese to plain touristic

Being-in-the-water in Běitóu does not necessarily mean that visitors will have access to an entirely local cultural immersion. Many resorts offer a hot spring experience on their premises and most of them lay claim to a Japanese background. Japan is a rich source of tourists for the area (Chavez 2012), and more and more Chinese are also attracted by such a foreign experience when they visit Taiwan. For a small fee, anyone who is not staying at the hotel can indulge in a crowded hot tub without meeting any locals. Interactions are formal, cold and polite. Long-stay tourists will acknowledge one another in the baths, but typically exhibit distanced and business-like behaviour. On those occasions, the Běitóu hot springs project an image that is common to standardized neoliberal touristic sites. Yet, as part of their touristic appeal, Běitóu's hot springs may represent an exotic experience for many foreigners.

However, the boundary between touristic and local facilities is not all that clear, as local users can flood any site at any moment. Twenty-four hour baths are a popular destination for youths after long nights of partying, which underlines the similarities with attraction parks or after-hours facilities. Indeed, the hot springs bring into play other users, other interactions and other perceptions of water; the presence of bathers in the hot springs is an essential component of the interactions that contribute to defining the situation and creating cultural knowledge and meaning.

Conclusion

The baths of Běitóu offer a glimpse into the knowledge and experience of being Taiwanese. The expected public dimension of aquatic public space conceals the fact that gathering around water is far more intimate than what goes on in regular public space, and might involve different cultural and religious principles. The spiritual and philosophical meaning of water has not been explored here as they have been elsewhere (Billette and Germain 2005; Naguib 2013), but without any doubt, the being-in-the-water of Běitóu brings into play a great deal of religious knowledge. The aim of the reflection here has been to underline the importance of interactions in (re)producing knowledge and culture, and the role of water in such interactions. Water appears not only as a material artefact, but as an extension of the human body through which verbal and non-verbal communication take place. Moreover, water is affected by the users' presence, their movements and their own body temperature. Through being-in-the-water in a social context such as the hot springs, this chapter concludes, the self enters into a unique consciousness that opens up particular channels of imbibing water knowledge.

Note

1 This Formosan stay was made possible thanks to a postdoctoral internship at the Chair in Taiwan Studies at the University of Ottawa (September 2013–August 2014), and at the Department of Geography of the National Taiwan University thanks to a fellowship of the Ministry of Foreign Affairs of the Republic of China (Taiwan) (September–November 2015).

References

Allio, F. (2000) 'Marcher, danser, jouer. La prestation des troupes processionnelles à Taiwan', *Études Mongoles et Sibériennes*, vol. 30–31, pp. 181–235.

Allen, J. (2006) 'Ambient power: Berlin's Potsdamer Platz and the seductive logic of public spaces', *Urban Studies*, vol. 43, no. 2, pp. 441–455.

Bateson, G. (1999 [1972]) *Steps to an Ecology of Mind*, University of Chicago Press, Chicago, IL.

Becker, H. S. (1983 [1965]) *Outsiders: Études de sociologie de la déviance*, Métailié, Paris, France.

Billette, A. and Germain, A. (2005) *Pratiques Municipales de Gestion de la Diversité Ethnoreligieuse à Montréal: Le Cas des Piscines Publiques: Étude Exploratoire*, INRS Urbanisation, Culture et Société, Montréal, Canada.

Boucher, N. (2013) 'Back to the future in Los Angeles: A critical review of history, and new indicators of the vitality of public spaces', in B. Momchedjikova (ed.) *Captured by the City: Perspectives in Urban Culture Studies*, Cambridge Scholars Publishing, Cambridge, UK, pp. 87–106.

Chang, H.-C. and Holt, R. G. (1991) 'Tourism as consciousness of struggle: Cultural representations of Taiwan', *Critical Studies in Media Communication*, vol. 8, no. 1, pp. 102–118.

Chavez, A. (2012) 'How can Japan help save the world? Be more Taiwanese', *The Japan Times*, 29 December 2012.

Clark, S. (1994) *Japan: A View from the Bath*, University of Hawai'i Press, Honolulu, HI.

Coleridge, S. T. (2000 [1817]) 'The dialectic of mind and nature', in L. Coupe (ed.) *The Green Studies Reader from Romanticism to Ecocriticism*, Routledge, London, pp. 21–22.

Coulson, N. (2013) 'Taipei, Water City', eRenlai, http://www.erenlai.com/en/focus/2013/taipei-city-of-water/item/5446-taipei-water-city.html, accessed 27 May 2016.

Cybriwsky, R. (1999) 'Changing patterns of urban public space: observations and assessments from the Tokyo and New York metropolitan areas', *Cities*, vol. 16, no. 4, pp. 223–231.

Davidson, J. W. (1903) *The Island of Formosa Past and Present*, Macmillan & Co., London.

Davidson, R. A. and Entrikin, J. N. (2005) 'The Los Angeles coast as a public place', *Geographical Review*, vol. 95, no. 4, pp. 578–593.

Deng, W. (2007) 'Using a revised importance – performance analysis approach: The case of Taiwanese hot springs tourism', *Tourism Management*, vol. 28, no. 5, pp. 1274–1284.

Douglass, M. (2000) 'Mega-urban regions and world city formation: Globalisation, the economic crisis and urban policy issues in Pacific Asia', *Urban Studies*, vol. 37, no. 12, pp. 2315–2355.

Eliasoph, N. and Lichterman, P. (2003) 'Culture in interaction', *American Journal of Sociology*, vol. 108, no. 4, pp. 735–794.

Eliav, Y. Z. (2000) 'The Roman bath as a Jewish institution: Another look at the encounter between Judaism and the Greco-Roman culture', *Journal for the Study of Judaism*, vol. 31, no. 4, pp. 416–54.

Eliav, Y. Z. (2010) 'Bathhouses as places of social and cultural interaction', in C. Hezser (ed.) *The Oxford Handbook of Jewish Daily Life in Roman Palestine*, Oxford University Press, Oxford, UK, pp. 605–22.

Erfurt-Cooper, P. and Cooper, M. (2009) *Health and Wellness Tourism: Spas and Hot Springs*, Channel View Publications, Salisbury, UK.

Goffman, E. (1973) *La mise en scène de la vie quotidienne. La présentation de soi*, Éditions de Minuit, Paris, France.

Heidegger, M. (1958) 'Bâtir, habiter, penser', in M. Heidegger (ed.) *Essais et Conférences*, Gallimard, Paris, France.

Heidegger, M. (1962) *Being and Time*, Harper & Row, New York.

Huang, L. and Kwok, R. Y.-W. (2011) 'Taipei's metropolitan developement: Dynamics of cross-strait political economy, globalization and national identity', in S. Hamnett and D. Forbes (eds) *Planning Asian Cities: Risks and Resilience*, Routledge, London, pp. 131–157.

Ingold, T. (2000) *The Perception of the Environment: Essays on Livelihood, Dwelling and Skill*, Routledge, London.

Iwata, N. and del Rio, V. (2004) 'The image of the waterfront in Rio de Janeiro urbanism and social representation of reality', *Journal of Planning, Education and Research*, vol. 24, no. 2, pp. 171–183.

Jacobs, A. (2010) 'Taiwan's steaming pools of paradise', *New York Times*, 16 March 2010.

Jennings, E. T. (2006) *Curing the Colonizers: Hydrotherapy, Climatology, and French Colonial Spas*, Duke University Press, Durham, NC and London.

Jordan, D. K. (2004) 'Pop in hell; representations of purgatory in Taiwan', in D. K. Jordan, A. D. Morris and M. L. Moskowitz (eds) *The Minor Arts of Daily Life. Popular Culture in Taiwan*, University of Hawai'i Press, Honolulu, HI.

Kwok, R. Y. W. (ed.) (2005) *Globalizing Taipei. The Political Economy of Spatial Developement*, Routledge, Oxford, UK.

Kwok, R. Y. W. and Hsu, J.-Y. (2005) 'Asian dragons, South China growth triangle, developmental governance and globalizing Taipei', in R. Y. W. Kwok (ed.) *Globalizing Taipei: The Political Economy of Spatial Developement*, Routledge, Oxford, UK.

Lahiri-Dutt, K. (ed.) (2006) *Fluid Bonds: Views on Gender and Water*, Stree, Kolkata, India.

Lai, P.-H. and Nepal, S. K. (2006) 'Local perspectives of ecotourism development in Tawushan Nature Reserve, Taiwan', *Tourism Management*, vol. 27, no. 6, pp. 1117–1129.

Laurier, E. and Philo, C. (2005) 'Cold shoulders and napkins handed: Gestures of responsibility', *Transactions of the Institute of British Geographers*, vol. 31, no. 2, pp. 193–207.

Lee, C.-F., Ou, W.-M. and Huang, H.-I. (2009) 'A study of destination attractiveness through domestic visitors' perspectives: The case of Taiwan's hot springs tourism sector', *Asia Pacific Journal of Tourism Research*, vol. 14, no. 1, pp. 17–38.

Lehmann, D., Tennant, M. T., Silva, D. T., Mcaullay, D., Lannigan, F., Coates, H. and Stanley, F. J. (2003) 'Benefits of swimming pools in two remote aboriginal communities in Western Australia: Intervention study', *British Medical Journal*, vol. 327, no. 7412, pp. 415–419.

Leroux, A. (2006) 'Poetry movements in Taiwan from the 1950s to the late 1970s: Breaks and continuities', *China Perspectives*, vol. 68, pp. 56–65.

Leybourne, M. and Gaynor, A. (eds) (2006) *Water: Histories, Cultures, Ecologies*, University of Western Australia Press, Crawley, Australia.

Lin, C.-J. (2004) 'The other woman in your home; social and racial discourses on "foreign maids" in Taiwan', in D. K. Jordan, A. D. Morris and M. L. Moskowitz (eds) *The Minor Arts of Daily Life: Popular Culture in Taiwan*, University of Hawai'i Press, Honolulu, HI.

Loughran, K. (2014) 'Parks for profit: The high line, growth machines, and the uneven development of urban public spaces', *City & Community*, vol. 13, no. 1, pp. 49–68.

Low, S. M. (2000) *On the Plaza: The Politics of Public Space and Culture*, University of Texas Press, Austin, TX.

Low, S. M. and Smith, N. (eds) (2006) *The Politics of Public Space*, Routledge, New York.

Marcotullio, P. J. (2003) 'Globalisation, urban form and environmental conditions in Asia-Pacific cities', *Urban Studies*, vol. 40, no. 2, pp. 219–247.

Mitchell, R. and Haddrill, K. (2004) 'From the bush to the beach: water safety in rural and remote New South Wales', *Australian Journal of Rural Health*, vol. 12, no. 6, pp. 246–250.

Mo, Y.-C. (2013) 'Beitou district rezoning to legalize hot springs', *Taipei Times*, 9 October 2013.

Morris, A. D. (2004) 'A history troubled and glorious', in D. K. Jordan, A. D. Morris and M. L. Moskowitz (eds) *The Minor Arts of Daily Life: Popular Culture in Taiwan*, University of Hawai'i Press, Honolulu, HI.

Mumford, L. (1964) *La Cité à Travers l'Histoire*, Éditions du Seuil, Paris, France.

Naguib, N. (2013) 'Aesthetics of a relationship: women and water', in J. E. Wagner (ed.) *The Social Life of Water*, Berghahn, New York.

Norcliffe, G., Bassett, K. and Hoare, T. (1996) 'The emergence of postmodernism on the urban waterfront; Geographical perspectives on changing relationships', *Journal of Transport Geography*, vol. 4, no. 2, pp. 123–134.

Richardson, M. (2003) 'Being-in-the-market versus being-in-the-plaza: Material culture and the construction of social reality in Spanish America', in S. M. Low and D. Lawrence-Zuniga (eds) *The Anthropology of Space and Place: Locating Culture*, Blackwell, Malden, MA.

Salpeter, D. (2013) 'Water in classical Chinese literature', eRenlai, http://www.erenlai.com/en/focus/2013/taipei-city-of-water/item/5434-water-in-classical-chinese-literature.html, accessed 3 July 2015.

Schafer, E. H. (1956) 'The development of bathing customs in ancient and medieval China and the history of the floriate clear palace', *Journal of the American Oriental Society*, vol. 76, no. 2, pp. 57–82.

Scott, S. (2009) 'Reclothing the Emperor: The swimming pool as a negotiated order', *Symbolic Interaction*, vol. 32, no. 2, pp. 123–145.

Sennett, R. (1976 [1974]) *The Fall of Public Man*, Faber & Faber, London.

Shan, X.-Z. (2014) 'The socio-demographic and spatial dynamics of green space use in Guangzhou, China', *Applied Geography*, vol. 51, pp. 26–34.

Shin, K.-S. (2010) *I'll Be Right There*, Other Press, New York.

Simon, S. (2004) 'From hidden kingdom to rainbow community; The making of gay and lesbian identity in Taiwan', in D. K. Jordan, A. D. Morris and M. L. Moskowitz (eds) *The Minor Arts of Daily Life: Popular Culture in Taiwan*, University of Hawai'i Press, Honolulu, HI.

Soldani, J. (2012) 'La fabrique d'une passion nationale: Une anthropologie du baseball à Taïwan', PhD thesis, Université Aix-Marseille, France.

Strang, V. (2004) *The Meaning of Water*, Berg, New York.

Strang, V. (2005) 'Common senses: Water, sensory experience and the generation of meaning', *Journal of Material Culture*, vol. 10, no. 1, pp. 92–210.

Strang, V. (2006) 'Aqua culture: The flow of cultural meanings in water', in M. Leybourne and A. Gaynor (eds) *Water: Histories, Cultures, Ecologies*, University of Western Australia Press, Crawley, Australia.

Strang, V. (2010) 'The Summoning of Dragons: Ancestral serpents and indigenous water rights in Australia and New Zealand', *Anthropology News*, vol. 51, no. 2, pp. 5–7.

Tabb, B. and Anderson, S. C. (2002) *Water, Leisure and Culture*, Berg, New York.

Taipei Hot Spring Association (2006) 'Brief history of Beitou', http://www.taipeisprings. org.tw/english/beitou/annals.htm, accessed 16 September 2014.

Tan, P. Y., Wang, J. and Sia, A. (2013) 'Perspectives on five decades of the urban greening of Singapore', *Cities*, vol. 32, pp. 24–32.

Thomas, W. I. (1990 [1923]) 'Définir la situation', in Y. Grafmeyer and I. Joseph (eds) *L'école de Chicago: Naissance de L'écologie Urbaine*, second edition, Champs Flammarion, Paris, France.

Thorpe, D. (2014) 'Key performance indicators for eco cities in the Asia-Pacific region', Sustainable Cities Collective, http://sustainablecitiescollective.com/david-thorpe/256536/key-performance-indicators-eco-cities-asia-pacific-region, accessed 8 January 2015.

Toussaint, S. (2008) 'Kimberley friction: Complex attachments to water-places in Northern Australia', *Oceania*, vol. 78, pp. 46–61.

Tsai, M.-C. (2006) 'Sociable resources and close relationships: intimate relatives and friends in Taiwan', *Journal of Social and Personal Relationships*, vol. 23, no. 1, pp. 151–169.

Tuan, Y.-F. (1982) *Segmented Worlds and Self: Group Life and Individual Consciousness*, University of Minnesota Press, Minneapolis, MI.

Wagner, J. E. (2013) *The Social Life of Water*, Berghahn, New York.

Warden, C. A. and Chen, J. F. (2009) 'When hot and noisy is good: Chinese values of renao and consumption metaphors', *Asia Pacific Journal of Marketing and Logistics*, vol. 21, no. 2, pp. 216–231.

Watson, S. (2006) *City Publics: The (dis)enchantments of Urban Encounters*, Routledge, London.

Winkin, Y. (2001) *Anthropologie de la Communication: de la Théorie au Terrain*, Éditions du Seuil, Paris, France.

Yang, C.-F. J., Gray, P. and Pope, H. G. (2005) 'Male body image in Taiwan versus the West: Yanggang Zhiqi meets the Adonis complex', *American Journal of Psychiatry*, vol. 162, pp. 263–269.

Yu, S.-D. (2004) 'Hot and noisy; Taiwan's night market culture', in D. K. Jordan, A. D. Morris and M. L. Moskowitz (eds) *The Minor Arts of Daily Life: Popular Culture in Taiwan*, University of Hawai'i Press, Honolulu, HI.

Yuet-Chau, A. and Moore, H. L. (2013) 'Actants Amassing (Aa)', in N. J. Long and H. L. Moore (eds) *Sociality: New Directions*, Berghahn Books, New York, pp. 133–155.

Zhang, H., Chen, B., Sun, Z. and Bao, Z. Y. (2013) 'Landscape perception and recreation needs in urban green space in Fuyang, Hangzhou, China', *Urban Forestry and Urban Greening*, vol. 12, no. 1, pp. 44–52.

Zitzmann, D. (2007) 'The Russian Banya under communism. Aleksandr Nikol'skii's Leningrad baths', *Osteuropa*, vol. 57, no. 1, pp. 97–112.

Zukin, S. (1998) 'Urban lifestyles: Diversity and standardization in spaces of consumption', *Urban Studies*, vol. 35, no. 5/6 pp. 825–839.

Zukin, S. (2006) 'L'espace public: expression de divisions', *Revue d'urbanisme*, vol. 350, September–October, pp. 67–69.

11 In the eye of the storm

Water in the cross-currents of consumerism, science and tradition in India

Neeraj Vedwan

Introduction

In the last decade, water has emerged as a political, economic and social flashpoint in India (Biswas 2016; Tiwari *et al.* 2009; Joy *et al.* 2008). Increased urbanization, industrialization and the expansion of commercial agriculture have added up to increase the pressure on water resources (Walker 2014). The state has responded to the intensifying crisis with a characteristic mixture of nonchalance and ideologically-motivated measures of dubious efficacy. One set of responses has entailed privatization or public-private partnerships, whose ostensible goal is to attract private investment into the water sector to promote efficiency.[1] This is particularly the case with hydropower, where the state has signed hundreds of memoranda of understanding (MOU) with private power companies (Seth 2014). Similar investment in water distribution, especially for domestic consumption, has been less forthcoming due in part to the grass-roots resistance which, although piecemeal and ad hoc, has nevertheless created uncertainty and deterred investors. The resistance is rooted in the apprehension that privatization will lead to price increases across the board, thereby making water even more inaccessible to the poor (Nair 2015).

A conspicuous feature of the debate surrounding water resources policy has been its embrace of the dichotomy between state-controlled, bureaucratic management and a pro-privitization and profit-driven philosophy.[2] The community-driven approach based upon grassroots empowerment has largely remained peripheral to the official discourse on water policy. For instance, Integrated Water Resources Management (IWRM), which enshrines participation as a cornerstone of its overarching philosophy, has been critiqued for not being radical enough in challenging the status quo (Biswas 2008; Ferreyra *et al.* 2008). Perfunctory participation has been the norm rather than exception in water resources management (Butterworth *et al.* 2010). The concept of a water user association, for instance, is a feature closely associated with IWRM to engage communities in the delivery of water resources by augmenting the state's efforts. The associations are consequently conceived as appendages to existing institutions, accepting the state's role as the allocator, and focusing almost solely on improving the efficiency of its efforts (Mollinga *et al.* 2006).

The efforts at involving communities in water resources management, in urban settings, have in most instances amounted to little more than 'responsibilising citizens' (Mosse 2008, p. 947) to take on a complex and difficult task that the state is unwilling to shoulder.

In this chapter, I focus on how political–economic preferences rooted in a particular neoliberal development discourse, shape the epistemological underpinnings of the regime(s) of water resources management that despite institutional and technical innovations, continue to dominate in India. What kinds of values, knowledge and perspectives does the entrenchment of a water resources management and philosophy that accords the highest value to technical and analytical expertise privilege (Preston 2004)? I draw upon a variety of data: surveys of a middle-class residential neighbourhood in Delhi, interviews with sugarcane farmers and media coverage of water issues, to specifically focus on the ways in which the ascendance of consumerism, along with the decline in environmental and social movements (Ray and Katzenstein 2005) has paved the way for an individualization of the water crisis.[3] The chapter concludes with a discussion of the emerging contradictions that characterize the role and meaning of water in people's lives and in policy discourses. In other words, despite the dominance of certain actors and metrics, the discourses of water are far from totalizing, and offer interesting possibilities for a broad questioning of the dominant paradigms of natural resource management and human–environment relations, and for creating new spaces for social and individual agency.

Consumerism, science and tradition in India

One of the most significant fallouts of the philosophical and practical stalemate in water resources policy-making – reflected in the public resistance, albeit sporadic, litigation and loss of credibility of water resources management institutions – has been the issue of access to domestic water. The state has clearly failed in its attempts, admittedly inadequate, to even provide a modicum of domestic water security to its most marginalized socio-economic groups (McKenzie and Ray 2009; Gandy 2004). The state has even failed to recognise and legitimise the myriad ways in which access to water is secured by the marginalized sections of society. Access to water, far from being a straightforward notion or simply a makeshift arrangement in the absence of de jure rights, can be conceptualized in a variety of different ways. On the one hand, lack of access can be seen to reflect lack of purchasing power in a marketplace that has commodified water. In contrast, Ranganathan and Balazs (2015) surmise that differential access is an outcome not so much of 'bundles of rights', but of 'bundles of power'. In the absence of a legal framework that would ensure water availability to all, the less fortunate sections of the population, especially the urban poor, are left to devise cultural-symbolic and micro-political strategies to enable access to water (Ribot and Peluso 2003). One proposed approach to redress the inequalities in access to drinking water is to deem access to drinking water a human right. However, Bakker (2007) describes the human rights approach to drinking water as being a continuation of the status quo, in as much as it implies the acceptance of an individualistic and perhaps market-based view of water. Focusing on the property rights regimes that govern access to and regulate drinking and irrigation water, could be a more effective non-state and non-market solution. From this perspective, conferring centrality on communities can assist in accomplishing important ethical and efficiency gains through recognition of their specific claims to water arising out of specific traditions and histories (Pahl-Wostl *et al.* 2007).[4]

The exclusion of local beliefs and experiences can have profoundly disenfranchising effects, besides lowering the likelihood of success of water management projects (Pahl-Wostl *et al.* 2011). As Mukhtarov and Gerlak (2014, p. 108) point out in their analysis of the failure of an IWRM project: 'By not engaging with the values, myths and images invoked in narratives advanced by the citizen groups, engineers and politicians ignored the roots of public anxiety and eventually failed to proceed with the project.'

The overarching water resources management philosophy, with its emphasis on monetisation, even co-opts extreme events for its purposes. Mehta (2001, 2005) describes the political ecology of drought and water scarcity in the states of Rajasthan and Gujarat and the use of extreme weather events – whether perceived or actual – to legitimise capital and technology intensive approaches to water management which would disproportionately benefit the urban and rural elites. Powerful actors, such as politicians and industrialists in these states, associated with institutions such as the Sardar Sarovar Project, have built a narrative of declining water supplies and threatened livelihoods to buttress their rationale for the commercial control of scarce water supplies in the name of development, a script that is being enacted widely in the country.

A turning point in the conceptualisation of water can be traced to the altered state-society relationship in the post-1991 period. The macro-economic changes in India following liberalisation led to a weakening of the developmental state and the enshrinement of the consumer as the engine of the economy. The concomitant changes in popular culture, facilitated by the mass media, positioned lifestyle and consumption as the bridges between individual aspiration and national well-being. The citizen-consumer – a concept that lies at the intersection of nationalism, consumerism and valorised technology – becomes the intended target, as well as agent of water resources management. The interplay of political economy of water and its epistemology is found in a cultural politics of water resources management, which pits city people against rural residents, middle class against the poor, and farmers against non-farmers – just to mention a few socio-economic fault lines (Mollinga 2010). The contestation over water between these disparate groups represents 'struggles over water [that] are simultaneously struggle[s] for power over symbolic representations and material resources' (Baviskar 2007, p. 1).

Consumerist ethos, which has assumed a hegemonic status among the middle class (Fernandes 2000), is associated with a view of water as simply a commodity whose availability should be governed by laws of demand and supply. Water, a public good, has therefore come to be treated by the state as a privilege affordable to a few, instead of being a universal entitlement.[5] This view has important implications for consumptive behaviours as well as for influencing new middle class identities that are taking shape in a globalizing India. The expectation that water should be available to those who can afford it in a big city like Delhi speaks simultaneously to a particular imaginary involving nationalism and consumerism,[6] as well as what has been termed the 'naturalization' of water (Swyngedouw 1997, p. 322). A common impulse connecting these phenomena is a form of biopolitics that defines control over nature as a metric of progress (Bakker 2012). The process of seeing water as a techno-monetary hybrid is part and parcel of the rise of the modern cityscapes.

Technical characteristics of water resources management affect the level and depth of public engagement. While science is ostensibly being deployed by the state in the service of rational water resources management for the benefit of all, the effect is to accentuate a top-down and capital-intensive model of water resources management that works, if at all, through the exclusion of specific areas and segments of the population. Mosse (1999, p. 947) describes one particular instance of the nexus between technology and privatization in the Indian context:

> [T]he water engineers and frontline staff who put the corporatisation of water into practice also tend to equate the citizenry with private, paying customers. In these terms, the slum-dwelling poor are marginalised as non-paying, non-deserving, difficult customers.

The dichotomies characteristic of Indian social and political life – rural-urban, Gandhian-capitalistic, caste binaries, subsistence-commercial – are superimposed on water resources management to create hierarchical and repressive structures. In general, larger, capital-intensive and technologically complex systems are more amenable to centralized control. However, even highly dispersed forms of irrigation, such as the water tanks, have been infiltrated by broader political systems and influence since the pre-colonial times.[7] Technology, as in space or weapons technology, becomes an expression of nationalism, a development clearly manifested in the preference and adulation for 'mega' projects like river-interlinking, which would transfer water from surplus river basins to deficit areas.

Swyngedouw (1997) has termed this technology-mediated transformation of water, from an erratic and even mystical gift to a controllable commodity, as its naturalization. Water – its control, sharing, flow and use – is driven by power, while power itself is exercised through water. The reimagination of the concept of citizenship in India, in exclusive and competitive ways, is linked to emerging environmental subjectivities. For the citizen-consumer, the view of the environment as a mere accessory to a modern and cosmopolitan lifestyle has led to paradoxical developments. For instance, technologies such as household water filtration, while affording urban residents access to clean water, which is now seen as a health commodity, also enable insulation from the larger social ecologies. In this regard, micro-technologies, such as those used in household water filtration, as well as macro-technologies involved in dam construction, for instance, have the same effect of promoting an instrumental relationship to water and, ultimately, alienation from it.

As described earlier, one of the key social changes that accompanied the neo-liberal reforms of 1990s was the transformation of middle-class sensibilities and the associated changes in the media landscape. In this chapter, I attempt to delineate the discursive similarities, derived from the hegemonic status of neoliberal beliefs among the middle class, between three seemingly different cases involving water. In the first, the middle-class values and perspectives related to drinking water in Delhi are explored, whereas in the second, a recent controversy in India pertaining to the use of water for maintaining cricket grounds is examined. In both cases, related to drinking water and agricultural water, a preference for market-based solutions, technical fixes and an affinity for 'efficiency' are evident. However, both the middle class and the media is largely dismissive of the redistributive stance of the state, and fails to view the issue of access to water through the lens of rights, entitlements or as abdication by the state of its basic responsibility. In contrast to these dominant narratives, the third case involves sugarcane farmers who describe the role of sugarcane in their lives in terms that underscore how culture and history have moulded the multi-dimensional relationship between people and the land. These farmers' stories are a world removed from the controversy surrounding the role of sugarcane in creating India's water crisis. Whereas the debates about sugarcane's adverse effects on water are framed in terms of metrics – like litres of water consumed per kilogram of sugar produced – farmers describe sugarcane in terms of supporting their precarious livelihoods and, in the long-term, transforming their identities.

A pluralistic approach to water resources policy would require that these suppressed and neglected voices and images also be accorded recognition and significance. The mutual constitution of water and society – themselves hybrid entities – has been variously captured through the concepts of 'hydrosocial cycle' (Linton and Budds 2014, p. 178) and 'water worlds' (Barnes and Alatout 2012, p. 483). These concepts consider water not as a singular, a priori, entity, but as a category that is, constituted in active engagement with simultaneously biophysical, social, economic, cultural and political forces. The meanings and symbolism of water, therefore, are produced in a dialectical relationship with the technological, economic and socio-political realities, which are configured by water in an historically contingent way. In other words, water is an inherently socio-natural category, manifested in a wide array of ways, with the result that 'different kinds of waters are realized in different hydrosocial assemblages' (Linton and Budds 2014, p. 175). That the relationship between water and society is forged at the level of their constitutive forces enables a non-deterministic and dynamic understanding of the shifting roles and meaning of water.

'We are reducing people to beggars: Water is not free'

One of the main initiatives promised by Arvind Kejriwal, the insurgent political candidate who became the chief minister of Delhi in 2015, was to provide 20,000 litres of free water each year to every family. It is a widely-known fact that the vast majority living in India's capital city reside in unauthorized settlements and lack reliable access to water. This electoral promise was an interesting turn in a long-running debate on water that had emphasized rationalizing the consumption of water. This meant that appropriate pricing of water was seen as the key to ensuring its sustainable use. I decided to pursue this issue with a set of middle-class respondents (n = 85) in a solidly middle-class neighbourhood in north-western Delhi, who were almost uniformly concerned about the situation prevailing with drinking water (i.e. scarcity and quality). Most of the sampled residents work for the private sector or for the government, though approximately 20 per cent own small businesses. The residents receive two hours of municipal water supply daily, one hour each in the morning and evening. However, in the summer it is not uncommon for residents to go without water for two or three days at a stretch. During these times, water is delivered by tankers and is inadequate to meet daily needs. Another aspect of the water supply that concerns residents is the quality. Most residents consider tap water to be of poor quality, and have installed filtration systems to purify water before consumption (Vedwan 2015).

When asked about water problems, most of the respondents mentioned waste, political problems and poor management as the biggest issues. Unauthorized settlements (or 'squatter colonies') were often blamed for siphoning off municipal water supplies. Water shortage was not seen as a structural issue. This is despite approximately a quarter of all households in Delhi lacking access to piped water supply. Delhi has seen the biggest demographic increase of any metropolitan city in the country, and the water availability has worsened because of reliance on neighbouring states for water supplies, which themselves are facing increased pressure from economic and population changes (Khandekar and Vasudeva 2013).[8] A 35-year-old corporate employee, whose family includes two children and his wife, said:

> People in this country are not used to paying for anything. It takes money to bring water to the tap, so what is the Chief Minister doing promising free water? The price of water is too low, but instead of increasing it we have [political] gimmicks. This is making people into beggars, even people who can afford to pay. How is this going to solve our water problems?

Another respondent, a retired government employee, mentioned a recent newspaper report (Bhatnagar 2010) that blamed leaky pipes for losing up to 30 per cent of the water flowing through them. Further, 'population increase has made it impossible to meet everyone's water needs', he said. People who have money are not affected by shortages, and 'this has always been the case'. Upon being asked what should be done to provide everyone with enough water to meet their basic needs, he said people must do the best that they can, and the government cannot take care of every issue under the sun.

Water was mentioned by several respondents, along with electricity, as a key amenity that had influenced their decision to live in the neighbourhood. The apartment complex studied is known in the area to have a fairly reliable supply of water and power. One resident, a software engineer who commutes to the Delhi suburb of Gurgaon, noted:

> We plan to move to Gurgaon in the next four or five years. The commute takes up a lot of my time and we have been thinking of renting an apartment there. One of the big problems in Gurgaon is that water supply is quite uncertain. Although they have much better quality apartments there, they are expensive. And it's not only rent, but power bills can also be very high because most apartment complexes have diesel-powered generator sets that supply power during the frequent load-shedding. Power and water are not as big of a problem here. But this is the capital of the country and like any big city abroad we should have 24/7 water and power. These should not even be issues.

The lens being used to diagnose water problems is primarily commercial and transactional, wherein the problem is a lack of purchasing power and obsolete technology (as in leaky pipes). What is interesting in this formulation is the scale and scope of the problem. In contrast to rural areas where water and power shortages have a long history of triggering public demonstrations and mass protests, the individualized and consumerist view of water in the city reduces it to a private concern.[9]

The relative anonymity and impersonality of social relations, coupled with segregation along class lines, makes it easy for middle-class respondents to treat water problems as resulting from flaws in people unlike them. A certain moralization of the water resources discourse is the result, which envisions a disinterested regime of technology and prices to be administered by managers as an antidote to – in their view – a debased and hopelessly populist politics. In the moral-semiotic universe of the middle-class residents, words like equity and social justice have been replaced with a profoundly asocial vocabulary centred on the individual as the agent as well as target of policy. The roots of this material-semiotic regime can be traced back to the economic liberalization of the 1990s, which has produced a diverse array of interests, with varying symbolic capital, vying for control over water resources. For instance, as will be clear from the discussion in the remainder of the chapter, in the current sociopolitical context the case for continuing the utilization of water for Indian Premier League (IPL) cricket matches is more effectively made than mounting a defence on behalf of sugarcane growers.

Sugarcane cultivation and India's water crisis

Sugarcane is an important cash-crop in the western, north-western and southern regions of India. Sugarcane farmers constitute a relatively prosperous and vocal section of the farming community in India. A result of the political clout of sugarcane growers has been a comparatively generous policy of price support provided by the provincial and federal governments, which has buttressed farmers' incomes in this sector (Sukhantkar 2014). A corollary of the political influence enjoyed by the sugarcane farmers has been that they have remained immune to the increasing criticism of the negative impact of this crop on water (Kuber 2016). Sugarcane is one of the most water-intensive crops grown in India, perhaps only second to rice, so the shifting discourse about its impact on water resources represents an interesting turn.

In the last decade or so, the politics surrounding sugarcane has been preoccupied with the issue of support price (Bhatt 2014). Farmers and their interest groups continually lobby for a higher government support price, which determines the price that the farmers get for their sugarcane from the sugar mill, be it private or public. More recently, however, the water-intensive nature of sugarcane has come under greater scrutiny (Shrivastava *et al.* 2011). To better understand the perspectives of sugarcane farmers, interviews were conducted in June and July 2015 with 15 individuals who have been engaged in growing sugarcane for decades. These interviews were conducted in a village in western Uttar Pradesh, a major sugarcane-producing region in India. The goal of conducting these interviews was to elicit the sugarcane growers' perspectives on the material and symbolic significance of sugarcane in their lives.

Western Uttar Pradesh, a region comprising over 20 districts, and the neighbouring states of Punjab and Haryana are known as the agricultural heartland of India. In western Uttar Pradesh, large parts of the region lack access to canals and river waters, and are therefore dependent on tubewell irrigation. Before the Green Revolution spread to these parts in the 1970s, agriculture was rain fed, which highly restricted sugarcane cultivation. The situation changed with the initiation of government efforts to promote groundwater irrigation, and the provision of state subsidies that led to a tremendous increase in the number of privately owned tubewells (Jewitt and Baker 2007). One of the consequences of the enhanced irrigation capacity was the ascension of sugarcane as the primary cash crop in the region (Siddiqui 1997). At the time, scant attention was paid to the long-term consequences of growing sugarcane for groundwater levels. In the past two decades, groundwater levels have plummeted in the area; while in the past, wells averaged a depth of 30 to 40 feet, a farmer now has to dig down to almost 150 feet to get groundwater. Many existing wells have to be deepened on a regular basis so that they can continue to provide water, and operating the wells has become an expensive business.

For farmers, sugarcane is neither the domain of fat-cat sugar barons, nor is it a catalyst for environmental disaster. In my interviews, an account of sugarcane emerged that emphasized its transformative effect on the lives of rural inhabitants, which would not have been possible without the package of technological and economic changes encapsulated in the Green Revolution. Unlike the media discourse and that of middle-class urban residents, which emphasizes the need to rationalise and economise water use, a different narrative rooted in the history of rural deprivation, memory of marginality and the aspiration for self-sufficiency, undergirds farmers' understanding of sugarcane – and by extension water – and its place in their lives. A 73-year-old farmer explained the importance of irrigation for their livelihoods:

> Sugarcane requires a lot of water, but why single out sugarcane when all crops need water? When there was no water [pre-Green Revolution] we could not even grow crops on our entire farm. Of the 30 bighas [six acres] we have, we could only raise crops on 15-18 bighas; the rest had no water. It was after we dug a tubewell that we were able to farm all of our land. Our village and the neighbouring villages were so poor that no outsider wanted to marry their daughter into our families. Wheat was so scarce that we used to feed it to our guests, while we ate mostly *chana* (gram). Now even the poorest person has no shortage of wheat to eat.

The sugarcane farmers recognize the centrality of water to their lives. One of their biggest complaints was about inadequate electricity, which on average is available for only about six hours a day. During the period in which these interviews were conducted, the daytime temperature routinely exceeded 43 degrees Celsius, yet the farmers wanted electricity for watering their parched fields rather than for other uses, for example, fans. When the electricity came on late in the night, sometimes at 3am, the whole neighbourhood would wake up and buzz with activity. Farmers would scamper off to their fields, located at a mile or more from their homes, to turn on their tubewells. On the issue of water and sugarcane, a 68-year-old farmer explained:

> Before sugarcane we had to struggle for everything. The first thing we did after we installed a tubewell was to grow more sugarcane. Earlier when we had to lift water out of wells with bullocks, we raised cane on only 2-3 bighas, but now we grow it on 25 bighas. We complain about the late payment of our dues by the sugar mills, but all said and done, sugarcane still gives the best returns. Tell me what will we do if we do not grow sugarcane? There is no market for vegetables. We cannot grow rice here, and wheat cannot compare with sugarcane in profits. It is sugarcane that sent my sons to college and now two of them have government jobs. People even celebrate birthdays today and spend more money on them than they used to on weddings. This is all due to money from sugarcane.

Sugarcane is deeply embedded in the lives of the farmers who raise it, and has virtually become a part of their identity. The recent agrarian history of western Uttar Pradesh has revolved around the issues of minimum support price (MSP) for sugarcane, arrears owed to farmers by the sugar mills and related issues that underpin the rural economy. Even the farmers' movements in the area, like the one led by the Bharatiya Kisan Union (BKU) in the 1990s, have sought to seamlessly weave together specific economic demands such as lower irrigation prices, remunerative MSP, with caste assertion (Gill 1994). A main goal of the BKU movement was to ensure access to water for the sugarcane growers, a demand that was clearly reflected in the campaign to waive off farmers' outstanding electricity dues.[10] The relationship between irrigation water, sugarcane and caste identity for farmers was forged at a specific politically historical moment in India's history when the 'urban bias'[11] of economic planning began to be noticed and resisted by a new class of capitalist farmers created by the Green Revolution (Lindberg 1994).

The sugarcane farmers' narratives were completely missing from the debate that raged in the media in early 2016 about the adverse impact of sugarcane cultivation on water resources. In this story, the primary villain, sugarcane, was portrayed as a commodity that solely benefits the class of rich politician-proprietors of sugar mills (often referred to, in the media, as the 'sugar mafia').

The Indian Premier League (IPL) versus sugarcane: A new 'urban bias' in action?

Far removed from the socio-political world of sugarcane farmers in western Uttar Pradesh, the recent controversy surrounding the IPL offers an interesting vantage point into the tussle between competing representations and epistemologies of water resources management. The IPL, which began in 2008, has inaugurated a period of uninhibited blending of cricket and entertainment in India. IPL games, with their carnivalesque atmosphere, present an unabashed celebration of consumerism in the service of a reconfigured nationalism (Mehta 2009). The game tickets, which are priced to limit access to the well-heeled middle-class, have come to symbolize the aspirational credo of the emerging consuming classes. An integral part of the games is the garish display by groups of scantily clad female cheerleaders, ostensibly to lend a 'fun' and 'international' dimension to domestic cricket. Cricket, which was traditionally associated with a certain sartorial and cultural stiffness, owing to its upper-class origins, has been reincarnated in a consumerist avatar, and as a vehicle for a masculine identity rooted in national-sexual politics (Tambe and Tambe 2010).

Although this transformation of a sport, known for being a 'gentleman's game', into a glamorous and high voltage visual spectacle is often described as democratization, the apotheosis of the commercial aspect is unmistakable. The auction of the IPL players and the astronomical fees offered to them, especially the stars, are featured prominently in the print and electronic media, and imparts a casino like aura to the sport. The complete transformation of the game of cricket is best captured by the centrality accorded to capital and profit by the owners, managers and players. In this sense, it is of piece with the liberalization of the Indian economy in the 1990s, which led to the unleashing of market forces and the rollback of the state from the economic sphere.

Cricket represents the first sport in India to which the logic of the market has been applied. Those who claim the IPL represents a democratization of cricket, selectively draw on the examples of players from non-metropolitan backgrounds that have received lucrative contracts.

How does the story of the commercialization of cricket intersect with water problems in India? In 2016, protests and outrage erupted when an IPL game was scheduled in Mumbai, the capital of the drought-stricken state of Maharashtra. Maharashtra was in the third year of a protracted drought that led to mass outmigration, rural destitution and even suicides (*The Tribune* 2016). Drought-affected villages lacked drinking water, which was delivered erratically by state-owned water tankers. In such an atmosphere, the holding of a cricket game in Mumbai triggered controversy, for it required the use of prodigious quantities of water to maintain the playing field. A non-governmental organization (NGO) filed a public interest petition in Maharashtra High Court, challenging the planned use of at least 6 million litres of water in maintaining the cricket pitches for the 20-odd games to be played in the state. The court in Mumbai ruled that in view of the prevailing drought situation, expending large volumes of water in holding a game was wrong, and the IPL games, therefore, should be shifted to venues out of the state (Bhasin 2016). 'People in Marathwada [a region of the state severely affected by the drought] do not get water for three to four days. This is a criminal waste', observed the court in its ruling (Vishal 2016).

An immediate storm of criticism followed, and the court was accused of 'conspicuous sanctimoniousness', and of 'acting like a moral police' (Vardarajan 2016). Charges of economic illiteracy were levelled and calculations were presented showing the miniscule impact of water use in cricket grounds on the overall water availability in the state. The NGO Loksatta, which filed the petition to shift the venue of the games, was criticized for peddling 'juvenile morality' (Bhalla 2016). Politicians were blamed for pursuing irrational policies that had exacerbated, if not created, the drought. A particular agricultural policy, namely promoting sugarcane cultivation, came in for a great deal of criticism. The media debate on the desirability and feasibility of restricting IPL cricket can be seen to have been conducted entirely in what has been described as the 'prescriptive epistemic form' (Mukhtarov and Gerlak 2014), wherein the technical and quantifiable attributes of water resources management come to be emphasized. The highly circumscribed debate played out within the boundaries determined by considerations of efficiency and environmental impact – all of which were laid down in strictly measurable terms. This is the framework that led to direct comparisons between IPL and sugarcane – the main cash crop in Maharashtra on which millions of farming livelihoods depend (Jain 2016). The editorial of the *Financial Express* noted:

> Each kilogram of sugar requires 2,068 litres of water which translates to 2 million litres of water per tonne of sugar. So, if the water usage numbers for IPL are correct, we just need to reduce Maharashtra's sugar production

by 3 tonnes! Last year, Maharashtra produced 10 million tonnes of sugar, so that's a drop in the ocean.

(Jain 2016)

Gone is the broad public compact, elevated to a patriotic calling, which prevailed up until the 1980s, in favour of agricultural production to achieve national self-sufficiency in food production. Now sugarcane emerges as the villain of the piece, whose insatiable thirst for water is apparently being ignored, while the focus, unfairly, has shifted on to the IPL. This rhetorical move is as much about the fraying of a national consensus in favour of agriculture, as it is a protest against the perceived stigmatization of cricket, especially commercial cricket, which is regularly hailed in the middle-class centric media as a new vehicle for social mobility, cultural renewal, as well as national self-assertion (Spaaij *et al.* 2015).

Sugarcane, as described earlier, is indeed a water-intensive crop. But there is divergence in expert opinion about the actual impact of sugarcane on water. For instance, Damodaran (2016) counters the claim that sugarcane is responsible for the water woes of Maharashtra thus:

> Bashers of sugarcane will tell us how it takes 2,000-odd litres of water to produce one kg of sugar. But they won't say that this water is consumed over 12 months, or that it goes towards production of fodder, electricity and alcohol as well. And if one were to also add that the mills themselves consume no additional water or electricity – they are surplus in both – it would virtually give a lie to the perception of sugarcane being a water-guzzler.

The role of sugar barons – namely owners of sugar mills – and their political clout has also been widely commented upon. In fact, in Maharashtra it is common for any politician of any standing to own a sugar mill or two. Owning sugar mills is not only a gateway to profits; it also provides an opportunity to cultivate political patronage in an area where sugarcane is the backbone of the rural economy.

The media discourse on sugarcane versus cricket, as described above, is peculiar not only for what it highlights but also for its omissions and silences. It completely neglects the social dimension of sugarcane and the anchoring role it plays in the life of the farmers and others dependent on it. The policy discussions and controversies take place in a rarefied realm where prices, quotas and efficiency are discussed in a disembodied manner, with virtually no regard to the array of cascading impacts on the millions of livelihoods at stake.

The annoyance and indeed incomprehension expressed by many commentators at the ruling of the Mumbai High Court was a particularly telling moment in the controversy. It represented the uncritical and completely unreflexive denial of even the possibility of a non-economic valuation of water. What was neglected in the rush to criticize the judicial ruling was the fact that the court's judgment was a nod in the direction of the marginalized and largely voiceless masses, particularly in the rural areas, for whom the IPL represents perhaps nothing more than an extravagant and insulting spectacle from which they are excluded. Symbolic as a Court's decision is bound to be, nonetheless it is perhaps of great import to those who have been relegated to the margins of the economic battlefield shaped by globalization and neo-liberal principles. But to the detractors of the Court's ruling, the only metric that mattered was economic and quantitative, and the very intangibility of the ruling marked it as irrational and whimsical. The pitting of the IPL against sugarcane was the latest incarnation of the battle between a moribund agriculture sector and a resurgent sports economy, a proxy, it can be argued, for the neo-liberal social order. Whereas agriculture is a spent force in terms of its ability to shape the middle-class identity, the latter, especially cricket, occupies the prized real estate at the intersection of globalization and nationalism. In another bygone era, the controversy might have been labelled as the showdown between the ideas of Bharat and India, but in the current iteration it seems like yet another nail in the ceremonial burial of a long-dead idea.

Conclusion

The policy and discourse pertaining to water resources in India is heavily skewed by the dominance of technocratic and commercial considerations. In cases as disparate as urban attitudes to drinking water and the controversy surrounding the sugarcane crop, the discussion is framed in narrow terms, excluding cultural, normative and practical aspects of water resources. The middle-class respondents demonstrate to a large extent the internalization of the hegemonic status achieved by the prescriptive episteme, which due to its articulation by the most influential sections of society has achieved a near unassailable status. An exclusive focus on the economic aspect of water resources excludes non-commodifiable aspects of water, which is reflected in the marginalization of specific voices and concerns, namely those of the rural, women and the poor. What is perhaps required, therefore, is a conception of 'social nature [which] regards nature itself as constituted and reconstituted through socio-natural processes, both materially and discursively' (Budds 2009, p. 19). The links between technical-material conditions and epistemologies of water management must be recognized. Following Stuart (2007), water, especially in urban contexts, must be de-fetishized from a commodity to an integral part of the fabric of everyday lives so that the full complement of its meanings can be restored.

Both the middle-class and farmers' values pertaining to water reflect a certain territorialisation of specific socio-political positions and histories (Swyngedouw 1999). As the economic role of agriculture has shrunk, sugarcane growers and their interests have become provincialized, while the neo-liberal, consumer-driven economic model has been inscribed onto a much larger, national even global, canvas giving it the gloss of universality (Thiel 2010). New forms of 'hydrosocial territories' (Boelens et al. 2016, p. 2) have been mapped out with their own constellation of technologies, economic rationale and institutions, animated by appropriate 'epistemological belief systems, political hierarchies and naturalizing discourses' (ibid. p. 2). A first step to tackling India's water crisis could consist of recognizing the plurality in conceptions of water(s). Only when it is acknowledged that these contested ideas come wrapped up in their own specific modes of apprehension and seemingly logical ways of problem definition and resolution, can the task of deciphering meanings, symbols and interests begin (Swyngedouw 2006; Strang 2004). In this regard an expanded notion of interest – not merely economic utility – could allow us to relate perceptions and values to the different spatial scales at which identities operate. Water values, instead of being treated as givens, could more fruitfully be considered as the outcomes of positionalities which help to relate them to political-economic forces, while allowing for a more dynamic interplay of agency and structure (Ioris 2012). In this schema, the relational, even oppositional, nature of values provides a connection between the broad regulatory powers of the state and the unfolding of subjectivities.

Notes

1   Goldman (2007) has argued that a broad institutional consensus about the global water crisis among players such as the World Bank, has led to a concerted push toward privatization of water distribution.
2   Beginning in the 1990s, international institutions such as the International Monetary Fund began a worldwide campaign, especially in developing countries, to aggressively promote public-private partnerships (PPP) to improve the distribution of domestic water, ostensibly to enhance efficiency (Tati 2005). The experience with PPP has been mixed at best and, faced with tariff hikes and no improvement in infrastructure, a large number of cities that had implemented PPPs have opted for "re-municipalisation," that is, reverting the control of water infrastructure back to the government (Nair 2015).
3   Ray and Katzenstein (2005) describe the changing character of social movements in India in the last five decades. From the class-based mobilization of the mid-twentieth century and other mass movements that were aimed at broad social justice for marginalized caste, linguistic and ethnic groups, social movements since the 1990s, in tandem with the neoliberal economic reforms, have become fragmented and have increasingly taken on a more right-wing flavour. Broadly speaking the state has relinquished much of the socio-economic space to the markets, with the result that poverty alleviation – a core function of the post-colonial state – has almost been entirely abandoned in favour of a trickle-down economic policy.
4   Truly empowering communities and enabling them to exercise control over water resources is unlikely to happen in the absence of deep institutional reforms. The status quo definitely militates against such a transfer of power from the state to communities. Subramaniam (2014) documents the state's attempts to co-opt common property resources such as rivers and *johads*, revived through community efforts, by granting exclusive rights over them to private businesses.
5   Yogendra Yadav, *Swaraj Abhiyan* founder and well-known public personality, who recently undertook a *padyatra* in support of the drought-stricken people of Latur in Maharashtra, noted the emergence of a thriving market for drinking water in the area. Water is available to those able to purchase it but for others it is a case of 'water, water everywhere, but not a drop for free' (Yadav 2016).

6 The idea of 'global city' that has gained currency in India in the last decade is animated by the fusion of nationalism with consumerism. Dupont (2011) has described the envisioning and execution of this concept in Delhi which has had disastrous consequences for the poor. The idea found its perfect expression in the Commonwealth games organized in Delhi in 2010, which provided the stage for the nationalist-consumerist middle-class fantasies to play out. The emphasis on aesthetics, which was reflected in the efforts to create a 'world-class' infrastructure, was put into practice through the expulsion of the homeless population. The personal and the political came together in the social-aesthetic engineering, effected at a huge expense to the taxpayer, to fulfil the middle-class aspirations of global citizenship. This was also presented as simultaneously a nationalistic act of building 'brand India' (Kaur 2012).

7 Mosse (1999) in his study of the contemporary policy discourse that emphasizes revival of traditional tank management in South India, while blaming the colonial state for the decline of local institutions, describes the narrative as a product of the state's inability to manage the numerous irrigation tanks spread throughout the region. The picture of irrigation tanks maintained by autonomous village communities has been a convenient fiction fostered by the state from the pre-colonial times to delegate responsibility while obscuring the nexus between the wider political structures and the local socio-political relations.

8 At a macro-level India has 16% of the world's population but only 4% of its freshwater resources. The per-capita freshwater availability in India has been declining since the 1970s and is expected to deteriorate further in the future. Gupta and Deshpande (2004) describe a situation in which the total water availability by 2050 will be inadequate to meet the expected demand.

9 It is worth noting that on issues such as garbage collection, crime, and encroachment, the residents of these middle-class neighbourhoods often use institutions like Resident Welfare Associations (RWA) to lobby the state to take favourable action. Chronic water problems do not invite similar action, most probably owing to the increasing tendency to view water as a commodity (and not a public good).

10 One of the most successful and largest sit-ins organized by the BKU in Delhi to press for the socio-economic demands of sugarcane growers was triggered by the deaths of several farmers, protesting high power tariffs, in an incident of police firing in the town of Shamli in 1987 (Dhangare 2015).

11 The leaders of farmers' movements have made claims of 'urban bias' in India's economic planning almost since the inception of the republic. This is owing to several factors, such as the state's push toward heavy industries in the 1950–1960s, domination of planning institutions by urban and upper-caste intellectuals, and more broadly, a privileging of urbanization and industrialization at the expense of rural revival. Sharad Joshi, a farmer leader of the 1980–1990s, used the evocative 'India vs. Bharat' binary to portray the conflict between these two competing visions of India (Lenneberg 1988, p. 450).

References

Bakker, K. (2007) 'The "commons" versus the "commodity": Alter-globalization, anti-privatization and the human right to water in the global south', *Antipode*, vol. 39, no. 3, pp. 430–455.

Bakker, K. (2012) 'Water: Political, biopolitical, material', *Social Studies of Science*, vol. 42, no. 4, pp. 616–623.

Barnes, J. and Alatout, S. (2012) 'Water worlds: Introduction to the special issue', *Social Studies of Science*, vol. 42, no. 4, pp. 483–488.

Baviskar, A. (2007) 'Introduction. Waterscapes: The cultural politics of a natural resource', in A. Baviskar (ed.) *Waterscapes: The Cultural Politics of a Natural Resource*, Permanent Black, Ranikhet, India.

Bhalla, S. (2016) 'No proof required: Fadnavis everywhere, and no water to drink', *Indian Express*, 9 April 2016, http://indianexpress.com/article/opinion/columns/maharashtra-drought-crisis-ipl-fadnavis-everywhere-and-no-water-to-drink/, accessed 15 April 2016.

Bhasin, R. (2016) 'Maharashtra drought: Bombay High Court suggests shifting IPL matches to other states', *Indian Express*, 12 April 2016, http://indianexpress.com/article/india/india-news-india/ipl-matches-maharashtra-people-bombay-high-court-maharashtra-cricket-association-water-crisis/, accessed 27 May 2016.

Bhatnagar, G. (2010) '40 percent of water supply gets wasted', *The Hindu*, 6 January 2010, http://www.thehindu.com/news/cities/Delhi/40-per-cent-of-water-supply-gets-wasted-study/article76718.ece, accessed 27 May 2016.

Bhatt, V. (2014) 'The bitter battle over sugarcane prices', *Tehelka*, 20 September 2014, http://www.tehelka.com/2014/09/the-bitter-battle-over-sugarcane-prices/, accessed 30 May 2016.

Biswas, A. (2008) 'Integrated water resources management: Is it working', *International Journal of Water Resources Management*, vol. 24, no. 1, pp. 5–22.

Biswas, S. (2016) 'Is India facing its worst-ever water crisis?', *BBC*, 27 March 2016, http://www.bbc.com/news/world-asia-india-35888535, accessed 17 June 2016.

Boelens, R., Hoogesteger, J., Swyngedouw, E., Vos, J. and Wester, P. (2016) 'Hydrosocial territories: a political ecology perspective', *Water International*, vol. 41, no. 1, pp. 1–14.

Budds, J. (2009) 'Contested H2O: Science, policy and politics in water resources management in Chile', *Geoforum*, vol. 40, pp. 418–430.

Butterworth, J., Warner, J., Moriarty, P., Smits, S. and Batchelor, C. (2010) 'Finding practical approaches to Integrated Water Resources Management', *Water Alternatives*, vol. 3, no. 1, pp. 68–81.

Damodaran, H. (2016) 'In fact: Sugarcane can't be blamed for Marathwada drought woes', *Indian Express*, 15 April 2016, http://indianexpress.com/article/explained/marathwada-drought-maharashtra-water-crisis-in-fact-why-sugarcane-cant-be-blamed-for-marathwada-drought-woes-2754087/, accessed 2 June 2016.

Dhangare, D. (2015) *Populism and Power: Farmers' Movements in Western India, 1980-2014*, Routledge, Delhi, India.

Dupont, V. (2011) 'The dream of Delhi as a global city', *International Journal of Urban and Regional Research*, vol. 35, no. 3, pp. 533–554.

Fernandes, L. (2000) 'Restructuring the new middle class in liberalizing India', *Comparative Studies of South Asia, Africa, and the Middle East*, vol. 20, no. 1 and 2, pp. 88–104.

Ferreyra, C., de Loe, R. and Kreutzweiser, R. (2008) 'Imagined communities, contested watersheds: Challenges to integrated water resources management in agricultural areas', *Journal of Rural Studies*, vol. 24, pp. 304–321.

Gandy, M. (2004) 'Rethinking urban metabolism: Water, space and the modern city', *City*, vol. 8, no. 3, pp. 363–379.

Gill, S. (1994) 'The farmers' movement and agrarian change in the Green Revolution belt of North-West India', *Journal of Peasant Studies*, vol. 21, no. 3–4, pp. 195–211.

Goldman, M. (2007) '"Water for all!" policy became hegemonic: The power of the World Bank and its transnational policy networks', *Geoforum*, vol. 38, no. 5, pp. 386–400.

Gupta, S. and Deshpande, R. (2004) 'Water for India in 2050: First-order assessment of available options', *Current Science*, vol. 86, no. 9, pp. 1216–1225.

Ioris, A. (2012) 'The positioned construction of water values: Pluralism, positionality and praxis', *Environmental Values*, vol. 21, no. 2, pp. 143–162.

Jain, S. (2016) 'IPL vs Sugarcane: That's really the equation in Maharashtra', *Financial Express*, 8 April 2016, http://www.financialexpress.com/fe-columnist/

ipl-vs-sugarcane-thats-really-the-equation-in-maharashtra/233581/, accessed 10 May 2016.

Jewitt, S. and Baker, K. (2007) 'The Green Revolution re-assessed: Insider perspectives on agrarian change in Bulandshahr District, Western Uttar Pradesh, India', *Geoforum*, vol. 38, no. 1, pp. 73–89.

Joy, K. J., Paranjape, S., Gujja, B., Goud, V. and Vispute, S. (eds) (2008) *Water Conflicts in India: A Million Revolts in the Making?* Routledge, New Delhi, India.

Kaur, R. (2012) 'Nation's two bodies: rethinking the idea of 'new' India and its other', *Third World Quarterly*, vol. 33, no. 4, pp. 603–621.

Khandekar, N. and Vasudeva, A. (2013) 'Leaking pipes, dirty water, and ailing DJB', *Hindustan Times*, 20 April 2013, http://www.hindustantimes.com/delhi/leaking-pipes-dirty-water-and-an-ailing-djb/story-EhvHyty3goZ4jawfHul3uN.html, accessed 27 May 2016.

Kuber, G. (2016) 'A bitter sugar story', *Indian Express*, 19 April 2016, http://indianexpress.com/article/opinion/columns/maharashtra-drought-latur-marathwada-sugarcane-a-bitter-sugar-story-2759592/, accessed 28 May 2016.

Lenneberg, C. (1988) 'Sharad Joshi and the farmers: The middle peasant lives!', *Pacific Affairs*, vol. 61, no. 3, pp. 446–464.

Lindberg, S. (1994) 'New farmers' movements in India as structural response and collective identity formation: The case of the Shethkari Sangathana and the BKU', *Journal of Peasant Studies*, vol. 21, no. 3–5, pp. 95–125.

Linton, J. and Budds, J. (2014) 'The hydrosocial cycle: Defining and mobilizing a relational-dialectical approach to water,' *Geoforum*, vol. 57, p. 170–180.

McKenzie, D. and Ray, I. (2009) 'Urban water supply in India: Status, reform options and possible lessons', *Water Policy*, vol. 11, pp. 442–460.

Mehta, L. (2001) 'The manufacture of popular perceptions of scarcity: Dams and water-related narratives in Gujarat, India', *World Development*, vol. 29, no. 12, pp. 2025–2041.

Mehta, L. (2005) *The Politics and Poetics of Water. Naturalizing Scarcity in Western India*, Orient Longman, New Delhi, India.

Mehta, N. (2009) 'Batting for the flag: *Cricket, television and globalization in India*', *Sport* in Society: Cultures, Commerce, Media, Politics, vol. 12, no. 4–5, pp. 579–599.

Mollinga, P., Dixit, A., Athukorala, A. (eds) (2006) *Integrated Water Resources Management: Global Theory, Emerging Practice and Local Needs*, Sage, New Delhi, India.

Mollinga, P. (2010) 'The material conditions of a polarized discourse: Clamours and silences in critical analysis of agricultural water use in India', *Journal of Agrarian Change*, vol. 10, no. 3, pp. 414–436.

Mosse, D. (2008) 'Epilogue: The cultural politics of water: a comparative perspective', *Journal of Southern African Studies*, vol. 34, no. 4, pp. 939–948.

Mosse, D. (1999) 'Colonial and contemporary ideologies of "community management": The case of tank irrigation development in South India', *Modern Asian Studies*, vol. 33, no. 2, pp. 303–338.

Mukhtarov, F. and Gerlak, A. K. (2014) 'Epistemic forms of integrated water resources management: Towards knowledge versatility', *Policy Sciences*, vol. 47, no. 2, pp. 101–120.

Nair, S. (2015) 'Privatisation of water supply: the muddy picture', *Indian Express*, 25 December 2015, http://indianexpress.com/article/explained/privatisation-of-urban-water-supply-the-muddy-picture/, accessed 10 May 2016.

Pahl Wostl, C., Craps, M., Dewulf, A., Mostert, F., Tahara, D. and Taillieu, T. (2007) 'Social learning and water resources management', *Ecology and Society*, vol. 12, no. 2, article 5, available from: http://www.ecologyandsociety.org/vol12/iss2/art5/

Pahl-Wostl, C., Jeffrey, P., Isendahl, N. and Brugnach, M. (2011) 'Maturing the new water management paradigm: Progressing from aspiration to practice', *Water Resources Management*, vol. 25, no. 3, pp. 837–856.

Preston, T. (2004) 'Environmental values, pluralism, and stability', *Ethics, Place and Environment*, vol. 7, no. 1–2, pp. 73–83.

Ranganathan, M. and Balazs, C. (2015) 'Water marginalization at the urban fringe: Environmental justice and urban political ecology across the North–South divide', *Urban Geography*, vol. 36, no. 3, pp. 403–423.

Ray, R. and Katzenstein, F. (eds) (2005) *Social Movements in India: Poverty, Power and Politics*, Rowman and Littlefield, New York.

Ribot, J. and Peluso, N. (2003) 'A theory of access', *Rural Sociology*, vol. 68, no. 2, pp. 153–181.

Seth, B. L. (2014) 'India's run-of-river hydro: Ill-defined, under-studied – and growing at a fast clip', *International Rivers*, 18 March 2014, https://www.internationalrivers.org/resources/india%E2%80%99s-run-of-river-hydro-ill-defined-under-studied-%E2%80%93-and-growing-at-a-fast-clip-8268, accessed 17 June 2016.

Shrivastava, A. K, Srivastava, A. K. and Solomon, S. (2011) 'Sustaining sugarcane productivity under depleted water resources', *Current Science*, vol. 101, no. 6, pp. 748–754.

Siddiqui, K. (1997) 'Credit and marketing of sugarcane', *Social Scientist*, vol. 25, no. 1/2, pp. 62–93.

Spaaij, R., Farquharson, K., Marjoribanks, T. (2015) 'Sport and social inequalities', *Sociology Compass*, vol. 9, no. 5, pp. 400–411.

Strang, V. (2004) *The Meaning of Water*, Berg Publishers, Oxford, UK.

Stuart, N. (2007) 'Technology and epistemology: Environmental mentalities and urban water usage', *Environmental Values*, vol. 16, no. 2, pp. 417–431.

Subramaniam, M. (2014) 'Neoliberalism and water rights: The case of India', *Current Sociology*, vol. 62, no. 3, pp. 393–411.

Sukhantkar, S. (2014) 'The sweet spot', *Indian Express*, 14 November 2014, http://indian-express.com/article/opinion/columns/the-sweet-spot/, accessed 27 May 2016.

Swyngedouw, E. (1997) 'Power, nature, and the city: The conquest of water and the political ecology of urbanization in Guayaquil, Ecuador: 1880–1990', *Environment and Planning*, vol. 29, no. 2, pp. 311–332.

Swyngedouw, E. (1999) 'Modernity and hybridity: Nature, regeneracionismo, and the production of the Spanish waterscape, 1890–1930', *Annals of the Association of American Geographers*, vol. 89, no. 3, pp. 443–465.

Swyngedouw, E. (2006) 'Circulations and metabolisms: (Hybrid) natures and (cyborg) cities', *Science as Culture*, vol. 15, no. 2, pp. 105–121.

Tambe, A. and Tambe, S. (2010) 'Cheerleaders in the Indian Premier League', *Economic and Political Weekly*, vol. 45, no. 36, pp. 18–21.

Tati, G. (2005) 'Public-private partnership (PPP) and water supply provision in urban Africa: The experience of Congo-Brazzaville', *Development in Practice*, vol. 15, no. 3–4, pp. 316–324.

The Tribune (2016) 'Maharashtra drought: Is cricket more important than people, HC asks BCCI', 6 April 2016, http://www.tribuneindia.com/news/sport/maharashtra-drought-is-cricket-more-important-than-people-hc-asks-bcci/218598.html, accessed 2 June 2016.

Thiel, A. (2010) 'Constructing a strategic, national resource: European policies and the up-scaling of water services in the Algarve, Portugal', *Environmental Management*, vol. 46, no. 1, pp. 44–59.

Tiwari, V. M., Wahr, J. and Swenson, S. (2009) 'Dwindling groundwater resources in Northern India, from satellite gravity observations', *Geophysical Research Letters*, vol. 36, no. 18, doi:http://dx.doi.org/10.1029/2009GL039401.

Vardarajan, T. (2016) 'The Bombay High Court wants a free hit in the IPL. But where's the no ball?', *The Wire*, 16 April 2016, http://thewire.in/29919/the-bombay-high-court-wants-a-free-hit-in-the-ipl-but-wheres-the-no-ball/, accessed 10 May 2016.

Vedwan, N. (2015) 'Water crisis in India: Cross-currents of tradition, nationalism and globalization', *Anthropology Now*, vol. 7, no. 1, pp. 29–38.

Vishal, R. (2016) 'IPL 9: Sense prevails in Bombay high court's verdict to shift matches out of Maharashtra', *India*, 6 April 2016, http://www.india.com/sports/ipl-9-sense-prevails-in-bombay-high-courts-verdict-to-shift-matches-out-of-maharashtra-1088146/, accessed 5 June 2016.

Walker, G. (2014) 'A theoretical walk through the political economy of urban water resources management', *Geography Compass*, vol. 8, no. 5, pp. 336–350.

Yadav, Y. (2016) 'Drought in India: There's water everywhere in Latur, but not a drop of it's free', *First Post*, 30 May 2016, http://www.firstpost.com/india/drought-in-india-theres-water-everywhere-in-latur-but-not-a-drop-of-its-free-2803922.html, accessed 2 June 2016.

12 Balinese wet rice agriculture in transition

Water knowledge between a sentient ecology and the pursuit of development

Lea Stepan

1 Introduction

Modernizing processes in Bali, Indonesia, have led to heavy pressure on local water resources. Agricultural intensification known as the 'Green Revolution', and the extensive expansion of mass tourism, became part of a distinct political programme for national economic growth under the authoritarian presidency of General Suharto in 1966–1998, which led to dramatic changes in local water use (Hardjono 1991; Lansing 1987, 1991). At present Bali struggles with water shortages, especially in the densely-populated tourist areas of the southern plains, as well as with polluted streams and water bodies (e.g. Cole and Browne 2015). Water in the natural environment, which is central to Bali's irrigated wet rice economy, is also an integral part of ritual activities and is worshipped for its vitalizing and purifying qualities (Geertz 1972; Hooykaas 1964; Lansing 1991; Ottino 2000).

Competing interests between state-induced development strategies and local perceptions of water have generated contestations and conflicts over resource management and access (Cole 2012; Cole and Browne 2015; Waldner 1998). The managerial aspect of water and the potential of the irrigation cooperatives to cope with agricultural transformations in Bali has received much attention (Jha and Schoenfelder 2011; Lorenzen 2015; Lorenzen *et al.* 2011; MacRae 2005, 2011; MacRae and Arthawiguna 2011; Strauß 2011). But beyond the managerial aspect, local perceptions/reactions in the context of contested water resources are understudied. To analyse the competing perceptions of water under influences of globalisation and modernisation, Appadurai's notion of 'scapes', defined as 'deeply perspectival constructs, inflected by the historical, linguistic, and political situatedness of different sorts of actors' (1996 p. 33), is useful. Through the framework of a 'waterscape', diverse paradigms can be grasped and analytically framed as disjointed and, at the same time, as bearing interconnected meanings of water.

The study draws on fieldwork conducted in 2014 and 2015 among local peasants in Wongaya Betan (Penebel, Tabanan), a highland village at the foot of Mount Batukaru. The example is presented to explore the ways in which rice farming villagers cope with the influences of economic policies of agricultural intensification or as Reuter (2011, p. 59) puts it, with the political shift 'towards universal de facto adoption of capitalism'. This process is not assumed to be limited or divided into separate knowledge domains, for example, local farmers or state politics, but attention is focused at the village level. Dialectics between paradigms of state-induced policies and local perceptions of the environment shape a selective discourse in the context of environmental transformations. Rather than looking at apparently opposing positions between local and global 'knowledge', judging the authenticity of 'modern' agriculture, or romanticizing the 'traditional', this chapter aims to identify the circulating knowledge fragments which have been adapted or rejected by local villagers and under what circumstances. What is addressed is not so much the question of whose knowledge defines water related environmental problems but how the interplay of knowledges shapes meanings and values at the local level. The villagers find themselves caught between shifting paradigms of state-backed agricultural intensification and aspirations for economic growth. Rather than being seduced by an abstract desire for accumulation of wealth and an eclectic adoption of capitalist premises, the negotiations are characterized by selective and transformative processes of development (see section 4). These merge with national and international agendas of preservation (United Nations Educational, Scientific and Cultural Organization – UNESCO) and so-called sustainable resource management (e.g. organic farming, see section 5) embedded in a spiritual framework of Hindu-Balinese cosmology. Finally, the practical limitations of initiatives such as those of UNESCO are discussed in relation to Bali's current water issues.

Through the combination of recent approaches of relational perceptions, which Rival calls 'perceptionism' (Rival 2014), with agriculture as a classic field of materialist anthropological research, new light is shed on modes of knowing and interacting with water and the environment. On the one hand, water becomes a hermeneutic lens to understand how villagers come to cope with epistemologies of modernisation. On the other hand, by turning attention to sentient relations and interactions with water in the context of political shifts, the meaning of water as a co-constitutive part of an enlivened environment is reworked against the hegemony of techno-scientific categories of water as 'object' and natural/material resource.

2 Wet rice cultivation and local ecology in Bali

Irrigated wet rice terraces are one of Bali's prominent images. Based on a sophisticated irrigation technology, the island's watersheds are divided into several wet rice irrigation territories. These are adjusted to the geomorphology and the flow of water through the rutted volcanic landscape from the mountains to the sea. Larger weirs split up the watershed into ever smaller sub-areas to direct the water through canals and tunnels into blocks of irrigated terraces managed by local farmer associations, called *subak*; the size of irrigated farmland determines the precise allocation for each water inlet.[1] Bali's customary irrigation associations embrace not only local governance and technical hydrology but also embody social and ecological principles embedded into a Hindu-Balinese geo-cosmology (Geertz 1972; Lansing 1991, 2006). The cosmological and ecological aspects of water in rice cultivation were deliberately sidelined between 1966 and 1998 in the economic politics of the 'New Order' era (Lansing 1991; Pedersen and Dharmiasih 2015).

To highlight the multiple dimensions and dialectics of engagements with water in the context of Balinese irrigation practices, it is useful to recapitulate Lansing's (1987) influential analysis of the relationship between cosmology, ecology and hydrological allocation. Lansing drew attention to the interlocking cycles of ritual practice, ecology and wet rice irrigation practices. These are embedded into relational perceptions of water along the cosmic axis of powerful tensions between mountain and sea (*kaja* – towards the mountain, *kelod* – towards the sea). He elaborates in his later work (Lansing 1991), that each direction is related to different perceptions of qualities of water. Whereas upstream water is connected with vitality and fertility, downstream water is perceived as cleansing and purifying. Upstream divisions of the water system are most often marked by a temple, a weir shrine or a field shrine. Their position is meant to exert influence over the physical and social components of the irrigation system and of the terraced ecosystem. In the light of the above-mentioned relational context this might not only be understood in terms of resource management but also through the transformative character of these cosmically powerful places.

Although Lansing's work on irrigation in Bali did not remain uncontested (e.g. Hauser-Schäublin 2003; Helmreich 1999; Lansing and de Vet 2012; Schulte-Nordholt 2011), he pointed out that knowledge practices of traditional wet rice cultivation embody a form of local ecology that is sustained by its embeddedness into a web of economic, social and cosmological relations. Epistemologies of water in Bali must therefore acknowledge the multiple dimensions of practical engagements; practices which include not only skilled technical and engineering know how but embody different layers of meaning encompassing and addressing social, economic, spiritual and ecological knowledge. These water practices seem to be increasingly challenged by modernisation processes underway since the second half of the twentieth century; in particular, the long-term effects of agricultural intensification and ever-increasing demands for freshwater by the growing tourism industry (Cole 2012).

From 1999, after the collapse of Suharto's authoritarian New Order regime, the political setting was characterized by a strengthening of legal and institutional plurality, and a decentralized democracy. These socio-political changes opened up possibilities for active entrepreneurship and creative negotiations (Lorenzen and Roth 2015). As a result of technological, demographic, economic and ecological changes generated through the strong economy of the tourism sector, concerns over a loss of local tradition led to a call 'to preserve Balinese culture'. Efforts to strengthen customary institutions like the Balinese wet rice irrigation associations (*subak*) led to their incorporation into national and regional development plans (Pedersen and Dharmiasih 2015). The next section clarifies how perceptions of water and knowing the environment are analysed in this chapter.

3 Perceptions of water: A sentient waterscape

The flow of water in Balinese irrigation is perceived as a relational axis between mountain (*kaja*) and sea (*kelod*). Rather than being an abstract and determinable space, the spatial directions *kaja* and *kelod* are considered as relative power-fields of a divine presence, which permanently influences human beings through its different qualities. The *kaja-kelod*-axis is an area of tensions in a relational, enlivened environment that requires appropriate human behaviour and from which orientation and meaning is gathered (Hornbacher 2005, p. 402). Rather than reflecting a cognitive representation model (a model *for* 'reality'– Geertz 1973, p. 93), it ties human existence into a web of open relations and interactions with the spiritually enlivened environment that are characterized by tensions and uncertainty. As the term 'animism' has a difficult history in anthropology, such conceptualisations of the environment are currently reworked to overcome the epistemological divide between the 'subject' and an outside 'object'- world (see e.g. Bird-David 1999; Bird-David and Naveh 2008; Descola 2013 [2005]; Ingold 2000, 2011). The essential difference to prior notions of animism is to avoid the idea of 'infusion of spirit into substance, or agency into materiality' but to acknowledge the dynamic, transformative potential of the entire field of relations within which beings of all kinds, more or less person-like or thing-like, continually bring one another into existence' (Ingold 2011, p. 68). Thus, a theoretical approach to water knowledge in Bali needs to move beyond constructivist paradigms.

Numerous attempts have been made to overcome the Cartesian divide between the material/object and the subject (e.g. Bennett 2010; Coole and Frost 2010) or the artificially constructed dichotomy of natural and cultural environments (Lahiri-Dutt, this volume). This debate has indeed a long history but its persistence highlights the necessity to question representational models and engage with alternative ways to analytically do justice to such relations. Ingold's (2000) concept of the 'dwelling perspective' offers an approach to deal with these relations, not only in terms of a methodological appropriateness of theories but also with regard to epistemological questions. In reworking Uexküll's influential semiotic *Umwelt* concept (1956 [1934]), the dwelling perspective is theorized as an active and sentient engagement within the lived-in environment 'based in feeling, consisting in the skills, sensitivities and orientations that have developed through long experience of conducting one's life in a particular environment' (Ingold 2000, p. 25). In relation to epistemologies of water, this approach emphasizes that water is not to be understood as materiality with added cultural meaning. It is not just imbued with meaning in a sense of *attaching* meaning like an additional layer onto a common notion of water as abstract materiality, and thus as an externalized 'natural' element, for example, in a positivist–reductionist sense as H$_2$O (on this, see also Linton 2010). Rather, meaning emerges from the relational context of the agent's sentient and skilled engagement with water in its context of practice. This means that the water derives its significance through meaning that is '*gathered* from the waterscape itself' (Ingold 2000, p. 192, emphasis in original). Ingold describes this engagement and the knowledge within the lived–in environment as a 'sentient ecology'.

Other than traditional ecological knowledge, which for example Berkes (1999, p. 8) defined 'as a cumulative body of knowledge, practice, and belief, evolving by adaptive processes', Ingold's approach escapes not only the reification or romanticising of a 'body of knowledge' but points to skilled practices and sensory perceptions within a lived–in environment. His approach prioritises the idea of an endless 'becoming' through encounters in history (Ingold 2011). Practices and techniques, which are thereby developed in the lived–in environment, are embedded in a matrix of social relationships. This is in stark contrast to the common notion of technology being an external aspect of human sociality which disconnects technical and social relations. Technology rendered as opposition to society sets "the epistemological conditions for society's control over nature by maximizing the distance between them" (Ingold 2000, p. 314). Instead of controlling the environment, skills and technical practices of a sentient ecology seek to draw the environment into the nexus of social relations; they rely on mutualism and reveal relations within it. Irrigation practices, in this sense, are not just a matter of hydrologic control but emerge from sensory techniques and dynamic mutual connections with the environment. The dimension of the sensory experience within the environment and the meanings that are gathered from it has particular significance for epistemologies of water in Bali (see also Ottino 2000, p. 55) and the way in which modernisation processes are negotiated.

Ingold's approach gains further relevance with regard to local hermeneutics and perceptions of the environment along the cosmic axis between mountain and sea (*kaja*-kelod). The differentiation of the previously described qualities (upstream–downstream) does not correspond to a hierarchical valuation. Instead of constituting polar opposites representing moral dichotomies (good-evil) or spiritual dichotomies (sacred-profane/pure-impure), Hornbacher (2005, p. 401) shows that the complementary cosmic power fields express transformation processes. The geocentric spatial orientation along *kaja-kelod* reinforces and raises awareness about circular transformation processes like that of the life cycle or the flow of water. This becomes evident with regard to the perceptions of the Crater Lakes and the sea. The sea, which is associated with the place of death, dirt and resolution, is at the same time the place of origin of life, epitomized in the freshwater crater lakes on top of the mountains.[2] The divine powers of the Crater Lake and the sea become an entity brought into being, for example, through agricultural practices and ritual activities embodied in water. In analogy with an anthropomorphic conception of the lived–in environment, water or bodily fluids are likewise perceived as the vitalizing energy manifest in, and flowing through, the body. The steady flow of water in the canals of an irrigation network is thus considered to secure the vitality of both the cosmos and the body. Accordingly, standing or blocked fluids cause malfunctions of the micro- and the macro-cosmic flow of life, for example, bodily diseases or putrid farmland (Ottino 2000; Weck 1986). *Kaja-kelod* as relational space of embodied knowledge practices is therefore central for relations with water in general and for agriculture in particular. Thus, the waterscape appears not only as a socio-cultural and material product of meaning but is itself meaningful in dialectic, sensed interactions.

Appadurai's notion of 'scapes' strengthens in turn the political interconnectedness that Ingold seems to fall a bit short on. Balinese images of society and self-perceptions have been strongly influenced by externally produced images and ideals of culture and society (e.g. Vickers 1989), which led to a 'preoccupation with Bali as an orderly and harmonious "cultural" rather than political society' (Lorenzen and Roth 2015, p. 102). The analytical approach to perceptions of the environment has therefore to acknowledge the mutual constitution of the political and sentient relationality. While Ingold (2011) offers an alternative way to understand perceptions of the environment in terms of relations, encounters and materiality, his approach remains weak with regard to political entanglements and restraints. Just as political trajectories, self-perceptions and images of society are mutually co-constitutive, so too are perceptions of the environment. With reference to Ingold's 'sentient ecology' and Appadurai's 'scapes', the Balinese irrigation system might therefore be understood as a 'sentient waterscape'.

4 Green Revolution and agro-cosmologies in Wongaya Betan

Tabanan district, which is also known as the 'rice bowl of Bali', had particular relevance for the state-imposed programme for the intensification of rice production in the 1970s and 1980s. Although the negative impacts of Green Revolution practices in Asia are well documented (e.g. Baghel 2014; Lansing 1995; Scott 1985; Shiva 2002, 2006, 2009), agricultural techno-scientific intensification strategies are still prevalent, though now the scientific paradigm has shifted from 'efficiency' to 'sustainability'. The modernisation paradigm of the 1970s marked a shift to 'develop' the country by means of economic growth (Aditjondro and Kowalewski 1994), more precisely by a constructed need to shift to surplus production of cash crops to avoid dependency on food imports under the pressure of a fast-growing population (Hardjono 1991). Such politics embody an epistemology of seeing technology, nature, culture and society as informed by a socially constructed scarcity that can be 'cured' by technical development under the paradigm of economic growth (Yapa 1996). In this vein, national self-sufficiency of rice production and affordability of basic commodities was given priority over sustainability of long-term rural ecology and the peasants' well-being. Technical innovations for irrigation (e.g. the

Bali Irrigation Project or BIP) and harvest increase through the introduction of high yielding seeds, combined with petrochemical fertilizers and pesticides (such as the government programme *Bimbingan Massal* – BIMAS, which translates as 'massive guidance') were the techniques to facilitate this 'development'.

The *subak* Wongaya Betan is located in the district of Tabanan, close to Mount Batukaru at the upstream end of the watershed Yeh Hoo. Agricultural intensification practices started in Wongaya Betan later than in other parts of the island, around 1980 because of the interior location of the village.[3] Yet the experienced impacts were paradigmatic: rising production costs but steady crop sale prices, declining soil conditions and diminishing water supply, all of which were experienced from 1990 onwards. MacRae aptly described the situation of Balinese wet rice farmers as: 'A contradictory picture of economic unsustainability and awareness of a need for change combined with reluctance to change' (2011, p. 74). Trapped by an 'entrepreneurial capitalism' that guided techno-scientific changes in agriculture and agricultural research which 'serves the *needs* of capital by responding to the *demands* of farmers' (Levins and Lewontin 1985, p. 214), rice farming and irrigation appear as pure economic reflex of commodity production, embedded in the relations of the capitalist market system and the power of knowledge out of the science labs. Water, in the context of these innovations, functions as an abstracted material part of a human dominated nature. It is rendered measurable and calculable (see also Linton 2008; Scott 1998). The productivity of the waterscape is then considered solely in terms of generating economic value; its social, ecological or cosmological value, and different modes of knowing become sidelined or epistemologically separated. Given that these transformations do not happen in a cultural vacuum, this study shows how the introduction of new resources, technologies and the shift in value were adapted by local villagers.

Even though the state-imposed paradigms of economic growth during the authoritarian 'New Order' regime usually left few options for public resistance, the imposition of capitalist values did not lead to a blind adoption of them. Nevertheless, it might not be understood as a form of resistance in the sense of a self-conscious political opposition to a capitalist economy, but rather as 'culturally subversive' active engagement led by the villagers' specific 'principles of existence' (Sahlins 2005, p. 28). Due to their proximity to Mount Batukaru, the sacred dwelling place of the gods and divine ancestors, the *subak* has certain responsibilities to interact with its most upstream source (*kaja*), the sacred Crater Lake Tamblingan located near the summit. In exchange for ensuring the flow of water, for fertility and divine blessings, the goddess at the Crater Lake requests to be ritually 'paid' with a share of the harvest of a local rice crop variety (*Padi Bali*). Thus, the farmers continued to plant the local rice crop varieties, once a year, during the main planting season (*kerta masa*). For the second harvest, each farmer can choose a preferred crop (mostly the new varieties which function as cash crops). According to them, they do not dare alter their cropping schedule, even though the extended growth period of the local crop allows only for two harvests per year, whereas elsewhere up to three are possible.

When a neighbouring *subak* replaced their local crops with new varieties, they experienced three complete harvest failures. Perceived as the divine punishment for disobeying their human duties in favour of economic gain, this was seen as evidence of the need to modify Green Revolution practices; to adapt them to local responsibilities of exchange with their co-dwellers in the enlivened environment – the sentient waterscape. Even if villagers are tempted by economic aspirations, as illustrated by the experiences of the neighbouring subak, the incorporation of new resources does not necessarily signify a shift in values or a modernized understanding of well-being (Sahlins 2005, p. 38). Instead, in this particular village, the accumulation of abstract wealth through the cultivation of cash crops, and the adoption of private surplus production in wet rice agriculture, was negotiated with regard to the sensed responsibility towards their land as a purveyor of well-being, the gods and the divine ancestors.

The argument of a re-negotiated economic value of water relates to the perception of water in the web of cosmic powerful tensions, which corresponds to a highly localized ritual tradition and responsibilities. These responsibilities within the lived-in environment shed light on the ritual relationship with the Crater Lake Tamblingan as the ultimate representation of well-being and source of life. The lake marks a central cosmic hub for circular transformation processes like the flow of water and the life cycle. Thus, although the ritual obligations also embody economic objectives, like a blessed rice harvest and prosperity, the ritual interactions at the Crater Lake reach beyond epistemologies of the *homo oeconomicus*. This view of economy aims at the welfare of the *oikòs* and the *polis*[4] (see Sahlins 2005 p. 23). The self-conscious embeddedness in an interactive relationship, thus, also refers to matters of contingency of (human) existence and ontology.

Even though all the *subaks* in Tabanan are ritually connected to Lake Tamblingan as their ultimate upstream irrigation source, almost all *subaks* completely replaced the local rice crop (*padi tahun*) with the new varieties (*padi baru*). Only a few *subaks* at the foot of Mount Batukaru continued to plant the local varieties, considered as their ancestors' heritage. Beyond the mere preservation of the highly-valued crops, this development also creates ritual-economic dependencies, as the local rice is an important part of the offerings in larger temple ceremonies, and sold for a considerably higher price. Thus, locality and the developed sense of distributed specific responsibilities towards the flow of water and the environment remain essential. Yet, epistemologies of economic gain and the blanket technology package of Green Revolution practices are neither completely rejected nor entirely adopted. They are negotiated in reference to deviating discourses of economy and thereby produce hybrid modes of production. This is in line with Hornbacher's (2005, p. 59) note on transformations of perceptions of agriculture and production. She argues, Balinese rice farming activities differed from a perception of agriculture as a private production based on self-interest and instead related the practice to the entirety of human existence. Working in the fields was perceived as service to the deities manifested in the elementary parts of rice farming (earth, water, rice-plant). Rather than pursuing self-interests, this kind of work related to responsibilities towards the whole community. Yet, Hornbacher (2005) suggests that with Bali's differentiating economy, a distinction between community-based activities and private production is becoming increasingly prevalent.

The planting cycles of the *subaks* in the same irrigation network are aligned to form cooperation networks that share the water between upstream and downstream farmers and which generate bioecologically important fallow periods for the fields (see also Lansing 1987). The spiritually articulated responsibility of this particular farmers group to plant the local varieties and thereby prolong their own planting cycles thus incorporates knowledge of economic resource management into a cosmology of sharing practices (see Bird-David 1992). It ensures the flow of water for irrigation also during the dry season throughout the whole watershed.

This resonates with the ideas of conservation or community-based natural resource management in international environmentalism discourses. But it is precisely this knowledge that is often structurally and politically neglected in larger development projects (see Hill, this volume, about farmer managed irrigation systems). Nonetheless, it is important to avoid ecological romanticism of ideas such as animism[5] and spiritual ecological knowledge; these conceptualisations are not necessarily concerned with the idea of conservation *per se*. Bird-David and Naveh (2008), for instance, found for the Nayaka of South India, that the basic concern lies in striving for a good relationship with their co-dwellers in the shared lived-in environment. In this sense, even though practices do sometimes embody ideals of community-based natural resource management, these might not necessarily be perceived by locals as a primary commitment to environmental or natural resource conservation.

But as international environmentalism has deeply influenced the contemporary conceptualizations of customary resource management, this offers 'a new way to model, idealize, and justify communal land and forest tenure' (Davidson and Henley 2007, p. 8). In addition to discourses of international environmentalism, international debates on indigeneity traverse discourses of culture, custom and tradition, local knowledge and religion that characterize many of the societal movements that followed the collapse of the New Order regime.

It is noticeable that the sensed responsibility towards the spiritually enlivened environment was referred to with the religious term '*bhisama*'. This term, recently introduced by the leaders of the Indonesian Hindu Society (PHDI), refers to religious regulations to control potential labile issues and suggests an 'apparently unambiguous identification of religion, culture and environment' (Warren 2012, p. 300). The PHDI is a newly established body of so-called religious experts to politically legitimate Balinese 'beliefs' as orthodox religion in confrontation with a strong Muslim majority in Indonesian state politics. After decades of government control, these political negotiations also include a re-appropriation of religion and traditional practices (Picard 1997, 2011). This indicates an apparent move to adopt religious rhetoric in a political discourse, also in the context of modernized agriculture. Pedersen and Darmiasih (2015) discuss a recent tendency of an 'enchantment' of wet rice cultivation, that contrasts with the previous ideology under the presidency of Suharto until 1998 which had sidelined this dimension. Yet, in the present political setting concerned with preservation of 'traditional culture', it is precisely this dimension that gains particular rhetorical relevance to legitimate water knowledge and practices.

5 Organic farming and sustainable development

In terms of profitability, the contemporary rice farming economy is barely able to compete with the growing tourism economy. As a result, water resources are increasingly under pressure of being overused for tourist facilities. High costs of the envisioned modern life-style urge farmers to sell their land for tourist properties. Since 2012, international initiatives like that of the United Nations Educational, Scientific and Cultural Organization (UNESCO), include the preservation of the wet rice cultivation tradition through their inclusion in the world heritage list. Further means to address environmental degradation, especially the poisoning of water bodies by agrochemicals, are 'sustainable resource management' and 'natural conservation' through organic farming initiatives. Most of the numerous small-scale organic farming projects spread over the island were initiated by groups consisting of expatriates and locals with alternative visions for Bali's agriculture, influenced by global health and environmental movements (MacRae 2005). But the initiative in Wongaya Betan is conducted as cooperation between local villagers, international academics and government extension workers, and thus under the umbrella of the national policy for sustainable development in agriculture. A closer look at the underlying meanings of organic farming reveals divergent notions of sustainability between state-led formulations and local perspectives.

The government-supported shift to organic farming marks a turn in politics and academia towards an evaluation of the impacts of the Green Revolution modernisation processes under the paradigms of preservation, environmental conservation, sustainable development and sustainable resource management. This shift, initiated in the 1980s, goes hand in hand with a wave of awareness over a man-made environmental crisis and the emergence of transnational environmental organisations (Ariesusanty 2011; Peet and Watts 1996). But Escobar's (1996) critical stance towards concepts of sustainability points to a cautious reflection on such policies as they often reinforce a logic of command-control and seek an epistemological reconciliation of economic growth and ecology. Accordingly, 'nature' became substituted in the post-Second World War ecological discourse by 'environment' and an urban-industrial perspective as an exclusive system considering only those aspects that keep it functioning. Water, land and seeds in agriculture – in this perspective epitomized as nature – appear as a mere appendage to the environment, requiring proper management through regulation, technology and research. Escobar concludes that such an agenda accepts the scarcity of resources and reduces ecology to higher forms of efficiency, with only human seen to be playing an active role. What is at stake under sustainable development strategies are therefore concerns over the impacts of environmental degradation on the potential of economic growth, rather than the consequences of the latter on the environment. The agenda thus refers less to the sustainability of the environment than to sustainable economic gains (Escobar 1996: 52).

To address the long-term effects of agrochemical supplies on the environment, some *subak* members of Wongaya Betan conducted a pilot project for organic farming in cooperation with the Governmental Research and Assessment Centre for Technology and Agriculture (BPTP, see also Fox 2012; MacRae 2011). The project is acknowledged as an example for successful implementation of organic farming practices in reaction to the ecological challenges of what has come to be called 'conventional' agriculture, in other words, the blanket technology package and input responsive agriculture of the Green Revolution (e.g. Lorenzen 2015; Roth 2014). The local initiative is part of a larger national policy by the Indonesian Government of general revitalization objectives and particularly of the ambitious policy 'Go Organic 2010'.

The six key objectives of the agricultural revitalization programme of the Indonesian government (RPPK, 2005) explicitly state the amelioration of negative impacts on the environment, and the necessity of a sustainable resource management through appropriate regulations, technology and research. Yet, priorities are still set on sustainability of national food security and defeat of yield stagnation (Las *et al.* 2006, p. 175). Thus, while Escobar's critique is quite relevant to the present case, his analysis operates on a conceptual macro level, and needs to be reworked with regard to meshed knowledges at the local level.

The pilot project for sustainable organic farming in Wongaya Betan started in 2004. With the expert knowledge of extension workers from the BPTP and the local expertise of some farmers, organic fertilizer in Wongaya Betan was produced and applied on the irrigated wet rice fields. To create humus, waste from livestock manure was recycled, and mixed with rice husk and natural bacteria. Combined with a fraction of the recommended dose of urea, production levels increased and were kept steady. Apart from structural economic struggles, healthy soil condition due to revitalized microorganism and predator-prey circles in the fields, higher harvest yields and larger grains were achieved. These resulted also in lesser pest attacks and efficient water-use because the increased biomass is able to retain water much longer. Farmers repeatedly emphasize the better harvest results which relates to the quality of the grains as well as to their marketing price.

Organic farming practices, albeit only small independent initiatives, were established not only in Wongaya Betan but also in other areas, especially around the famous tourist destination Ubud, Gianyar. Nevertheless, it is important to notice that the organic movement is still marginal in Bali. The present effort is, therefore, a showcase intended to encourage organic practices in the future. Those involved in the organic farming initiative in Wongaya

Betan combined different knowledge systems and aspirations creating synergies across and beyond their previous knowledge and engagement with their water-scape. Business projects with various products of organic and local rice varieties contribute to the economic success of the project which increased villagers' income. It opened a new channel for participation and incorporation of farmers into the economy of the tourism sector or agro-ecotourism (Sumiyati 2013). Against the backdrop of growing water demands and the concerns over global methane emissions, the BPTP research centre is constantly researching new technologies to improve harvest yields and cope with water related problems. This is used to justify the implementation of a sustainability policy based on a perception of a degrading environment and its declining productivity that can be cured by appropriate knowledge and a technology package.

Looking at the case from an analytical perspective that assumes a 'becoming' of meanings and relationships that are gathered from the waterscape through the mutual constitution between politics and sensed responsibilities, expands upon Escobar's (1996) useful criticism of the underlying epistemological premises of sustainable development. Concerning the marketing of their organic crops, the local peasants are still bound up in discourses of the imposed state-backed neoliberal market-system and local forms of community-based economy (see e.g. Lorenzen 2015; MacRae 2011; MacRae and Arthawiguna 2011). But at the same time, organic farming as a form of government supported sustainable development allows the villagers to actively adapt new technological knowl-edge in their framework of a sentient waterscape and negotiate modernisation paradigms and epistemologies of economy and ecology.

The application of the Green Revolution technology package did not result in a general epistemological adoption of seeing water and soil as abstract, exploitable natural resources nor to a rejection of economic gains through intensification. Water as an integral part of the landscape is interwoven into technical, ritual and social relations and meanings. As such, water is part of the complex network of relations between humans and land within an anthropo-morphic cosmology. This mode of knowing water incorporates it as a social and material co-agent (see also Krause and Strang 2016). Yet, this knowledge of the landscape does not exclude its economic usage. To be precise, water and land have previously been conceived as usable resources but the difference lies exactly in the sustainability aspect of the usage that emerges out of the inter-woven meanings and relations between humans and land. This accounts not only for Bali but has been recognized for other Austronesian areas (Jacka 2001; Stürzenhofecker 1998). Whereas expert knowledge on organic farming prac-tices promotes sustainability of natural resources through site specific ecosys-tem management, local peasants have emphasised the importance of sustaining the vitality of the earth, as life-giving and sustaining entity. It is notewor-thy that this refers to production but also to the anthropomorphic concep-tion of the landscape and thus the life cycle itself and not just to soil or water. Landscape refers to just this complex meshwork of economic usage, spiritual agents, relational meanings, qualities and responsibilities (Waldner 1998). Land

and water comprise therefore not only the dimension of a usable resource, but are embedded into the complex of the enlivened relationships with the environment.

Organic farming under the paradigm of sustainability can thus be seen as a form of reanimating or revitalizing the waterscape in the sentient environment which aims at the source of life, Goddess Mother Earth (*Ibu Pertiwi*). The villagers in Wongaya Betan emphasize this idea through the name of their organic farming project and peer-farmer education centre *Somya Pertiwi*, which means 'Benevolence of the Earth Goddess'. It adds to the perceived significance to sustain the vitality of the Goddess/Mother Earth in the project. However, to emphasise the adaption of knowledge fragments into the framework of local perceptions is not to emphasise mythology or cosmology as the infinite source for constant reiterations or invariably recurring meanings and values. Rather, it points to aspects that might give insights into meanings and culturally important perceptions of transformation processes.

However, such negotiations only became possible in the political setting that followed the fall of Suharto's authoritarian New Order regime and global concerns of environmental problems entering state politics. This allowed eventually for creative entrepreneurship in agribusiness and new realms for cooperation and negotiations. Thus, in the new political context a new space was opened for alternative perceptions of water and the practice of wet rice cultivation. Water knowledge is tied into the dialectics of diverse epistemologies between techno-scientific approaches of rice intensification, sustainable development, and the agrarian-cosmological order; all adapted in a framework of water knowledge in a sentient waterscape. Notwithstanding the creative potential to negotiate diverse knowledges in emerging situations, new political contexts also created limitations for a critical engagement with environmental challenges of modernisation.

6 Preserving local ecological wisdom

The case of Wongaya Betan shows that another issue on stage (in environmental policies and local responses) is the question of preservation and heritage. During the same period of the Green Revolution, under the political programme of President Suharto, tourism became the main sector for economic development and gained more importance than agriculture. Under the authoritarian developmentalist New Order regime, Bali Province was the target of a political agenda to foster further economic growth through revenues of mass tourism and real estate projects (Davidson and Henley 2007, pp. 9–11; Picard 1997). Early public discussions in Bali were mainly focused on concerns over a 'loss of culture' and questions of authenticity of material culture and artistic practices. Discussions over environmental degradation concerned primarily the diminishing quality of the tourism product: the experience of a unique Balinese culture (Picard 1997 p. 203).

Cultural preservation constitutes, therefore, a central agenda of the decentralized government after the fall of Suharto to sustain the cash cow – tourism.

Political concerns for cultural preservation (in the form of long-term plans or in the context of UNESCO world heritage projects) have been formalized and extended to include the agricultural sector. Yet, as Pedersen and Dharmiasih (2015) argue, the decentralized government continues the modernist project under shifted paradigms. In the aura of global environmentalism, the Balinese wet rice economy serves to present the image of local ecological wisdom on an international stage to its visitors. Media comments of the chairman of the Balinese branch of the Indonesian Tourism Industry Association confirm this view behind national preservation endeavours. A newspaper article about the nomination of the *subak* system as an UNESCO World Heritage Cultural Landscape, cited his reaction: 'It will drive Bali's image as a cultural tourism destination in the world. Many tourists will come to the area to see the tradition' (Erviani 2012). Thus, the heritage conservation aspect is enacted on the local scale as a marketing agenda to preserve Balinese traditional ecological wisdom as cultural capital to promote tourism. This is intended to once again serve the ultimate modernist project of economic growth.

Whereas aspirations for economic growth alone are not to be blamed, the dangerous point that must be recognized is that notwithstanding good intentions of heritage conservation initiatives and the fame of local ecological wisdom, presently they work at the expense of a critical engagement with pressing issues of water shortage and pollution. As Warren critically remarks, the increasingly polarizing discourse between religious and economic perceptions of the environment gains power as it traverses social strata and 'speak[s] to the everyday dialectics of identity and interests, meaning and practical decision-making ... but with so little practical consequence to date' (Warren 2012, p. 303). Instead, conceptualisations and relations concerning environmental responsibilities (as discussed in the previous sections) are politically veiled by the power of this preservation and heritage discourse.

7 Concluding remarks

Balinese water resources are entangled with cosmological perceptions of water as a part of a sentient waterscape, that is threatened by the neoliberal agendas and state interests, that see it solely as an economical resource. In recent years, local people's perceptions and experiences are further confronted with national and international initiatives of water conservation and environmental preservation. This creates new contexts for local negotiations of globally circulating concepts, such as 'organic farming', 'agro-eco-tourism', 'cultural heritage', 'sustainable development' and 'resource management'. The present waterscape in Bali is shaped by competing epistemologies of water that are tied into diverse practices based on different modes of engagement. By analysing transformations of water-related practices in the present case study with regard to their underlying epistemological frame, this chapter highlights how the relationship between state-induced modernisation paradigms and local perceptions shapes present negotiations, meanings and values. The discussion demonstrates that national and international initiatives for preservation, which are now expanded to wet rice cultivation, remain questionable with regard to their practical implications for water in Bali. As shown above, these are still caught up in paradigms of a modernist development to present Bali's visitors with the image of an ancient ecological wisdom. Yet, these concerns for preservation ultimately veil a critical ecological engagement with present environmental problems under the blanket package of a religious or heritage discourse. The methodological approach to water in agriculture, which framed it as a sentient waterscape, sheds light on modes of knowing, interacting and legitimating water practices. On the one hand, water becomes a hermeneutic lens to understand how villagers come to cope with epistemologies of modernisation. On the other hand, the meaning of water as a co-constitutive part of an alive environment has been reworked against the hegemony of techno-scientific categories of water as functional 'object' and natural/material resource. Water is not just attached with meanings, which are commonly declared as merely cultural constructions but acts at the interface of a sensed materiality and powerful agency.

Notes

1   The size of the irrigated rice fields varies considerably between less than 50 ha and over 400ha. Geomorphology and the availability of water determine the size.
2   In this example, the powers of Lake Tamblingan are most prominently connected with the sea temple Pekendungan, which is known for its apotropaic and cleansing powers. On the perception of Lake Batur and its association with the sea, see Lansing (2006)
3   Also, the comparatively more productive lowlands of the southern plains were prioritized for spreading and application of new seed varieties.
4   Sahlins refers here to Aristotle, who, in his view, treated this sense of economy first accordingly.
5   See section 3. *Perceptions of water* this chapter on the terminology of animism.

References

Aditjondro, G. and Kowalewski, D. (1994) 'Damning the dams in Indonesia: A test of competing perspectives', *Asian Survey*, vol. 34, no. 4, pp. 381–95.
Appadurai, A. (1996) *Modernity at Large. Cultural Dimensions of Globalization*, University of Minesota, Minneapolis, MN.
Ariesusanty, L. (2011) 'Indonesia: Country Report' *The World of Organic Agriculture. Statistics and Emerging Trends*, IFOAM, Bonn, and FiBL, Frick.
Baghel, R. (2014) *River Control in India: Spatial, Governmental and Subjective Dimensions*, Springer, Dordrecht, Heidelberg, New York and London.
Bennett, J. (2010) *Vibrant Matter: A Political Ecology of Things*, Duke University Press, Durham, NC.
Berkes, F. (1999) *Sacred Ecology. Traditional Ecological Knowledge and Resource Management*, Taylor and Francis, Philadelphia and London.
Bird-David, N. (1992) 'Beyond 'The Original Affluent Society': A culturalist reformulation [and Comments and Reply]', *Current Anthropology*, vol. 33, no. 1, pp. 25–47.

Bird-David, N. (1999) '"Animism" revisited: Personhood, environment, and relational epistemology', *Current Anthropology*, vol. 40, no. S1, pp. S67–S91.

Bird-David, N. and Naveh, D. (2008) 'Relational epistemology, immediacy, and conservation: Or, what do the Nayaka try to conserve?', *Journal for the Study of Religion, Nature & Culture*, vol. 2, no. 1, pp. 55–73.

Cole, S. (2012) 'A politcal ecology of water equity and tourism. A case study from Bali', *Annals of Tourism Research*, vol. 39, no. 2, pp. 1221–1241.

Cole, S. and Browne, M. (2015) 'Tourism and water inequity in Bali: A social-ecological systems analysis', *Human Ecology*, vol. 43, no. 3, pp. 439–450.

Coole, D. and Frost, S. (eds) (2010) *New Materialisms: Ontology, Agency, and Politics*, Duke University Press, Durham, NC.

Davidson, J. and Henley, D. (2007) *The Revival of Tradition in Indonesian Politics: The Deployment of Adat from Colonialism to Indigenism*, Routledge, London.

Descola, P. (2013 [2005]) *Beyond Nature and Culture*, University of Chicago Press, Chicago, IL.

Erviani, N. K. (2012) 'Bali's 'subak' farms to be named UNESCO World Heritage Site'. *The Jakarta Post*, http://www.thejakartapost.com/news/2012/05/22/bali-s-subak-farms-be-named-unesco-world-heritage-site.html, accessed 03 January 2016.

Escobar, A. (1996) 'Constructing nature. Elements for a poststructural politial ecology.', in R. Peet and M. Watts (eds) *Liberation Ecologies: Environment, Development, Social Movements*, Routledge, London.

Fox, K. (2012) 'Resilience in action. Adaptive governance for subak, rice terraces and water temples in Bali', Indonesia, PhD thesis, University of Arizona, Tucson, AZ.

Geertz, C. (1972) 'The wet and the dry: Traditional irrigation in Bali and Morocco', *Human Ecology*, vol. 1, no. 1, pp. 23–39.

Geertz, C. (1973) *The Interpretation of Cultures: Selected Essays*, Basic Books, New York.

Hardjono, J. M. (1991) *Indonesia: Resources, Ecology, and Environment*, Oxford University Press, Oxford, UK.

Hauser-Schäublin, B. (2003) 'The precolonial Balinese State reconsidered: A critical evaluation of theory construction on the relationship between irrigation, the State, and ritual', *Current Anthropology*, vol. 44, no. 2, pp. 153–181.

Helmreich, S. (1999) 'Digitizing "development". Balinese water temples, complexity and the politics of simulation', *Critique of Anthropology*, vol. 19, no. 3, pp. 249–265.

Hooykaas, C. (1964) *Āgama Tīrtha: Five Studies in Hindu-Balinese Religion*, Noord-Holland, Amsterdam, Netherlands.

Hornbacher, A. (2005) *Zuschreibung und Befremden: Postmoderne Repräsentationskrise und verkörpertes Wissen im balinesischen Tanz*, Reimer, Berlin, Germany.

Ingold, T. (2000) *The Perception of the Environment: Essays on Livelihood, Dwelling and Skill*, Routledge, London.

Ingold, T. (2011) *Being Alive: Essays on Movement, Knowledge and Description*, Routledge, London.

Jacka, J. (2001) 'Coca-Cola and Kola: Land, ancestors, and development', *Anthropology Today*, vol. 17, no. 4, pp. 3–8.

Jha, N. and Schoenfelder, J. W. (2011) 'Studies of the subak: New directions, new challenges', *Human Ecology*, vol. 39, no. 1, pp. 3–10.

Krause, F. and Strang, V. (2016) 'Thinking relationships through water', *Society & Natural Resources*, vol. 29, no. 6, pp. 633–638.

Lansing, J. S. (1987) 'Balinese "water temples" and the management of irrigation', *American Anthropologist*, vol. 89, no. 2, pp. 326–341.

Lansing, J. S. (1991) *Priests and Programmers. Technologies of Power in the Engineered Landscape of Bali*, Princeton University Press, Princeton and Oxford.

Lansing, J. S. (1995) *The Balinese*, Harcourt Brace, Fort Worth, TX.

Lansing, J. S. (2006) *Perfect Order: Recognizing Complexity in Bali*, Princeton University Press, Princeton, CT.

Lansing, S. and de Vet, T. (2012) 'The functional role of Balinese water temples: A response to critics', *Human Ecology*, vol. 40, no. 3, pp. 453–467.

Las, I., Subagyono, K. and Setiyanto, A. P. (2006) 'Isu dan pengelolaan lingkungan dalam revitalisasi pertanian [Environmental issues and management in agricultural revitalization]', *Indonesian Agricultural Research and Development Journal*, vol. 25, no. 3.

Levins, R. and Lewontin, R. C. (1985) *The Dialectical Biologist*, Harvard University Press, Cambridge, MA.

Linton, J. (2008) 'Is the hydrologic cycle sustainable? A historical-geographical critique of a modern concept', *Annals of the Association of American Geographers*, vol. 98, no. 3, pp. 630–649.

Linton, J. (2010) *What is Water? The History of a Modern Abstraction*, UBC Press, Vancouver, Canada.

Lorenzen, R. P. (2015) 'Disintegration, formalisation or reinvention? Contemplating the future of Balinese irrigated rice societies', *The Asia Pacific Journal of Anthropology*, vol. 16, no. 2, pp. 176–193.

Lorenzen, R. P., Lorenzen, S., Schoenfelder, J. W. and Jha, N. (2011) 'Changing realities—Perspectives on Balinese rice cultivation', *Human Ecology*, vol. 39, no. 1, pp. 29–42.

Lorenzen, R. P. and Roth, D. (2015) 'Paradise contested: Culture, politics and changing land and water use in Bali', *The Asia Pacific Journal of Anthropology*, vol. 16, no. 2, pp. 99–105.

Lorenzen, S. and Lorenzen, R. P. (2008) 'Institutionalizing the informal: Irrigation and government intervention in Bali', *Development*, vol. 51, no. 1, pp. 77–82.

MacRae, G. (2005) 'Growing rice after the bomb', *Critical Asian Studies*, vol. 37, no. 2, pp. 209–232.

MacRae, G. (2006) 'Banua or Negara? The Culture of land in the highlands of Bali', in T. A. Reuter (ed.) *Sharing the Earth, Dividing the Land. Land and Territory in the Austronesian world*, ANU Press, Canberra, Australia.

MacRae, G. (2011) 'Rice farming in Bali: organic production and marketing challenges', *Critical Asian Studies*, vol. 43, no. 1, pp. 69–92.

MacRae, G. and Arthawiguna, I. W. A. (2011) 'Sustainable agricultural development in Bali: Is the subak an obstacle, an agent or subject?', *Human Ecology*, vol. 39, no. 1, pp. 11–20.

Ottino, A. (2000) *The Universe Within: A Balinese Village Through its Ritual Practices*, Karthala, Paris, France.

Pedersen, L. and Dharmiasih, W. (2015) 'The enchantment of agriculture: State decentering and irrigated rice production in Bali', *The Asia Pacific Journal of Anthropology*, vol. 16, no. 2, pp. 141–156.

Peet, R. and Watts, M. (1996) 'Liberation ecology. Development, sustainability, and environment in an age of market triumphalism', in R. Peet & M. Watts (eds) *Liberation Ecologies: Environment, Development, Social Movements*, Routledge, London.

Picard, M. (1997) 'Cultural tourism, nation-building, and regional culture: The making of a Balinese identity', in M. Picard & R. E. Wood (eds) *Tourism, Ethnicity and the State in Asian and Pacific Societies*, University of Hawaii Press, Honolulu, HI.

Picard, M. (2011) 'Balinese religion in search of recognition: From Agama Hindu Bali to Agama Hindu (1945–1965)', *Bijdragen tot de Taal-, Land- en Volkenkunde*, vol. 167, no. 4, pp. 482–510.

Reuter, T. (2011) 'Understanding fortress Bali: The impact of democratization and religious revival in Indonesia', *Jurnal Kajian Bali*, vol. 1, no. 2, pp. 58–72.

Rival, L. (2014) 'Encountering nature through fieldwork: Expert knowledge, modes of reasoning, and local creativity', *Journal of the Royal Anthropological Institute*, vol. 20, no. 2, pp. 218–236.

Roth, D. (2014) 'Environmental sustainability and legal plurality in irrigation: the Balinese subak', *Current Opinion in Environmental Sustainability*, vol. 11, pp. 1–9.

Sahlins, M. (2005) 'The economics of Develop-man in the Pacific', in J. Robbins & H. Wardlow (eds) *The Making of Global and Local Modernities in Melaneisa. Humiliation, Transformation and the Nature of Cultural Change*, Ashgate, Hampshire, Burlington, UK.

Schulte-Nordholt, H. (2011) 'Dams and dynasty and the colonial transformation of Balinese irrigation society', *Human Ecology*, vol. 39, pp. 21–27.

Scott, J. C. (1985) *Weapons of the Weak : Everyday Forms of Peasant Resistance*, Yale University Press, New Haven, CT.

Shiva, V. (2002) *The Violence of the Green Revolution: Third World Agriculture, Ecology and Politics*, Zed Books, London.

Shiva, V. (2006) 'Staying alive. Women ecology, and development', in N. Haenn and R. Wilk (eds) *The Environment in Anthropology. A Reader in Ecology, Culture, and Sustainable Living*, New York University Press, New York and London.

Shiva, V. (2009) *Leben ohne Erdöl: Eine Wirtschaft von unten gegen die Krise von oben*, Rotpunktverlag, Zürich, Switzerland.

Strauß, S. (2011) 'Water conflicts among different user groups in South Bali, Indonesia', *Human Ecology*, vol. 39, no. 1, pp. 69–79.

Stürzenhofecker, G. (1998) *Times Enmeshed. Gender, Space, and History among the Duna of Papua New Guinea*, Stanford University Press, Stanford, CA.

Sumiyati, W., Windia, I. W. *et al.* (2013) 'Subak development programs to implement agro-ecotourism', Paper presented at the International Symposium on Agricultural and Biosystem Engineering (ISABE), University Gadjah Mada, Yogjakarta, Indonesia.

Uexküll, J. v. (1956 [1934]) *Streifzüge durch die Umwelten von Tieren und Menschen. Ein Bilderbuch unsichtbarer Welten*, Rowohlt, Hamburg, Germany.

Vickers, A. (1989) *Bali: A Paradise Created*, Periplus, Berkley and Signapore.

Waldner, R. (1998) *Bali - Touristentraum versus Lebensraum? Ökosystem und Kulturlandschaft unter dem Einfluss des internationalen Tourismus in Indonesien*, Lang, Bern, Switzerland.

Warren, C. (2012) 'Risk and the sacred: Environment, media and public opinion in Bali', *Oceania*, vol. 82, no. 3, pp. 294–307.

Weck, W. (1986) *Heilkunde und Volkstum auf Bali*, Bap Bali, Jakarta, Indonesia.

Yapa, L. (1996) 'Improved seeds and constructed scarcity', in R. Peet and M. Watts (eds) *Liberation Ecologies. Environment, Development and Social Movements*, Routledge, London and New York.

13 Water flows uphill to power

Hydraulic development discourse in Thailand and power relations surrounding kingship and state making

David J. H. Blake

Introduction

Water resources have been acknowledged as an important linking theme in Southeast Asian studies, reflected in multiple aspects of society, history, economy and culture (Bray 1994; Boomgaard 2007), leading some commentators to assert that human-water relationships, and not the much better-studied human-land relationships, are a stronger determinant factor in the region (Rigg 1992). Water and its control have been identified as a subject of research and theorization throughout the twentieth century, from every day, localized forms of contestation right up to overarching theories providing the basis for sociological explanations for differential types of state formation. An oft-cited example has been Karl Wittfogel's (1957) twin concepts of 'hydraulic society' and 'Oriental despotism', which were in part, a critical response to and elaboration of Marx's historical materialism notion of an 'Asiatic mode of production'. Simultaneously, a debate has simmered for decades regarding the compatibility of Wittfogel's hydraulic society concept to the early political formation and development paradigm of South and Southeast Asian nations, which has been most vibrant in the case of Bali (e.g. Geertz 1959; Hunt *et al.* 1976; Geertz 1981; Lansing 1991; Christie 1992; Schoenfelder 2004; Lansing *et al.* 2005; Nordholt 2011) and to a lesser extent, Java (e.g. Bray 1994; Christie 1992) and Sri Lanka (e.g. Leach 1959, 1980; Goldsmith and Hildyard 1984).

In the years following its publication, Wittfogel's hydraulic society hypothesis became the subject of vigorous academic arguments, often embedded in divisive Cold War-inspired political rhetoric (Peet 1985), including a flurry of critiques that essentially rejected and provided counter examples to the purported universal correlation between large-scale hydraulic control and totalitarian agro-managerial organization (see Mitchell 1973; Price 1994). According to Barker and Molle (2004, p. 8), 'the intensity of the intellectual debate in the post-war period has sent many researchers in a quest of hydraulic societies in Asia, which has not always been successful and convincing'. Pointing to parochial case studies across tropical and subtropical Asia, critics argued that certain extensive irrigation systems were not necessarily the product of a centralized, powerful and despotic state (e.g. Leach 1959; Geertz 1980; Leach 1980; Lansing 1991).[1] In the case of the ancient Khmer empire, Stott (1992, p. 53) stoutly rejects Wittfogel's theories and expresses little doubt that its hydraulic development paradigm was a result of indigenous ingenuity, confidently claiming, 'yet again we have a brilliant, farmer-level, South East Asian response to a distinctive set of environmental conditions. There is no need for a hydraulic hypothesis'. Going further still, Robbins (2004, p. 48) condemns Wittfogel's thesis outright as 'fundamentally flawed on both empirical and theoretical grounds', asserting that 'there is no evidence of an empirical association between large irrigation schemes and centralized authority either in contemporary cases or ancient periods'.

Acceptance and rejection of the hydraulic society hypothesis

A significant part of the debate pivots around the extent to which irrigation development in ancient states was considered a technological tool for socio-political domination imposed from above by despotic state leaders or autocratic kings, as opposed to the outcome of more organic processes encompassing indigenous technologies that emerged from below. In a study of archaeological records, Stargardt (1992, p.62) for instance, criticized the notion that 'the creation and maintenance of hydraulic works in Asia was a prime example of the despotism of a centralized court bureaucracy, with a monopoly of managerial as well as technical expertise, over the undifferentiated rural population', claiming it was unsupported by the evidence, even for Wittfogel's archetypal case of China. Some observers in the case of Bali have stressed the more instrumental aspects of irrigation development and societal power relations by ordinary people (e.g. Christie 1992; Schoenfelder 2004), while others have focussed more on the employment of symbolic and ideological motivations by powerful dynasties (Geertz 1981; Nordholt 2011).

At the same time, it has been noted how processes of socio-ecological transformation, kingship and state-making interact through hydraulic control, in which 'the symbolic and material form a single complex' (Lansing 1991). Likewise, in the context of Tamil Nadu state, Mosse contends that '[m]edieval kings and chiefs controlled and gifted water flows, creating landscapes which inscribed their rule into the hydrology and thus naturalized it' (2003, p. 4), leaving tangible legacies in contemporary waterscapes.

It has not been uncommon for academic authors to reject the hydraulic society hypothesis based on a shallow understanding or a misinterpretation of his contributions, maintains Price (1994, p. 193), who noted a marked tendency for anthropologists studying irrigation to simply overlook Wittfogel's work, 'or to merely cite it to dismiss it instantly as 'reductionistic', 'simplistic', or 'mechanical''. Also, referring to anthropological studies of water resources, Worster (1992, p. 30) maintained:

[O]ne of the most serious weaknesses in that literature, it must be said straight off, is that the modern experience with irrigation hardly appears in it. Nowhere do the ecological anthropologists – nor does Wittfogel for that matter – seem to realise that the link between water control and social power might occur in places other than the archaic cradles of civilization nor that the past hundred years have seen more irrigation development than all of previous history.

One other common characteristic of scholarly scepticism around hydraulic society's applicability to modern Southeast Asian nation-states seems to have been a marked reluctance to examine in detail the dynamics of contemporary rulers' agency in presiding over bureaucratically centralized states that promote a hydraulic engineering paradigm. Linked to this research lacuna is questioning the extent to which such rulers may embody particular characteristics originally ascribed by Wittfogel to an ideal hydraulic despot.

This chapter seeks to explore the relevance of the hydraulic society hypothesis in the context of modern Thailand, not so much by recording the nature and extent of the bureaucratic hold of state agencies over the water resources development paradigm, but by considering the role of the present monarch's[2] relationship with the sector, whether materially, discursively or symbolically. It contends that to understand the importance of water resources as a critical and contested resource in Thai society, one must examine the agency of uber-elite actors as much as farmers and every day water users, permitting epistemological insights into the interplay between knowledge, power and discourse. Such a perspective acknowledges Hart's (1989) contention that societal power struggles are invariably inter-connected with struggles concerning access to and control over natural resources. Water resources control, in particular, has consistently proven to be central in this regard (Mollinga 2008), with irrigation viewed as a socio-technical system providing unique opportunities to mould and discipline system users as a supreme 'technology of control', facilitating its governmentality (Foucault 1991). Before examining the evidence to link the reigning monarch with a controlling position of a hierarchical society partially ordered along hydraulic lines, the chapter first briefly revisits Wittfogel's hypothesis to establish what he said concerning the rulers directing the hydraulic paradigm.

Wittfogel's hydraulic society hypothesis revisited and its pertinence to Thailand

A defining characteristic of hydraulic societies identified by Wittfogel (1957) was the governing presence of an autocratic emperor, pharaoh or king (often revered as a semi-divine deity) who acted both as chief ideologist and arbiter of the state's hydraulic development apparatus. The ruler would be responsible for playing 'the decisive role in initiating, accomplishing, and perpetuating the major works of hydraulic economy', relying upon a coterie of *aides de camp* for its execution (Wittfogel 1957, p. 27). He referred to the existence of an organizational web for managing the hydraulic works covering the whole, or at least the 'dynamic core', of the nation. As a corollary, he posited, 'those who control this network are uniquely prepared to wield supreme political power' (Wittfogel 1957, p. 27). Popular revolution would be near impossible, and so even when a ruling dynasty died out or was overthrown by force, the new regime was unlikely to differ very much from the old. He emphasised how hydraulic states lacked appropriate processes of external control and internal balance and 'under such conditions there develops what may be called a *cumulative tendency of unchecked power*' (Wittfogel 1957, p. 106, emphasis in original).

In *Oriental Despotism: A Comparative Study of Total Power*, Wittfogel (1957) considered Thailand (or Siam, as it was known pre-1939) to be an atypical example of a hydraulic society, where the pre-modern state's hydraulic engineering ambitions was principally concerned with building navigation canals to transport rice surplus to the capital to meet the needs of the ruling elite, as well as the transport of troops and messengers to secure the wider kingdom.

Based on its hydraulic density, Wittfogel classified Siam as a so-called 'loose' hydraulic society,[3] characterized by a compact hydraulic core represented by Bangkok and the lower Chao Phraya delta, and an extensive, relatively undeveloped periphery. Beyond a brief description of pre-modern Thailand's context, Wittfogel provided scant evidence to justify his case for Thailand's inclusion as a hydraulic society, which tends to suggest that he was not overly familiar with its specific historical and hydraulic development context. Supporting Wittfogel's critical distinction between the dominance of hydraulic and hydro-agricultural forms of state formation, Ishii (1978) conceded that a monsoonal climate provides a less straightforward opportunity for would-be hydraulic rulers to exercise effective social control than is permitted in the arid or semi-arid environments of Asia where classical agro-managerial states and the grandest hydraulic structures first appeared and flourished. He classed pre-modern Thailand as a 'quasi-hydraulic society', where the state's control over irrigation technology was only partial. Examining the context of pre-nineteenth-century Thailand's water resources development paradigm, Brummelhuis (2007, p. 19) by contrast, dismisses the possibility that it could have been a hydraulic society, reasoning that there was no evidence of large-scale irrigation works, rice growing was a matter for individual households and there was 'not an inkling of a state controlling a large region politically via waterworks'.

Interestingly, the only serious examination of the applicability of Wittfogel's theory to Thailand's modern political economy and agricultural water management context,[4] appears to have been undertaken by Australian anthropologist Gehan Wijeyewardene (1973). Studying rural development trends in communities near Chiang Mai, Wijeyewardene maintained that the Thai bureaucracy pursued a strategy that further entrenched state-centric agro-managerialism, citing a 66-fold expansion of irrigated land area expedited under government schemes between 1907 and 1967[5] as evidence. This transition occurred under a variety of political regimes, from absolute monarchy pre-1932 to a series of weak military-civilian coalitions thereafter, with irrigation expansion efforts post-World War Two as US and Western development aid flooded in. Concurring with the findings of Riggs (1966),[6] he noted a progressive transfer of land and property from peasants into the hands of bureaucrats 'and the class they appear to dominate'. Wijeyewardene (1973, p. 101) linked the state's pre-eminence in water resources development to a nationalistic developmental mindset, observing that, '[I]n its official ideology, no doubt sincerely held, the bureaucracy is concerned with national development and security ... The irrigation programme is the *most spectacular manifestation of this ideology'* (Wijeyewardene 1973, p. 100, emphasis added). Conceding that while Thailand's irrigation development paradigm did not possess the 'overriding factors' referred to in Wittfogel's theory, he felt sufficiently confident to conclude, 'the country today is much more a hydraulic society than in the past' (Wijeyewardene 1973, p. 108). However, it should be noted, he dismissed the notion that Thailand was controlled by a hydraulic despot without elaboration.[7] The chapter now turns to examine the seminal role of the current monarch in the expansion and entrenchment of a modern hydraulic society in Thailand.

Tracing the Thai King's fascination with water resources control and renovation of a theocracy

The ninth monarch in the Chakri dynasty, king Bhumibol Adulyadej (literally meaning "Strength of the Land, Incomparable Power") has been closely associated with many aspects of water resources management throughout much of his long reign spanning over seven decades. He was born in 1927 in Boston, U.S.A., but spent much of his early childhood in Switzerland, making infrequent visits to Siam. one of the origins of Bhumibol's celebrated lifetime interest in water resources may be traced back to childhood lessons in Lausanne, with descriptions of the king learning the basics of soil and water relationships from Swiss tutors, a fact frequently invoked in popular biographies.

> "Some people wonder why I became interested in irrigation or forestry", His Majesty said in one of his speeches 30 years ago. "I remember that when I was 10 years old, a science teacher who is now dead taught me about soil conservation. We had to write: '*There must be forest on the mountain or the rain will erode the soil and damage the mountain surface.*' This is a fundamental fact of soil and forest conservation and of irrigation. If we fail to maintain the highland forest, we will have problems ranging from soil erosion to sedimentation in dams and in rivers. Both can lead to floods."
> (*Bangkok Post* 2006)

A boyhood curiosity in natural resources management and hydrology was later formalized during a brief period studying engineering at Lausanne University, apparently inspired by a German-trained irrigation engineer uncle.[8] But his nascent interests in water control were not brought to fruition in a practical sense until several years into his reign, under the despotic Field Marshall Sarit Thanarat regime in the late 1950s and early 60s (Chaloemtiarana 2007). After several years of thwarted ambition and limited sovereign power allowed under the anti-royalist government of Field Marshall Phibun Songkhram, the young king was subsequently given considerable opportunity to put into practice some of the hydraulic concepts he had earlier been taught in Switzerland. Sarit played a crucial role in restoring the weakened Chakri dynasty's weakened position since the 1932 coup and, in particular, elevating the sovereign's reputation as a sacral, virtuous *dhammaraja*[9] (Handley 2006; Jackson 2010). Sarit built his regime upon promoting indigenous principles of authority and social hierarchy, adopting a paternalistic style of rule supposedly adapted from the ancient Sukothai system of monarchical governance. He appointed staunch royalist M. R. Kukrit Pramoj as the regime's chief ideologue to help in defining new interpretations of a 'Thai-style democracy', to be built under a patronage system where 'the king is the head and the government and bureaucracy are its organs' (Hewison and Kengkij 2010, p. 188). As Kukrit once stressed, '[t]he king must be both God and human. It is the burden of the king to consider where the dividing line between the two is' (Pramoj 1983, cited in The National Identity Board, 2000). For Kukrit, the king's foremost political role was to provide a watchful eye over government in the best interests of the nation, thus acting as a traditional benevolent, righteous and paternalistic leader (Hewison and Kengkij 2010). According to Fong (2009, p. 688): 'The syncretism of *devaraja* and *dhammaraja* with a relational view metaphorically expressed as between father and children, or *pho-luk*, constituted Thailand's unwritten social contract between king and subjects.'

As has been argued by Leach (1959), in comparing the dominant Indianized model of acquisition of political authority by south Asian rulers over the last 2,500 years against that in China, was largely one of 'charismatic' over 'bureaucratic' type, and such a distinction could be said to hold true for Siam/Thailand. The former ruler generally 'achieved power by personal ascendancy and he held authority in the capacity of a living Buddha or as a Chakravartin'[10] (Leach 1959, p.4).

Constructing a staunchly military-royalist network at the centre, Sarit systematically set about promoting the monarchical institution, both as a beacon of loyalty for the mass populace to follow, but also as a way to bolster his personal legitimacy. Sarit would routinely refer to the 'Army of the King' and the 'government headed by the King', while the king in return anointed Sarit 'Defender of the Capital' on the day of the 1957 coup, notes Baker and Phongpaichit (2005, p. 177). Sarit scrapped a Land Act enacted by the previous government that was unpopular in palace circles, restored several defunct royal ceremonies[11] and encouraged an expansion of the king's role and participation in rural development activities. According to Chaloemtiarana (2007) and Krittian (2010), the revival of royal ceremonies contributed to a solidification of Sarit's political power while simultaneously bolstering monarchical authority that had been lost since the 1932 coup, making each interdependent. Krittian (2010) regarded royal ceremonies as part of a 'sophisticated political ideology machine' to entrench royal and military power. To quote Wittfogel's (1957, p. 92) words, '[T]he agromanagerial sovereign cemented his secular position by attaching to himself in one form or another of the symbols of supreme religious authority.'

Significantly, it was Sarit who reintroduced the practice of subjects crawling in front of royalty, a symbolic gesture of reverence that had been earlier banned under the absolute monarchy of King Chulalongkorn (Krittikarn 2010). According to Wittfogel (1957, p. 152), 'no symbol has expressed total submission as strikingly, and none has so consistently accompanied the spread of agrarian despotism, as has prostration.' He emphasised the proliferation of prostration in ancient state-centred hydraulic civilizations and how, despite rationalization, it has remained 'a symbol of abject submission' wherever it was practiced (Wittfogel 1957, p. 154). In Thailand, subjects are not only expected and encouraged to prostrate themselves in front of the king, but also before his children and closest advisors. The paper now considers some of the pathways by which the king has come to be widely referred to by state actors and royalist supporters as the 'Father of Water Resources Management'.

Emergence of the paternalistic high priest of hydraulic development

From the late 1950s, the king and queen embarked on increasing visits to remote regions of the country, attracting large crowds of awed villagers wherever they went. Indeed, a key component of the 'complex assembly' of cultural meanings comprising the so-called standard total view of the monarchy (or STVTM) referred to by Hewison (1997), is that the king is cast as an egalitarian 'development king', bringing 'modernity and progress to the country and works tirelessly for the welfare of his subjects' (Ivarsson and Isager 2010, p. 3). Ranked uppermost amongst the vocational pursuits sparked by these rural exposure tours were exploring agricultural projects to replace opium cultivation amongst highland populations in the north and the potential of irrigation technology introduction to transform small-scale farming (Baker and Phongpaichit 2005). The virtual parachuting in of a god-like king to solve the problems of peasant farmers, long ignored by the government, was both an opportunity and a threat to the bureaucratic polity's hitherto dominance (Riggs 1966). The multiple magico-divine (*saksit*) mystification elements constructed around Bhumibol's reign (Jackson 2010), only served to confirm his inestimable potency in divining matters hydro-meteorological for the good of the nation.

The king's ideas concerning water resources development were initially trialled at his own palaces and royal development centres, then later through his personal charity foundation[12] and so-called 'Royally-initiated Projects' (*khrong-gan an-nuang maa jaak pracha-damree*).[13] Initially these projects were supposedly dependent on the monarch's personal financial resources, they concentrated on providing social welfare for disaster relief and providing direct charity to the poor, plus a production unit for royal films and radio broadcasting projects (Public Relations Department 2000). As the ideological struggle against communism intensified during and after the Sarit regime, so the Royally-initiated Projects expanded their remit and geographical scope into new fields of development, assuring royal hegemony (Chanida 2007). According to one bureaucratic agency produced report justifying the projects, the king established them with the ultimate mission of alleviating a range of water-borne problems:

> His Royal Initiative Projects aim to relieve problems of water whether flood, drought or pollution aim to mitigate people's sufferings. He endeavours to study [on] how to tackle water resources development and management because he is certain that without the water-related problems that damage crop yields and with availability of sufficient water, the standard of living of those impoverished rural people could be better.
>
> (Department of Water Resources 2008)

His concepts and observations were readily offered as technical and policy advice to the government and public though up-country visits, speeches, exhibitions and hagiographical royal publications during the last two decades of the twentieth century (Handley 2006). As with the legendary Sinhalese kings noted by Leach (1959), Bhumibol's reputation has been systematically cemented by a vast bureaucratic machinery devoted to elevating his cult-like status as a great irrigation engineer and hydraulic innovator, rather than a conqueror or builder of cities. The king's vision was not solely focused on developing water resources capture, storage and distribution infrastructure on the ground, but was also turned to manipulating the capricious Thai weather, having previously determined that water scarcity resulting from 'uncertain rainfall' was *the* principle obstacle besetting Thailand's farming sector (National Identity Board 2000). Such official narratives suggest that his fascination with rainmaking dates to charitable trips to remote regions to present *kathin* robes to monks (see Gray 1991), that promoted national unity amongst ethnic groups and identified potential development projects.

A striking characteristic of state-sanctioned reports surrounding the king's role in artificial rainmaking is that they tend to stress his personal agency in driving the programme. In 1975, the king is reported to have established a 'Royal Rain-Making Office' under the supervision of the Permanent Secretary to the Ministry of Agriculture and Cooperatives (Ministry of Agriculture and Cooperatives n. d.). Official accounts stress the king's enthusiasm for and scientific knowledge of artificial rainmaking has accumulated over the years, evidenced by his 1999 'discovery' of a novel technique for achieving greater cloud density and increasing the areal extent of precipitation. The king named and patented this new cloud-seeding method as the 'Super Sandwich' technique, which reputedly has made Thailand 'the centre of tropical rainmaking activities in this region' (Thaiways Magazine 2011).

A review of various official reports reveals that both the palace itself and the wider government public relations machine has long recognized the immeasurable propaganda value to be derived from thoroughly milking the artificial rainmaking narrative, through conjuring the potent image of a scientifically astute monarch, privy to ancient knowledge and mysterious powers (Blake 2012). Building a popular mythology based around the 'Royal Rainmaker' moniker (i.e. divine creator of royal rain or *fon luang*)[14] has arguably been an indispensible elite device in legitimating a bureaucratic-military-royal dominance over almost the entire hydrological cycle. It could be argued that such paternalistic benevolence to some degree reduced climatic uncertainty and water scarcity for peripheral farmers, while imperceptibly integrating them into the core of Thai nationhood. For decades, no other political competitor could have hoped to match the king's comprehensive domination of the state's hydraulic development discourse, max-imising what has been termed by Wittfogel (1957, p. 133) the 'rulers' publicity optimum', subtly shaping public opinion towards a common-sense acceptance of the legitimacy of sovereign power. In short, through a complex network of royalist supporters, Bhumibol has succeeded in creating a comprehensive 'regime of truth' (Foucault 1991) that has proven remarkably resilient and unassailable by counter-narratives, protected as it is by strict defamation and lèse majesté laws, used alongside other instruments of state censorship (see Streckfuss, 2011).

It is instructive that one time Director-General of the Royal Irrigation Department (RID), Pramote Maiklad,[15] has claimed that the king is regarded as the 'Royal Father' of Thai water resources management due to his 'high intellect, capability and continuous devotion' to solving national water resources related problems (Maiklad 2002, p. 246). Perhaps more than any other honourific title, Pramote's label best encapsulates the exalted position in which he is held within Thai hydraulic society. The accolade was formally bestowed upon Bhumibol in 1997 by Prime Minister Banharn Silpa-Archa infamous for his godfather-esque control of the economy, including irrigation distribution, in his Central Plains constituency of Suphan Buri (Molle 2007). In a presentation speech by the prime minister, on the occasion of the king's 50th year on the throne, Banharn wholeheartedly endorsed to the nation the king's hydraulic brilliance:

> Your Majesty's interest in water is not only limited to help lessening the water shortage problem to ease the hardship of people, but also to maintain the quality and quantity of water in balance, a crucial factor for human survival. Management includes draining water from low-lying areas where floods stagnated, flood protection and mitigation, as well as waste water treatment. In our humble recognition of Your Majesty's Kindness and Competence on water resources development and management, I, on behalf of the Royal Thai Government and all the Thai people, beg for your gracious permission to offer to Your Majesty the epithet of "The Father of Water Resources Management"
>
> (Ministry of Agriculture and Cooperatives n. d., p.135).

Similar descriptions predominate in numerous state-sanctioned accounts (e.g. Amporn 1996; Kiattikomol 2000; National Identity Board 2000; Ehrlich 2011), reinforcing the notion that the king possesses supernatural talents in the field of water resources development that has enabled him to singularly identify the principle problems besetting the sector, while correctly prescribing the necessary solutions (invariably infrastructural and technological) for implementation by the bureaucracy.

Beyond internationally-recognised innovations in artificial rainfall techniques, some of Bhumibol's most vaunted domestic technological inventions have included creating designs for various water resources storage and flow control technologies (e.g. *'fai maew'* – a type of hill stream weir adapted from Northern ethnic groups; and the *'gaem ling'* – floodplain water storage reservoirs, mimicking the anecdotal storage of bananas in monkey's cheeks for later consumption); a patented floating paddlewheel aeration device for pollution mitigation (the *'Chaipattana* aerator'); bio-engineering using certain grass species for slope erosion control and capturing run-off; providing a design for Bangkok's flood defence strategy which has led to the royal seal of approval for numerous large-scale irrigation and floodwater control structures, such as the Pasak Cholasit Dam[16] (Handley 2006; Grossman and Faulder 2011).

If one accepts the basic premise that water resources development and management is inherently political by nature (Mollinga 2008; Molle, Mollinga *et al.* 2009), then the repeated claim of conservative royalists that the king is somehow able to remain 'above politics' in the political milieu of Thailand (e.g. Streckfuss 2011) becomes a questionable proposition. Not only has the king been intimately involved in multiple aspects of water resources governance across scales from micro-management to national policy and planning over many decades, he has also been an integral actor in the construction of the dominant problem and solution framework upon which water resources management policy discourse is based, arguably becoming its chief overseer and ideologue (Blake 2012, 2015). Moreover, it would not be unreasonable to argue that while the king has consistently acted in the manner of an executive director of the nation's water resources development paradigm, he has remained essentially protected from critical scrutiny of the socio-political and environmental outcomes of his visions and material interference, very much in the tradition of Wittfogel's autocratic rulers.

The hydraulic ruler's viziers and 'network monarchy'

Oriental Despotism discusses in considerable detail how power was concentrated in the hands of a single, absolute ruler in a hydraulic society and the profile of an archetypal sovereign. Wittfogel (1957, p. 305) stated:

> [I]n his person the ruler combines supreme operational authority and the many magic and mythical symbols that express the terrifying (and allegedly beneficial) qualities of the power apparatus he heads. Because of immaturity, weakness, or incompetence, he may share his operational supremacy with an aide: a regent, vizier, chancellor, or "prime minister". But the exalted power of these men does not usually last long. It rarely affects the symbols of supreme authority. And it vanishes as soon as the ruler is strong enough to realize the autocratic potential inherent in his position.

If one provisionally accepts that Thailand qualifies as a modern variant of a hydraulic society, does the above description reasonably reflect the king's rise to power and the concentration of hierarchical authority in the uber-elite around him over the past half century? Previous sections have addressed some of these elements, but the means by which he has attained supreme authority deserve closer attention. While it is widely recognised that the king commands strong authority and legitimacy derived from constitutional powers invested in him, the success of his rule has also drawn in no minor part from revived magical, mythical and sacral symbols, ceremonies and rites, referred to by Fong (2009) as constituting 'primordial simulacra'. However, such remarkably strong and socially entrenched expressions of operational supremacy could neither be attained nor maintained by the king alone, but has required a considerable coterie of close aides, courtiers, counsellors and sycophantic bureaucratic advisors, who have formed the next tier down from the monarch in exerting control over Thailand's ruling polity during the past six decades. It is these people who have ensured that the king's influence is felt both materially and metaphysically in even the most humble of rural households, as everyone depends on water resources for their life and livelihood. It is therefore worth examining a few of the key individuals that have assisted in securing a relatively static hydraulic regime, thus contributing to the monarch's elevated power beyond mere figurehead status.

I have already referred to the influential figures of Sarit, Kukrit and Pramote as key creators or sculptors of the Bhumibol hydraulic high priest legend and cult in Thai society. To provide a complimentary conceptual term to help understand this dynamic, I refer to McCargo's (2005) concept of 'network monarchy', which contends that Thai politics is best understood in terms of political networks. Further, between 1973 and 2001, the most influential network was centred on the palace. McCargo's theory argues that the monarchy is required to work through proxies to maintain its legitimacy, but it has never achieved outright domination, despite frequent and active interventions in the political process. It may help to elaborate on some of the powerful actors through which the king individually and the collective monarchical institution has worked and drawn strength, in particular the seminal role played by his closest and longest-serving proxy.

While Field Marshall Sarit Thanarat and his successors as past prime ministers, Field Marshall Thanom Kittikachorn, Thanin Kraivixien and (to a lesser extent) M. R. Kukrit Pramoj, may be perceived as key facilitators in promoting the king's extraordinary hydraulic bent to the nation, the strongest contender for the 'grand vizier' position since the late-1970s, has been the king's highest-placed chancellor, namely General Prem Tinsulanonda (Blake 2015). The first three figures were undoubtedly important statesmen and monarchical revivalists during the early and middle period of Bhumibol's reign, who forged a nationally revered persona around Bhumibol as a benevolent and paternalistic 'development king'. However, only General Prem has consistently lasted the course at the monarch's side over several decades, accepting royally appointed roles of commander-in-chief of the armed forces, regent, prime minister (March 1980–August 1988) and subsequently from 1998 to the present, President of the Privy Council.[17] McCargo (2005) argued that the king has trusted Prem almost unreservedly, viewing him as incorruptible and skilled as an alliance-builder and patronage wielder. From around 1980 up until the start of Thaksin Shinawatra's term as prime minister in 2001, 'Prem served effectively as Thailand's "director of human resources", masterminding appointments, transfers and promotions. Prem's power was never absolute, though it was always considerable' (McCargo 2005, p. 506).

During his tenure as prime minister, General Prem proved himself indispensable in ensuring the king's utopian visions about water resources management were positioned near the top of the nation's development agenda (Blake 2012), especially those targeted at the perennially sensitive and heavily securitised northeast region (Sneddon 2000; Baker and Pongpaichit 2005). Towards the last few years of his premiership, Prem was frequently seen fronting efforts to promote the ill-conceived Green Isaan Project as an exemplar of bureaucratic-military cooperation and monarchical benevolence (Guyot 1987). He helped to spread a popular perception amongst rural dwellers and the wider public that water was essentially a symbolic gift from the king.[18] The Green Isaan Project, perhaps more than any other previous regional development scheme, promised significant political power and benefits to whichever individual or group could ensure its successful implementation in the 'arid' and 'impoverished' northeast, thus fulfilling the king's stated wishes to see the region become a prosperous and fertile 'green zone' (*Bangkok Post* 1987b). With its implicit royal mandate, the Green Isaan Project ultimately became the target for an internecine power struggle between several strategic groups (cf. Evers and Benedikter 2009), once it had been established as a potent ideological vehicle for whomsoever could control the public discourse, thereby obscuring non-hydrological material benefits, such as project links to conditional foreign aid and the international arms trade (Hewison 1994; Molle, Floch et al. 2009). Prem, as a southerner, eventually had to concede the project's control to his ambitious protégé, General Chavalit Yongchaiyudh, who swapped his military garb for a politician's suit, founding a political party with strong roots in the Northeast and later briefly becoming the prime minister in 1997.

Prior to Prem's long tenure as the king's hydraulic grand vizier, the role in some ways had been pioneered during the late 1950s and early 60s by Field Marshall Sarit Thanarat. As already elaborated, Sarit was a shrewd leader who realised his political fortunes were dependent on closely allying himself to a monarchical revival and according to Handley (2006), in return the palace 'wholeheartedly' supported Sarit's toppling of the anti-royalist Field Marshall Phibun Songkhram in 1957. Following Sarit's premature death from liver cancer in 1963, the premiership passed to another royalist military dictator, Field Marshall Thanom Kittikachorn, who helped maintain his predecessor's rigid ideological system based on the indivisible triune of 'Nation, Religion, King' (Fong 2009) and processes of 'Thai-ization' of the periphery (Phatharathananunth

2006). Thanom continued to endorse a growing bureaucratic programme of pharaonic hydraulic development initiated under Sarit, with many major hydropower and irrigation dams commissioned during a long period of repressive military regimes, including the Bhumibol (1964), Ubon Ratana (1966), Sirindhorn (1971), Sirikit (1972) and Chulabhorn (1972) projects; each named in honour of a member of the royal family.[19]

Following Thanom's flight into exile during a temporary period of democratisation (October 1973 to October 1976), the aristocratic M. R. Kukrit Pramoj briefly assumed the prime minister's post (March 1975 to April 1976). While in a sense he represented a rather more liberal and culturally refined figure than the succession of brutal military dictators that preceded him, he was nevertheless, a lifelong propagandist[20] in abetting the monarchy's ascension to its present position (Connors 2008). During his ten-month tenure in power, Kukrit's government set in motion a massive irrigated agriculture stimulus programme, largely focused on the communal construction of small dams and reservoirs across the northeast, which bore the unmistakable hallmarks of the king's hydraulic and rural social engineering vision. These landscape features, known as '*fai Kukrit*',[21] are still visible today and although scarcely used for irrigation purposes, remain fixed as examples of forced participation in the collective folk memory and language of villagers that manually toiled to build them (Blake 2012).

Ultimately, however, as an elected premier, Kukrit was insufficiently autocratic, conservative or militaristic for either the palace's or the generals' wiles, for instance, by seeking to introduce populist measures such as land reform, farmer support programmes, enforce the minimum wage, decentralise government, abolish the Anti-Communist Act and repatriate 25,000 US troops stationed in Thailand, thus forcing his resignation in January 1976 (Handley 2006). For this and several other reasons, he fails to qualify as a model vizier, unlike Sarit, Thanom and Prem. One other noteworthy and still active close aide of the king who has been instrumental within the network monarchy in supporting royal hydraulic developmentalism is Dr Sumet Tantivejakul.[22] Sumet has acted as a tireless promoter of the king's water resources development projects, inventions and developmental philosophy over many decades. He has been a loyal right hand man, once describing Bhumibol as an 'environmental activist' in a book celebrating the king's lifetime achievements in balancing the development needs of society with that of the environment (see Handley 2006, p. 385).

The utopian promise provided by irrigation, termed '*chonla prathaan*' (literally meaning 'royally bestowed water') in the vernacular, has been a powerful material and symbolic means for organizing society, promised to the people by a benevolent monarch and delivered by a coterie of elite political allies, royal-backed institutions and technocratic bureaucracies.

Conclusion

This chapter has framed Thailand as representative of a modern variation of a hydraulic society, which is claimed to have developed a markedly centralized model of water resources control under a powerful, theocratic and paternalistic ruler. The observation that King Bhumibol's rule has been in large part legitimized and empowered through a systematic domination of the national hydraulic discourse is not in itself validation of a hydraulic society, but it is one of several indicators that would support the hypothesis. Validating Wijeyewardene's (1973) earlier observations, I have argued that over the twentieth century, the Thai state has steadily implemented a comprehensive irrigation development programme that has expanded from the centre to the periphery. This has not merely incorporated formerly marginal spaces like the Nam Songkhram Basin (see Blake 2012), but it has also further entrenched control by a military-bureaucratic-royalist network of interests over rural society using water resources, both instrumentally and symbolically. Irrigation development has been used as a key tool of state simplification, discipline and control, making legible complex socio-ecological waterscapes (cf. Scott 1998; Mosse 2003). The Thai state has achieved this through a mixture of consensual and coercive instruments and strategies at multiple spatial and temporal scales. This is not to infer that such a process of domination has been without contestation or resistance from below, as sub-altern individuals and groups have continuously struggled for improved participation, rights and access to a bundle of resources, including water.

Since the dictatorial Sarit regime reinvigorated the Thai monarchy, it has re-emerged as virtually untouchable institution at the centre of the national body politic, through a sacral kingship that has become an indispensible part of the public mythology of the modern Thai state. Although, the king's personal agency in controlling the water resources management discourse in Thailand has waned in recent years as his health has declined, the monarch still retains considerable symbolic power in presiding over the paradigm. His ubiquitous name, image, ideas and dictums are routinely invoked by hydraulic bureaucrats and powerful members of the network monarchy to justify numerous development projects and programmes nationwide. These may entail the construction of new hydraulic infrastructure or the defence of royalist network-directed schemes designed to tackle real or perceived crises, which effectively acts as a potent means to silence any public criticism (although a high degree of self-censorship naturally operates). While Thailand represents a less absolute regime than well-documented hydraulic society variants in earlier state formations in the great river basins of semi-arid and arid Asia, it nevertheless still qualifies on several counts. Indeed, with a military dictatorship installed in power since May 2014, marking a shift away from democratic aspirations towards a stricter paternalistic authoritarianism, there have been numerous arrests leading to the imprisonment of hundreds of ordinary citizens on charges of *lèse majesté*, alongside the enactment of and various other state-sanctioned humans rights abuses (Human Rights Watch 2015). Such hydraulic insights into power relations provide an intriguing modern context to the original concerns of Wittfogel regarding statecraft, kingship, hydraulics and despotism.

Notes

1 It should be noted that Leach (1959) does not reject Wittfogel's theory outright, but merely questions its historical relevance to the specific locality he studied, later conceding that 'the classical Sinhalese kingdom, with its capital at Anuradhapura, was a striking and characteristic example of what Wittfogel has called 'hydraulic civilization''' (Leach 1980, p. 92–93).

2 This chapter was written prior to the death of King Bhumibol Adulyadej on 13 October 2016.

3 A 'loose' hydraulic society was considered one in which the hydraulic agriculture, 'while lacking economic superiority, is sufficient to assure its leaders absolute organizational and political hegemony' (Wittfogel 1957, p. 166).

4 Cohen (1992) provides a critique of northern Thailand's classification as a HS during the nineteenth century, noting how much irrigated lands were under the control of an aristocracy and royalty, rather than the devolved, peasant-led irrigation management systems of popular imagination.

5 Tellingly, since Wijeyewardene's (1973) study, state irrigation coverage and annual expenditure by the RID on the sector has increased manifold (Turral 2008), even as Thailand's economic base has gradually shifted from an agrarian to a predominantly manufacturing and service industry orientation.

6 Riggs proposed that Thailand up to the 1960s was essentially run by a powerful military-bureaucratic complex, with little political space for participation from civil society or the private sector in national governance.

7 While Wijeyewardene was not specific about which 'overriding factors' he referred to, his outright rejection of a hydraulic despot is somewhat troubling, especially given the fact he conducted his fieldwork during a period of brutal military dictatorship and civil society suppression by state forces (see Baker and Phongpaichit 2005).

8 Bhumibol never completed the engineering degree course at Lausanne University, but following the mysterious death of his elder brother in June 1946, switched to studying political science and law, deemed more suitable disciplines in preparation for his unexpected ascension to the throne. The coronation took place in May 1950.

9 A *dhammaraja*, a word derived from the Pali language, is a monarch who upholds a royal code of conduct that emphasizes the major Buddhist precepts and the so-called Tenfold Practice or Duties of Kingship, believed to be a cornerstone of Thai kingship.

10 A *chakravartin* is an ancient Indian term used to refer to an ideal universal ruler, one who rules ethically and benevolently over the entire world (https://en.wikipedia.org/wiki/Chakravartin Accessed 28 December 2015). While there have not been any serious claims that Bhumibol is a universal ruler, as far as I am aware, it was noteworthy that during the king's 60th anniversary of accession to the throne in 2006, he was prominently proclaimed by state authorities as the 'King of Kings'.

11 One of the ceremonies revived under Sarit's tenure included the archaic Royal Ploughing Ceremony, a lavish combined Buddhist and Brahminical rite that had ceased in 1932. The king (and in recent years the Crown Prince) officiates at the ceremony held in May to ensure abundant rainfall and a bountiful rice harvest for the coming agricultural year. Fong (2009) describes it as an example of a primordialised royal ceremony used to reinforce the monarch's sacrality.

12 The Chaipattana Foundation was founded in 1988 to 'support the implementation of Royally-initiated and other development projects' in a way that does not replicate government projects and 'without political involvement' (Source: http://www.chaipat.or.th/chaipat_old/noframe/eng/ Accessed 26 April 2013).

13 These schemes are administered financially by a special agency established within the National Economic and Social Development Board, the Office of Royal Development Projects, which may call on the technical support of other state bureaucracies to advise on certain individual projects, such as the Royal Irrigation Department or Royal Forestry Department.

14 An anecdote provided by one of Bhumibol's closest aides, M. R. Butrie Viravaidhya, is recounted in Handley's (2006, p. 436–437) book, in which she claimed ('in all seriousness') that the king was a divine representation of Shiva, because he had the power to stop rain falling on a royal barge procession in 1996.

15 On mandatory retirement from his RID position, Pramote appeared to become an even closer personal confidante of the king and was elected a senator in Bangkok for one term and is presently the Chairman of the National Reform Council's (NRC) committee on natural resources and environment, under the military junta.

16 An illustration of this dam project is found on the reverse of the 1,000 baht denomination banknote.

17 The Privy Council of Thailand is a senior body of advisors close to the monarchy granted a range of powers and responsibilities under the constitution, with all eighteen members personally appointed by the king.

18 For example, at the launch of the Project in April 1987, the media showed military trucks converted to transport water for distribution in certain Northeastern villages with large banners down their sides declaring, '*Nam prathai jaak nai luang*' (i.e. 'From the generosity of the King') (Bangkok Post 1987a).

19 Curiously, the only exception to this series of large dams named after leading members of the royal family, was a chronological in naming a structure after the unpopular Crown Prince. It was not until 2001, that an existing hydropower dam in Kanchanaburi province was renamed Vajiralongkorn Dam, even though it was first commissioned in 1984.

20 He founded, owned and contributed to the popular newspaper, Siam Rath, which according to Handley (2006, p. 123), was used as a vehicle for furthering the interests of a remote monarchy through its reportage and essays, providing a space for 'consistent criticism of the king's rivals'.

21 It is instructive to note that the timing of the construction of '*fai Kukrit*' coincided with the building of similar utopian irrigation engineering works under totalitarian regimes in Vietnam (Evers and Benedikter 2009) and the Democratic Republic of Kampuchea (Bultmann 2012).

22 Dr Sumet is former Assistant Secretary-General of the National Economic and Social Development Board (NESDB) and current chairman of the royal 'quasi-NGO', the Chaipattana Foundation.

References

Amporn, S. (1996) *The Great King of Thailand*, One Asset Management Limited, Bangkok, Thailand.

Baker, C. and P. Phongpaichit (2005) *A History of Thailand*, Cambridge University Press, Cambridge, UK.

Bangkok Post (1987a) 'King sends water to the Northeast', 1 April 1987, p. 1.

Bangkok Post (1987b) 'Army set for Northeast 'green zone' project', 12 July 1987.

Bangkok Post (2006) 'Saviour of Nature', 19 May 2006, http://www.bangkokpost.com/60yrsthrone/saviour/index.html, accessed 21 December 2012.

Barker, R. and F. Molle (2004) 'Evolution of Irrigation in South and Southeast Asia', Comprehensive Assessment Research Report, Comprehensive Assessment of Water Management in Agriculture, IWMI, Colombo, Sri Lanka.

Blake, D. J. H. (2012) '*Irrigationalism*: The Politics and Ideology of Irrigation Development in the Nam Songkhram Basin, Northeast Thailand' PhD thesis, University of East Anglia, Norwich, UK,

Blake, D. J. H. (2015) 'King Bhumibol: The Symbolic "Father of Water Resources Management" and Hydraulic Development Discourse in Thailand', *Asian Studies Review*, vol. 39, no. 3, pp. 649–668.

Boomgaard, P. (2007) *A World of Water: Rain, Rivers and Seas in Southeast Asian Histories*, KITLV Press, Leiden, Netherlands.

Bray, F. (1994) *The Rice Economies: Technology and Development in Asian Societies*, University of California Press, Berkeley, CA.

Brummelhuis, H. T. (2005) *King of the Waters. Homan van der Heide and the origin of modern irrigation in Siam*, KITLV Press, Leiden, Netherlands.

Chaloemtiarana, T. (2007) *Thailand: The Politics of Despotic Paternalism*, Silkworm Books, Chiang Mai, Thailand.

Chanida, C. (2007) *'Khrong-gan an nueang ma jak phraratchadamri:gan-sathapana phraratcha amnat nam'* [The Royally-Initiated Projects: The Making of Royal Hegemony (B.E. 2494–2546)] Thammasat University, Bangkok, Thailand.

Christie, J. W. (1992) 'Water from the Ancestors: Irrigation in early Java and Bali' in J. Rigg (ed.) *The Gift of Water: Water Management, Cosmology, and the State in South East Asia*, School of Oriental and African Studies, University of London, London.

Cohen, P. T. (1992) 'Irrigation and the Northern Thai State in the Nineteenth Century' in G. Wijeyewardene and E. C. Chapman (eds) *Patterns and Illusions: Thai History and Thought*, The Richard Davis Fund and Department of Anthropology, Research School of Pacific Studies, The Australian National University, Canberra, Australia.

Connors, M. K. (2008) 'Article of Faith: The Failure of Royal Liberalism in Thailand' *Journal of Contemporary Asia*, Vol. 38, no. 1, pp. 143–165.

Connors, M. K. (2011) 'When the Walls Come Crumbling Down: The Monarchy and Thai-style Democracy', *Journal of Contemporary Asia*, vol. 41, no. 4, pp. 657–673.

Department of Water Resources (2008) *Annual Report 2008*. Department of Water Resources, Ministry of Natural Resources and Environment, Bangkok, Thailand.

Ehrlich, I. (2011) 'The King of Water' *Thailand Tatler*, Blue Mango Publishing Company Ltd., Bangkok, Thailand.

Evers, H. D. and S. Benedikter (2009) 'Hydraulic Bureaucracy in a Modern Hydraulic Society - Strategic Group Formation in the Mekong Delta, Vietnam', *Water Alternatives*, vol. 2, no. 3, pp. 416–439.

Fong, J. (2009) 'Sacred Nationalism: The Thai Monarchy and Primordial Nation Construction', *Journal of Contemporary Asia*, vol. 39, no. 4, pp. 673–696.

Foucault, M. (1991) 'Governmentality', in G. Burchell, C. Gordon and P. Miller (eds) *The Foucault Effect: Studies in Governmentality*, With two lectures by and an interview with Michel Foucault, University of Chicago Press, Chicago, IL.

Geertz, C. (1959) 'Form and Variation in Balinese Village Structure', *American Anthropologist*, vol. 61, no. 6, pp. 991–1012.

Geertz, C. (1980). 'Organization of the Balinese Subak' in E. W. Coward, Jr. (ed.), *Irrigation and Agricultural Development in Asia: Perspectives from the Social Sciences*, Cornell University Press, Ithaca and London.

Geertz, C. (1981) *Negara: The Theatre State in 19th Century Bali*, Princeton University Press, Princeton, NJ.

Gray, C. E. (1991) 'Hegemonic Images: Language and Silence in the Royal Thai Polity', *Man*, vol. 26, no. 1, pp. 43–65.

Goldsmith, E. and N. Hildyard (1984) 'Traditional irrigation in the dry zone of Sri Lanka', in E. Goldsmith and N. Hildyard (eds) *The Social and Environmental Effects of Large Dams*, vol. 1, Wadebridge Ecological Centre, Camelford, UK.

Grossman, N. and D. Faulder, (eds) (2011) *King Bhumibol Adulyadej: A Life's Work*, Editions Didier Millet, Bangkok, Thailand.

Guyot, E. (1987) *Isan Khieo - The Green Northeast*, The Institute of Current World Affairs, Hanover, NH

Handley, P. M. (2006) *The King Never Smiles*, Yale University Press, New Haven and London.

Hart, G. (1989) 'Agrarian Change in the Context of State Patronage', in G. Hart, A. Turton and B. White, *Agrarian Transformations: Local Processes and the State in Southeast Asia*, University of California Press, Berkeley and Los Angeles, CA.

Hewison, K. (1994) 'Greening of Isaan - more than just a pinch of salt', *Thai-Yunnan Project Newsletter*, vol. 24, p. 7.

Hewison, K. (1997) 'The Monarchy and Democratisation', in K. Hewison (ed.) *Political Change in Thailand. Democratisation and Participation*, Routledge, London and New York.

Hewison, K. and K. Kengkij (2010) 'Thai-Style Democracy: The Royalist Struggle for Thailand's Politics', in S. Ivarsson and L. Isager (eds) *Saying the Unsayable: Monarchy and Democracy in Thailand*, NIAS Press, Copenhagen, Denmark.

Human Rights Watch. (2015) *Thailand: UPR Submission 2015*. HRW's Submission to Human Right's Council. September 2015, https://www.hrw.org/news/2015/09/21/thailand-upr-submission-2015, accessed 28 December 2015.

Hunt, R. C., E. Hunt, G. Munir Ahmed, J. W. Bennett, R. K. Cleek, P. E. B. Coy, T. F. Glick, R. E. Lewis, B. B. MacLachlan, W. P. Mitchell, W. L. Partridge, B. J. Price, W. Roder, W. Steensberg, R. Wade and I. Wellman (1976) 'Canal Irrigation and Local Social Organization [and Comments and Reply]' *Current Anthropology*, vol. 17, no. 3, pp. 389–411.

Ishii, Y. (1978) 'History and Rice-Growing', in Y. Ishii (ed.) *Thailand: A Rice Growing Society*, The University of Hawaii, Honolulu, HI.

Ivarsson, S. and L. Isager (2010) 'Introduction: Challenging the Standard Total View of the Thai Monarchy' in S. Ivarsson and L. Isager (eds) *Saying the Unsayable: Monarchy and Democracy in Thailand*, NIAS Press, Copenhagen, Denmark.

Jackson, P. A. (2010) 'Virtual Divinity: A 21st-Century Discourse of Thai Royal Influence', in S. Ivarsson and L. Isager (eds) *Saying the Unsayable: Monarchy and Democracy in Thailand*, NIAS Press, Copenhagen, Denmark.

Kiattikomol, K. (2000) *His Majesty the Great Engineer*, The Engineering Institute of Thailand, Bangkok, Thailand.

Krittian, H. (2010) 'Post-Coup Royalist Groups: Re-inventing Military and Ideological Power' in S. Ivarsson and L. Isager (eds) *Saying the Unsayable: Monarchy and Democracy in Thailand*, NIAS Press, Copenhagen, Denmark.

Krittikarn, S. (2010) 'Entertainment Nationalism: The Royal Gaze and the Gaze at the Royals', in S. Ivarsson and L. Isager (eds) *Saying the Unsayable: Monarchy and Democracy in Thailand*, NIAS Press, Copenhagen, Denmark.

Lansing, S. J. (1991) *Priests and Programmers: Technologies of power in the engineered landscape of Bali*, Princeton University Press, Princeton, NJ.

Lansing, S. J., L. Pedersen and B. Hauser-Schaublin (2005) 'On Irrigation and the Balinese State', *Current Anthropology*, vol. 46, no. 2, pp. 305–308.

Leach, E. R. (1959) 'Hydraulic Society in Ceylon', *Past and Present*, vol. 15, April, pp. 21–26.

Leach, E. R. (1980) 'Village Irrigation in the Dry Zone of Sri Lanka', in E. W. Coward, Jr., *Irrigation and Agricultural Development in Asia: Perspectives from the Social Sciences*, Cornell University Press, Ithaca and London.

Maiklad, P. (2002) 'His Majesty the King: Royal Father of Thailand's Water Resource Management', *The Journal of the Royal Institute of Thailand*, vol. 27, Special Issue, December, pp. 232–246.

McCargo, D. (2005) 'Network Monarchy and Legitimacy Crises in Thailand', *The Pacific Review*, vol. 18, no. 4, pp. 499–519.

Ministry of Agriculture and Cooperatives (No date) 'His Majesty the King and Agricultural Development in Thailand', The Ministry of Agriculture and Cooperatives, Bangkok

Mitchell, W. P. (1973) 'The Hydraulic Hypothesis: A Reappraisal' *Current Anthropology*, vol. 14, no. 5, pp. 532–534.

Molle, F. (2007) 'Scales and Power in River Basin management: The Chao Phraya River in Thailand', *The Geographical Journal*, vol. 173, no. 4, pp. 358–373.

Molle, F., P. Floch, B. Promphakping and D. J. H. Blake (2009) 'The "Greening of Isaan": Politics, Ideology, and Irrigation Development in the Northeast of Thailand', in F. F. Molle, T. Foran and M. Käkönen (eds) *Contested Waterscapes in the Mekong Region: Hydropower, Livelihoods and Governance*, Earthscan, London.

Molle, F., P. P. Mollinga and P. Wester (2009) 'Hydraulic Bureaucracies and the Hydraulic Mission: Flows of Water, Flows of Power', *Water Alternatives*, vol. 2, no. 3, pp. 328–349.

Mollinga, P. P. (2008) 'Water, Politics and Development: Framing a Political Sociology of Water Resources Management', *Water Alternatives*, vol. 1, no. 1, pp. 7–23.

Mosse, D. (2003) *The Rule of Water: Statecraft, Ecology and Collective Action in South India*, Oxford University Press, New Delhi, India.

National Identity Board (2000) *King Bhumibol: Strength of the Land*, National Identity Office, Office of the Prime Minister, Bangkok, Thailand.

Nordholt, H. S. (2011) 'Dams and Dynasty, and the Colonial Transformation of Balinese Irrigation Management', *Human Ecology*, vol. 39, pp. 21–27.

Peet, R. (1985) 'Introduction to the Life and Thought of Karl Wittfogel', *Antipode*, vol. 17, no. 1, pp. 3–21.

Phatharathananunth, S. (2006) *Civil Society and Democratization: Social Movements in Northeast Thailand*, NIAS Press, Copenhagen, Denmark.

Price, D. H. (1994) 'Wittfogel's Neglected Hydraulic/Hydroagricultural Distinction', *Journal of Anthropological Research*, vol. 50, no. 2, pp. 187–204.

Public Relations Department (2000) *The Thai Monarchy*, The Public Relations Department, Office of the Prime Minister, Bangkok, Thailand.

Rigg, J. (1992) 'The Gift of Water', in J. Rigg (ed.) *The Gift of Water: Water Management, Cosmology and the State in Southeast Asia*, School of Oriental and African Studies, University of London, London.

Riggs, F. W. (1966) *Thailand: The Modernization of a Bureaucratic Polity*, East-West Center Press, Honululu, HI.

Robbins, P. (2004) *Political Ecology: A Critical Introduction*, Blackwell Publishing, Malden, MA.

Schoenfelder, J. W. (2004) 'New Dramas for the Theatre State: the Shifting Roles of Ideological Power Sources in Balinese Polities', *World Archaeology*, vol. 36, no. 3, pp. 399–415.

Scott, J.C. (1998) Seeing Like a State: How Certain Schemes to Improve the Human Condition have Failed, Yale University Press, New Haven & London.

Sneddon, C. S. (2000) 'Altered Rivers: Socio-Ecological Transformations, Water Conflicts and the State in Northeast Thailand', PhD thesis, University of Minnesota, Ann Arbor.

Stargardt, J. (1992) 'Water for Courts or Countryside: Archaeological Evidence from Burma and Thailand Reviewed, in J. Rigg (ed.) *The Gift of Water: Water Management, Cosmology and the State in South East Asia School of Oriental and African Studies*, University of London, London.

Stott, P. (1992) 'Angkor: Shifting the Hydraulic Paradigm', in J. Rigg (ed.) *The Gift of Water: Water Management, Cosmology and the State in South East Asia*, School of Oriental and African Studies, London.

Streckfuss, D. (2011) *Truth on Trial in Thailand:* Defamation, *Treason, and Lese-Majeste*, Routledge, London and New York.

Thaiways Magazine. (2011) 'Our Great King. Awards Presented to H.M. King Bhumibol', http://www.thaiwaysmagazine.com/king/awarda_presented.html, accessed 20 October 2011.

Turral, H. (2008) 'Policies and Strategic Planning for the Thailand Irrigation Sector Reform Programme', *Final Technical Report*, The Food and Agriculture Organization of the United Nations and the Royal Irrigation Department, Bangkok, Thailand.

Wijeyewardene, G. (1973) 'Hydraulic Society in Contemporary Thailand?' in R. Ho and E. C. Chapman (eds) *Studies of Contemporary Thailand*, Research School of Pacific Studies, Australia National University, Canberra, Australia.

Wittfogel, K. A. (1957) *Oriental Despotism: A Comparative Study of Total Power*, Vintage Books, New Haven and London.

Worster, D. (1992) *Rivers of Empire: Water, Aridity and the Growth of the American West*, Oxford University Press, New York and Oxford.

14 Waterscapes in transition

Past and present reshaping of sacred water places in Banaras[1]

Vera Lazzaretti

Introduction

This chapter explores the urban waterscape of Varanasi (Uttar Pradesh, India) and crucial moments of its transition. In particular, it focuses on the city's 'minor' water places, which have up to now received only fleeting attention in studies about the urban landscape and cultural practices linked to the city's space. Together with the river Gaṅgā and its tributaries, Varuṇā and Assī, minor water bodies constitute the urban waterscape and clearly testify to the multiple fragmentation of water, its knowledges and practices. These places include wells, ponds, rivulets, canals and urban lakes, which are looked after, worshipped, exploited, narrated, contested, or even forgotten by multiple local actors with various expectations and conceptions of water and its spaces. As argued in this chapter, they constitute privileged locations in which to explore the interaction of diverse epistemologies of water, as well as questioning dynamics of development of the city's landscape.

In order to describe the changing status and role of water places in Banaras (as the city is called by its inhabitants) and trace the trajectories of an urban waterscape in transition, the chapter details the case of a central yet somehow today hidden sacred well; this is the Jñānavāpī (Well of Knowledge) which is located at the core of the city's religious life and next to the temple of the patron deity, Kāśī Viśvanāth. By documenting the fragmented histories and sources about the well, as well as drawing on my own fieldwork around this water place, the chapter will highlight the peculiarities and priorities of the various actors involved in the past reshaping and ongoing redefinition of such a well. Additionally, a tentative picture of more general tendencies of development for the city's various sacred water places will be introduced.

The perspective of water will be adopted as a privileged and alternative lens through which to address Banaras in its *thickness*, and thus enrich understanding of the complexity of its space and the dynamics of its constant reshaping. Through this analysis water, indeed, will be seen as a distinct and promising epistemic object, which allows examinations of otherwise supposedly diverse sources, fields and epistemologies by directing our attention to their confluences. The water perspective, for example, pushes to consider the inherent overlapping of 'sacred' and 'profane' as an overlooked motif through which to further question the construction of a supposedly monolithic sacred city such as Banaras, as well as to develop a critical analysis of the current discourses about its heritage.

The redefinition of water places during crucial phases of transitions, on which the chapter focuses, indeed seems to result from the interactions of diverse views on water, which as a simplification could be labelled as 'sacred' and 'profane'. As detailed below, a first crucial phase of transition of the urban waterscape in the nineteenth century coincides for example with a mutual support of, on the one hand, the religious revitalization of an Hindū sacred geography promoted by the Brahman local élite and regional patrons, and on the other, the needs of sanitation, development and aesthetic gaze of the colonial administration. Furthermore, today's projects and discourses about the city's 'heritageization', which however seem to exclude a serious consideration of the urban waterscape, are constructed by the interactions of a supposedly secular mandate and new visions of religious consumption.

In the following section the city of Banaras and its waterscape will be introduced and the context and theoretical background of the present chapter will be framed. After that, the development of the city's waterscape will be tentatively retraced by drawing on various sources, whose integrative use will be seen as necessary when addressing water. Crucial transitions and the overlapping of epistemologies linked to both 'sacred' and 'profane' understandings of water during colonial times will be highlighted. The following section details the case of the Well of Knowledge and its history of ascent and decline, as well as discussing its current situation. It will be shown how this case illustrates transitions and paths to sacralization at a specific sacred water place, while also raising questions about the repositioning and the future of waterscapes in current 'beautification' projects linked to heritage discourses.

Framing Banaras through the lens of water

Banaras is a well-known pilgrimage destination of North India. As shown by numerous works (especially Freitag 1989, 2008; Bakker 1996; Dalmia 1997; Gaenszle and Gengnagel 2008; Desai 2007; Dodson 2012), the myth of a supposedly ancient, unchanging and exclusively Hindū sacred city[2] has been projected, reshaped and sustained by a multiplicity of voices and representations of its landscape. These include indigenous eulogistic literature,[3] colonial reports, visual depictions of its sacred space and the tourist industry, as well as academic works and a variety of discourses rethinking the city as a 'national heritage'. The latter are nowadays particularly pervasive due to a new wave of interest stemming from the current Prime Minister's cultural agenda and focus on Banaras' 'beautification' and development.[4]

Water in Banaras has been addressed by various scholars and from different angles: apart from the expected consistent production about the river Gaṅgā and related topics of pollution, sacredness, mythology and environment (Darrian 1978; Alley 2002; Madan 2005; Doron *et al.* 2014), the contributions include studies on specific water architectures in the city (Rötzer 1989; Jalais 2014); surveys of the urban waterscape in general (Singh 1994); and specific understandings of groups of workers linked to water (Schütte 2008; Doron 2013). Water has also been peripherally addressed by studies of specific social groups and their leisure and sport activities (Kumar 1988; Alter 1992). Another contribution drew attention to and summed up the pervasiveness of water in Banaras as an element through which local actors construct their ideas of time and space (Derné 1998). Further significant contributions about sacred water were made within the field of religious cartography (for example, Gutschow 2006; Gengnagel 2011), and water in Banaras is currently addressed by an ongoing project on 'Changing Sacred Waterscape in Banaras' at the University of Heidelberg.

The element water itself seems to have been, however, considered only peripherally in analyses of the urban sacred landscape and its historical evolution; in fact, if single historical shrines and the search for their 'original' locations have received much attention in major works (for example, Eck 1983; Bakker 1996), specific water places, their transitions and links to the wider history of the urban territory have been often overlooked. This is rather surprising if we think that while dismantlement and re-use of building material from previous shrines is a common pattern in Indian architectural history, water sources and their locations, if not fixed, are at least less movable than stones. Water places indeed represent more stable and land-anchored realities; their mutable although almost unmovable locations, as well as memory about them, persist in pointing out those very places where water was most likely found and worshipped at first, before they became the seats of major gods and the places for urban settlements. Furthermore, to my knowledge, and excluding the few works mentioned above, extensive studies about actual perceptions, consumption, uses, symbolic understandings and narratives of specific water places (both sacred and civic) in Banaras are lacking. Adopting the perspective of water, instead, challenges the dichotomies often attached to this renowned city, such as those of sacred and profane, or antiquity and modernity.

The variety of water forms and architectures in the urban landscapes of South Asia indeed testifies to the inherent overlapping of different spheres of human life linked to and realized through water, as well as to the various knowledges and uses of water, both in everyday life and in ritual practices. The actual construction of wells, reservoirs and water monuments is well documented. Sources on water places and architecture include inscriptions from the first millennium and treatises on architecture from the second millennium. Both types of evidence testify to the construction of water places as a highly meritorious and costly practice, sometimes associated with donations and rewards to Brahmans (Jain–Neubauer 1981). The practical function of supplying water for human beings, animals and fields seem to always overlap with the social and charitable dimension. On the one hand, shrines usually emerged next to pre-existing wells which were needed to supply water for worship and rituals. The term *tīrtha*, which is connected at first to the act of crossing rivers and, later, to the metaphor of passing through the human shore to reach the divine (Kramrisch 1946; Eck 1981), is used to indicate firstly the water places itself and, as a consequence, the nearby shrine (Jain-Neubauer 1981). On the other hand, water and its architecture are at the same time linked to mundane and pragmatic activities, such as retreat, irrigation, fertility and cooling (Jain-Neubauer 1981) and indeed they are places where dichotomies collapse and boundaries between sacred and profane dissolve (Dehejia 2009). Architectural texts also inform us about diffused knowledge about stone and minerals in building material of wells, which ameliorate water and make it precious for fertility, growth, nutrition, healing and embellishment purposes (Jain-Neubauer 1981). Moreover, since the emergence of pan-Indian sacred places and the evolution of the concept of *tīrtha*, water places can be seen as representatives of the merging of Brahmanical and 'folk' traditions (Eck 1981). Water worship, indeed, before being absorbed by major religious traditions of not only South Asia, seems to be related to previous aboriginal practices, sprits of places, natural deities and local knowledges of land and territory.[5]

Water is a pervasive element in Banaras: the city lies on the west bank of the Gaṅgā where the river turns into a bight that makes the water flow 'upstream', as often reiterated by its inhabitants and devotees to prove the divine character of the city. The Gaṅgā, with its diverse and changing landscapes, marks the flow of seasonal, monthly and daily rhythms. Apart from the expanding river, the city seems to have been cyclically affected in the past, and partially still today, by a sort of tide; ponds, rivulets and sometimes low lanes of the inner city were filled with water during rains, giving the image of a wet, flooded, amphibious city, which was and still sometimes even today is compared by travellers with Venice (Zara 2011); Banaras is imagined as a floating fish. Visiting the city during the monsoon is indeed a revealing experience: underground temples, which are considered to be the more ancient remains of the first urban strata, are flooded from presumed mythic connections with the river, the major ponds of the city overflow, and near the barrier formed by the *ghāṭ*-s[6] an impression of immediate vicinity of the growing river is palpable.

The presence of a fragmented, multiple, pulsing waterscape, whose specific places are scattered and need to be discovered in hidden or central locations of the city, makes the water's perspective particularly fruitful for further explorations of such an urban landscape and of the various engagements that inhabitants establish with what has been called a 'sentient waterscape' (Stepan, this volume).

Banaras' waterscape and its evolution

Water monuments and architectures in Banaras include a variety of ponds, called *kuṇḍ*-s, *pokharā*-s and *tālāb*-s, whose main function was to collect rainwater and drainage and direct them through the *nālā*-s, the rivulets of the city, into the Gaṅgā and its tributaries (Singh 1955). Other main typologies of water places include a variety of wells, named as *kūp*-s, *kuāṃ*-s or *vāpī*-s; these comprise both original civic constructions realized during *mughal* times and the British period (Rötzer 1989) and, as the case of the Well of Knowledge testifies, presumably ancient larger sacred water bodies, which have been lately reduced into smaller wells.

The variety of typologies, terminologies and practices of such a waterscape is documented by local scholars in their extensive cataloguing works about the city's sacred geography: compilers of modern glorifications and guides to the city's various *tīrtha*-s registered an enormous variety of names and locations of urban sacred water places, as they were surveyed during the nineteenth to twentieth centuries (Sukul 1977; Vyas 1987). These local surveys, which also acted as pilgrims' guides, testify to an incredible diversification of not only rituals and terminologies, but also of uses and qualities connected to the specific water places; water at a certain well is, for example, considered particularly apt for fertility purposes, while another source provides healing water for stomach diseases. Furthermore, auspicious moments of the calendar for when to bathe (*snān karnā*) and drink at these various places are recorded. Today, knowledges linked to water in these urban places are preserved in feeble memories of old inhabitants of the various historical areas, and silently celebrated on specific occasions by small groups of informed citizens, passing pilgrims or curious researchers.

A history of such a waterscape is difficult to trace, as long as the presence of water intersects various and, in a way, opposite sources, which indeed need to be considered together to approach the multifaceted nature of water. As I will document, clear evidence of transitions in the urban waterscape is only available starting from the colonial period; it is thus necessary to integrate these sources with other material, which is however, less conventional when addressing water-related problems. For example, local glorifications can be interrogated as historical sources by following the approach outlined by Ginzburg (2000), in which sources are read 'against the grain', thus questioning and sometimes even going against the intentions of their producers. Sacred texts and in particular local glorifications will be seen as the products of new historical and religious contexts and thus as means of resettlement of past heritage and previous geographies (Smith 2007, p. 2; Acri and Pinkney 2014), as well as the urban waterscape. The integrative use of different sources is anyway necessary in our analysis; in fact, only by addressing fragments of evidence, collected by various actors with sometimes contrasting purposes, can we trace a tentative history of Banaras' waterscape.

Studies in urban geography about the ancient history of Banaras indicate that the very first settlements grew both along the well-drained ridge that existed along the Gaṅgā, from the Varuna to the Assī rivulets, as well as in the interior land by taking advantages of tanks which supplied water and fertile ground (Singh 1955). Textual and material evidence testifies that ancient sacred places were also connected to water bodies, mainly named *kuṇḍa*-s and *vāpī*-s, which apparently belong to the first stratum of the urban fabric. For example, the only sanctuary of Banaras mentioned by the first description and catalogue of sacred places and pilgrimages, the *Tīrthayātrāparvan* (around III sec. C.E.) of the Mahābhārata, is Vṛṣadhvaja (MBh 3.82.69), which is connected to Kapiladhārā, a bathing pool, also named by the text. From the Gupta period the commercial city developed into a sacred centre, nevertheless heterogeneous, as testified by material evidence (Bakker and Isaacson 2004) and developed southwards, taking advantage of other natural ponds.

Local glorifications address urban water bodies as divine breeding grounds since at least the seventh century, most likely collecting and adapting pre-existing traditions centred on the various water places. Glorifications culminate with the naming and description of dozens of sacred wells and ponds in the thirteenth to fourteenth century *Kāśīkhaṇḍa*, a section of the Skandapurāṇa, which is the most extensive glorification about the city.[7] This text, as I will detail in the case of Jñānavāpī, greatly contributed to reinventing the city's sacred geography and, as far as the waterscape goes, it represents an extensive collection of local water places: their names, divine origins, qualities and locations are established and fixed by this eulogistic compendium, which is still today an authoritative and pervasive textual frame for any discourse about the urban sacred geography.

A great number of the major water monuments were, however, either reconditioned or constructed anew starting from the seventeenth up to the nineteenth century. In this period patrons from various regional powers, led by Marathas and Bengalis, funded the construction of riverfront palaces, *ghāṭ*-s, temples and rest houses for pilgrims in a sort of call for the resettlement of the city's sacred geography and ancient traditions (Couté and Léger 1989; Freitag 1989; Gutschow 2006). Other pre-existing water places were instead equipped with more solid and sometimes monumental structures, or reshaped according to new directions and uses of the city's space (Singh 1955), as also testified by the case of Jñānavāpī.

It is, however, from colonial time that we can distinguish more clearly the development of the city's waterscape during the nineteenth and twentieth centuries; maps, charts, drawings and surveys, realized with both cataloguing and intervening intentions, address water as an environmental and urban issue in a growing and developing city, while also keeping an eye on local beliefs. In this period, the overlap of different epistemologies connected to water become more evident.

Drastic changes happened between 1822 and 1880, when major works on the urban fabric were undertaken by the colonial administration; in this period, some of the major water places were drained, equipped with underground channels leading to the river and reduced to smaller tanks (Singh 1955). The map *City of Banarus*, surveyed in 1822 by James Prinsep, Assay Mint Master in Banaras (1820–1830), testifies to the state of the urban waterscape before major drain works were undertaken (Prinsep 2009).

Rich patrons such as regional sovereigns, leading citizens and wealthy pilgrims gave extensive pious donations for the Hindū cause; however, public works for the improvement of infrastructure were more difficult to fund and undertake due to the lack of a coordinating body. Some public works, such as the repair of river banks, *ghāṭ*-s, streets and the digging and cleaning of drains were undertaken by the local administrations randomly, while for only a few years a Committee for Local Improvement (1823–1829), of which Prinsep was the secretary, was in charge of more systematic planning and implementation (Medhasananda 2002). Under the Committee works for the reclamation of the Mandākinī and Matsyodarī were realized, together with the paving of a few lanes in the central area of the city. However, a Municipal Board was instituted only in 1867 and systematic public sewage works started later (Singh 1955).

The Committee records testify to caution and pragmatic concerns right at the start of the activities. Apart from the projection of a period of preliminary surveys, the Committee's first project was to repair the so-called 'Madho Rai kī masjid' at Pañcagaṅgā Ghāṭ and its minaret, even though, as stated 'the Committee are aware that the general object set forth in the Resolution of Government do not particularize the repair of religious worship' (Minute Book of the Committee, as quoted in Medhasananda 2002, p. 430). However, the repair of a shrine was justified as its 'position, structure and maintenance are ultimately connected with the safety, convenience and ornament of a town' (ibid.). The same record establishes that to prevent jealousy between Hindūs and Muslims the Committee should as well fund at the same time some Hindū shrine; the proposal was then that of sponsoring works at the Nāg Kuāṃ, one of the city's step-wells where an annual festival takes place, in order 'to preserve it to further dilapidation' (ibid.).

As is well known, the colonial administration's aspiration of egalitarian treatment of Hindūs and Muslims, its effects on the relationship between them and the modes of negotiating the 'public space' between the communities, local élites and colonial rulers have been already widely analyzed by numerous works (Cohn 1987; Bayly 1985, 1992; Pandey 1990; Desai 2007). In this context I seek to underline, however, that the Committee's attitude towards the city and the projects for the improvement of infrastructures and embellishment of symbolic 'ornamental' buildings met, and happened to support, the coeval 'resurrection' (Desai 2007) of the Hindū urban geography pursued by regional and local patrons as far as water places were concerned.

By looking at the overlap of urban civic space and sacred geography, one can better grasp the mutual interactions of diverse epistemologies at work in this period: development of streets and infrastructures, as well as the reclamation of land from large water bodies and the reshaping of water monuments, seem to have been necessary to respond both to the growth of population and to the increase of spatial ritual practices, as well as managing religious crowds. Local pilgrimages and religious festivals were developing and evolving together with the urban space. Indeed, religious activities contributed to the need for development of the urban fabric and to the enrichment of the city: pilgrims coming by land had to pay a tax to the government and their donations also provided benefits to sacred specialists, Brahman families and monasteries (Bayly 1999).

In this picture, major water bodies, such as large tanks and *nālā*-s, were directly affected by the urban development projects which involved their drainage, sanitation and coverage for hygienic purposes, as well as for the creation of a new public sphere: by changing them into underground channels for the dirty water, in fact, their reduced versions dwelled at the centre of newly established parks, such as Mandākinī, Beniā, Matsyodarī, created from the reclaimed land.

Some minor sacred water places must as well have been affected by the reshaping of the city, which was pursued on the one hand for a colonial ideal of improvement and urban development, as well as by an attitude of sporadic equity towards different religious communities, as testified by the case of Nāg Kuāṃ. On the other hand, as I will detail in the case described in the next section, the materialization of the imagined sacred landscape in the urban fabric also called for the restructuring of crucial water places, thus changing water uses; indeed, this resulted in the definition of the symbolic and ritual roles of water, as well as in the production of knowledge about it.

Defining, directing, reshaping water: The case of Jñānavāpī

The Jñānavāpī occupies today a paradoxical position in the urban landscape of Banaras: the well is central and hidden at the same time. On the one hand, it lies next to the temple of the patron deity of the city, Kāśī Viśvanāth, which is the principle destination for the many pilgrims visiting the city. The area, which comprises also the Gyān Vāpī mosque, has become a sort of unique cordoned-off compound under a security plan, after the Ayodhya events (1992) and subsequent claims by Hindutva leaders over Viśvanāth temple, and thus occupies a central role in narratives and media as a contested space. On the other hand, the well itself and its surrounding space occupy, both in these narratives (Lazzaretti, in preparation) and as far as religious activities in the area are concerned, a peripheral position. The well itself and the nearby minor shrines are today scarcely addressed by passing devotees who are, instead, directed in orderly lines towards Kāśī Viśvanāth; as I have claimed elsewhere, the temple is being transformed into the only acceptable and accessible destination in this huge compound, which comprises not only a variety of minor shrines, the well and the mosque, but also a series of previously civic lanes and spaces which were absorbed by the security project (Lazzaretti 2015).

The Jñānavāpī occupied, at least in the last two centuries, an important role in the city's religious life as the centre of the practice of pilgrimage in this pilgrimage city: it is the place where still today pilgrims are supposed to pronounce their vow (*saṅkalp lenā*) when commencing a pilgrimage circuit, and 'leave' the ending formula (*saṅkalp chornā*) to sanction their accomplishment, once they come back from any urban circumambulations. This role is, however, today obstructed by the security rules and barriers, which consistently affect the practices of local urban pilgrimage (Ibid.); furthermore, the evolution of the well in history and in the present are almost absent in scholarship about Banaras, despite its crucial position and role.

Despite the nearby shrine of Kāśī Viśvanāth, whose fame spread presumably at the time of the *Kāśīkhaṇḍa*, and whose current version was built at the end of the eighteenth century, the Well of Knowledge has been identified as the water place (*vāpī*) near which the first known main sanctuary of the city, Avimukta, dwelled, as mentioned by texts previous to the twelfth century[8] (Eck 1983; Bakker 1996). Textual evidence is not enough to prove the existence of the Well of Knowledge and its location; however, it is highly probable that if the notable ancient sanctuary of Avimukta existed and was located in the area under investigation (Bakker and Isaacson 2004), it was most likely attached to a water place; in fact, as detailed above, the presence of water was considered necessary for and preliminary to the construction of a shrine.

During the various constructions and reconstructions of the urban landscape (Bakker 1996; Bakker and Isaacson 2004), the well apparently maintained its location. It took, however, the *Kāśīkhaṇḍa*, with its immense reprojection of the city's sacred geography and previous mythological material (Smith 2007), to give a first extensive description of the Well of Knowledge. Jñānavāpī now acquired its name and was anchored to the city's new geography, as well as to the superimposing mythology of Śiva as a supreme, all-encompassing great god, and his inalienable connection to the city. The text narrates the mythological origins and special qualities of the water of Jñānavāpī (KKh 33.1–52 and 34) in an effort to resettle the previously existing divine landscape centred on Avimukta, as a newly conceived sacredscape, centred instead on Viśvanāth.

The Jñānavāpī is positioned by the text in the Nirvanamaṇḍapa or Muktimaṇḍapa, the 'Pavilion of Liberation' (KKh 79.56–68). This is described as a superb and powerful sacred area, and as the place in which pilgrims should start the Antargṛhayātrā, a pilgrimage within the inner area of the city (KKh 100.77), thus anchoring the area around the well to the practice of pilgrimage. The well is identified as the water form of Viśvanāth himself; as the assigned name Jñānavāpī reveals, this water is seen as bestowing knowledge, wisdom and liberation. The special power of Jñānavāpī's water seems to be even more meaningful in the city in which, according to glorifications, *mokṣa*, the final goal which frees human beings from rebirth, is easily accessible. Not only is this specific water considered to be a way to obtain wisdom and liberation but, as described by the text, it should also be used to perform a series of rituals in this *tīrtha*: in particular, for rituals for ancestors, for holy ablutions, for drinking while fasting, to be touched for piety and to make offerings to the *liṅga* and other deities in the area (KKh 33.31–50).

A later text, the *Kāśīrahasya* of the sixteenth century, further amplifies the importance of the area where the Jñānavāpī is located and its role in the city's geography. The pavilion where the well was positioned by the KKh and the surrounding area are here definitely anchored to the most important layer of the city's sacred geography: the pavilion is described as the place where pilgrims have to perform the initial and final rituals of the Pañcakrośīyātrā (KR 10.10–19), the most notable urban procession, which circumambulates the entire sacred field of Banaras (Singh 2002; Gengnagel 2011). The scheme of initial and final rituals to be performed in the pavilion of the Jñānavāpī seems to have been later employed as a rule for other local circuits, which flourished in nineteenth- and twentieth-century Banaras.

That the practice of local urban circuits became a major and profitable activity in these centuries, as also testified by the colonial interest in religious gatherings and urban pilgrimage routes, becomes clear when addressing the transitions at the Well of Knowledge in the nineteenth century. The area around the well received the attention of both the Committee for Local Improvement and of patrons from other regions, who participated actively in reshaping the space and, in doing so, defined the access to and uses of the surrounding space and its water. In his minutes of 1829, at the time of the abolition of the Committee, Prinsep talks about the 'ornamenting of the Gyan Bapee' and the paving of Viśvanāth *galī*, as works just accomplished (quoted in Medhasananda 2002, p. 437). The structure of the pavilion of the Well of Knowledge was, in fact, erected most likely in 1828, with the sponsorship of Baija Bai, widow of Daulat Rao Sindhia of Gwalior (Sherring 1868). At that time, or shortly after, what was before a pond with steps (*vāpī-vāv-bāvri-bāoli* all derive from the same Sanskrit root *vap*) was covered, and its water confined to the smaller well which is visible today and contained in a marble carved hexagonal enclosure. The size of the previous structure can still be guessed from the small grating positioned on the ground at one corner of the pavilion which covers the now underground stairs. According to Kedarnath Vyas, current incumbent of the area and author of one of the more notable contemporary pilgrimage guides (Vyas 1987), and whose family has been in charge of the well's activities and pilgrims' inflow and offerings for at least the last two centuries, the reshaping was done in order to give pilgrims more space for the performances of rituals and for rest. In the same period, his ancestors, and in particular a great devotee of the well, a certain Rukmini Devi, who was the wife of an uncle of Kedarnath's grandfather, requested and obtained from the colonial government a declaration that the step-well and its water could not be used by Muslims as an ablution place any more (personal communication with Kedarnath Vyas, March 2014 and January 2015).

Apparently, the Jñānavāpī was reduced and reconceived from being a multifunctional, shared and unconfined water place, to being a notable, now beautifully carved and crowded sacred place for Hindū pilgrims in a time when the practice of urban pilgrimage was flourishing and in a moment of reification of the city's sacred geography. Diverse actors' interests and epistemologies overlapped at Jñānavāpī at that time and merged in the reconfiguration of the geography of pilgrimage in the city.

On the one hand, the colonial administration, as probably suggested by the Committee, facilitated and encouraged the restructuring of the well with the intent of promoting cleanliness, sanitation, order and beauty for a growing central space. The frenetic activities at the well, and the consequent need of managing crowds, are testified by Sherring (1868), who describes the well as incredibly stinky and filthy due to the constant offerings of flowers and fruits in the water. On the other hand, the Hindū patrons who funded the monumentalization of the well's space were actively working for the establishment of a tangible Hindū heritage and for the consequent estrangement of the Muslim community, in a period when the whole area around the temple of Viśvanāth was being narrated and shaped as a highly contested space (Lazzaretti, in preparation).

Various minds with different purposes contributed to the reshaping of acceptable practices linked to water at the well, such as rituals linked to pilgrimage, worship of the surrounding deities and consuming of sacred water as divine essence; indeed, they helped to diffuse knowledge about it and construct its new fame. The Vyas family, who acquired and moved in the building near the well in the 1880s in order to be able to be nearer to the Jñānavāpī, as their 'working' place (personal communication with Kedarnath Vyas, January 2015), presumably played a crucial role in diffusing a new epistemology of the Jñānavāpī, drawing on the one hand on extensive knowledge of the local textual tradition and their own spatial knowledge; and on the other hand, on the approval of the colonial administration and the support of influential regional personalities.

Due to crowd fervour and the intense devotional activity, the well was later equipped with iron bars across the top to prevent the occasional suicides of liberation seekers; lately also a cloth has been spread over the top to prevent offerings from being dropped into the water. The tendency to direct, close and protect divine water seems to increase along with the specialization of the well as a pilgrimage confluence. Indeed, the well occupied a crucial role in the city's religious life until the last century, both as part of the daily visits to the patron deity and the surrounding shrines (Eck 1983) and as a crucial ritual space for the performance of pilgrimages, which was nonetheless sanctioned by members of the local royal élite, the Kāśī Nareś of Ramnagar, who still come once a year to perform starting rituals for the Pañcakrośīyātrā (personal communication with Kedarnath Vyas, March 2013).

Drastic changes to the well and further difficulties in accessing its water happened in recent times. As mentioned above, and detailed elsewhere (Lazzaretti 2015), the security plan started after the Ayodhya events is progressively transforming the area into a transitory, neglected and semi-empty space, from which local circuits divert. The Kāśī Viśvanāth Trust, which has been in charge of the Viśvanāth temple since 1983, is now trying to purchase also the land next to the well from the Vyas family, in order to realize the project of a unified compound with rest houses for pilgrims and a traditional learning institution (personal communication with P. N. Dubey, Additional Chief Officer of the Kāśī Viśvanāth Trust, March 2013). Apart from evident consequences that such a 'theme park' (where the theme on stage would be 'Hindū dharma') would have on the mosque and the Muslim community, it is not clear how the well and its water would be newly reconceived by such a project in what is another crucial moment of transition, not only for the Jñānavāpī, but for the urban waterscape itself.

Apart from the bombastic stress on the Gaṅgā river as a symbol of national identity and the periodical (mostly ineffective) campaigns launched against its pollution, water indeed seems to be left behind in current heritage discourses; in particular, the various water places of the city do not seem nowadays to be part of the selected fragments of culture which are shaped to constitute heritage (Bendix 2009). Peripheral water places in the city are, in fact, in a state of neglect and as media once in a while report, speculators have managed to have many ponds filled up with garbage and sold as building sites. The national campaign for the 'beautification' of the city's heritage concentrates instead on other major attractions, such as (only part of) the riverfront and related ceremonies on the river (Gaṅgā *ārtī*).

As for Jñānavāpī, local discourses insist on the declaration of Kāśī Viśvanāth as heritage while the Kāśī Viśvanāth Trust, supported by regional and central government, projects the reshaping of the whole area according to new standards of aesthetic religious consumption, which are, however, presented as part of the secularization of the compound.[9] These projects seem to totally exclude from the area the already confined Muslim minority, as well as discard previously constructed heritage spaces, such as the Jñānavāpī. Such water places, which had been once selected and shaped as part of the city's heritage by the mutual support of diverse epistemologies, become now peripheral and useless in the new ongoing 'heritageization' process.

Epistemologies of water in Banaras: Provisional conclusions

The city's waterscape has been invested with diverse meanings according to various narrating voices and myths. Indeed, it has been in the past the site for the overlapping of multiple and contrasting epistemologies whose promoting actors have sometimes managed to dialogue about water and because of water. Water places have been the objects of ameliorating plans by a colonial administration driven by an epistemology of objectification, order and beauty. Water, indeed, at that time was felt to be crucial in the reshaping of urban landscape and development projects, and water places played a role in informing distinct epistemologies.

At the same time, in fact, the symbolic power of some chosen water places, instead of being rejected, was reinforced by epistemologies of the sacred, sustained by powerful actors and sanctioned by ritual practices; indeed, these places have been described as inextricably linked to the concept and the status of the urban territory as a sacred ford, shaped by the Brahman élites. In particular, the need to define, canalize and direct water somehow emerged as a consequence of diverse demands and views about water.

Recent transitions at Jñānavāpī, which up to now include mystification, disuse and illegal occupation of space, appear instead to be accidental side-effects of the ongoing 'beautification' and 'heritageization' processes which, together with the security mandate, affect the central area, as well as the city in general. Dialogue about the city's multiple water places is now silently and passively rejected, while fragments of culture shaped in the past as heritages are discarded to the advantage of more attractive and lucrative locations, and pushed into oblivion by louder discourses.

More generally, it emerges that dialogues between distinct knowledge systems only happen among or are silenced by powerful actors, whose epistemologies are used to define the shapes and lead the transition (or the misuse) of an urban waterscape. The variety of water places themselves, however, stores deeper strata of local beliefs and practices which are meaningful to a multiplicity of inhabitants and define their sense of belonging, as well as their knowledges about specific locations in the city. These deeper strata are barely accessible from the surface and related epistemologies are hardly represented, if not openly rejected, both in transition phases and in reshaping projects. This is what we need to investigate further to bring the multiplicities of water, and its potentials, to the central stage.

Notes

1 I wish to thank the participants of the workshop 'Epistemologies of Water in Asia' held in Heidelberg in December 2014 for fruitful exchanges and stimulating discussions and, in particular, Jörg Gengnagel for inviting me in pursuing the water path. I deeply thank Kedarnath Vyas ji for the hours spent together talking and remembering about names, qualities and deities linked to water places in Banaras.
2 For a more extensive analysis on the academic literature about Banaras and the history of the 'deconstruction' of its myth see Lazzaretti 2013.
3 With glorification or eulogist literature I refer to the genre of *māhātmya-*s or *sthalapurāṇa-*s; these are sections of the *Purāṇa-*s, the ancient recitations of the Brahmans, which often circulate among sacred specialists and devotees independently. Glorifications are centred on specific sacred objects, such as sacred centres, temples, water places, holy grounds or areas. The city of Varanasi has a rich and varied eulogistic literature, which has been most widely analyzed by Bakker and Isaacson, 2004.
4 See the bibliography for a collection of recent articles about the topic.
5 An extensive, although rather old-fashioned, catalogue of water worship is found in Masani 1918.
6 The term *ghāṭ* indicates the series of steps leading to a water body, in this case a river. It is a diffused water architecture found all over South Asia. Banaras' *ghāṭ-*s form a sort of barrier which protects the city from the seasonal swell of the river during monsoon. For an extensive study about the *ghāṭ-*s of Banaras see Jalais 2014.
7 On the *Kāśīkhaṇḍa*, its dating and content see Hazra 1985; Bakker 1996; Bakker and Isaacson 2004; Smith 2007.
8 The *Tīrthavivecanakāṇḍa* of Lakṣmīdara, a compendium of previous glorification material, quotes a *Liṅgapurāṇa* which describes Avimukta as dwelling to the north of a water place (*Devasya dakṣiṇe bhāge vāpī tiṣṭhati śobhanā*, TVK, p. 109–110); this has been thought of as Jñānavāpī. Avimukta in fact is said to have been located in the area currently occupied by the mosque.
9 On the appropriation of secularism and uses of its language by politicized religion and the Hindū right see Prakash 2007.

Sources

Kāśīkhaṇḍa (*Skanda Purāṇa*), G. V. Tagare, ed. Motilal Banarsidass, Delhi, 1996.
Kāśīrahasya, Jagdīś Nārāyaṇ Dūbe, ed. Ādarśa Prakāśan Mandir, Vārāṇasī, 1984.

The *Mahābhārata*, J. A. van Buitenen, ed. University of Chicago Press, Chicago, 1973.
Tīrthavivecanakāṇḍa, Bhaṭṭaśrīlaksmīdharaviracite Kṛtyakalpatarau aṣṭamo bhāgaḥ, K. V. Rangaswami Aiyangar, ed., Gaekwad's Oriental Series 98, Baroda, 1942.

Media

Minimum selection of recent articles from the *Times of India* at http://timesofindia.indiatimes.com/city/varanasi:
'At Kāśī Vishwanath Temple, Modi offers 1,001 lotuses to "baba"', 14/05/2014.
'Demand of declaring KVT as national monument', 08/02/2015.
'Facelift of Dashaswamedh ghāṭ soon', 08/02/2015.
'Issue to declare KVT a national heritage: Locals take out procession', 16/02/2015.
'Modi to spend 100% MPLADS fund in Kāśī', 24/02/2015.
'HRIDAY at the heart of Varanasi's dev', 01/03/2015.
'Restore facilities, but preserve Kāśī's heritage', 01/03/2015.
'City "mahashamshans" getting facelift', 10/03/2015.
'Water-ATMs for tourists, pilgrims in Varanasi soon', 11/03/2015.
'Tourism ministry, industry join hands for facelift of Kāśī', 15/03/2015.

References

Acri, A. and Pinkney, A. M. (2014) 'Reorienting the Past: Performances of Hindū Textual Heritage in Contemporary India', *International Journal of Hindū Studies*, vol. 17, no. 3, pp. 223–230.
Asad, T. (2003) *Formations of the Secular: Christianity, Islam and Modernity*, Stanford University Press, Stanford.
Alley, K. D. (2002) *On the Banks of the Gaṅgā: When Wastewater Meets a Sacred River*, University of Michigan Press, Ann Arbor.
Alter, J. S. (1992) *The Wrestler's Body: Identity and Ideology in North India*. University of California Press, Berkeley.
Bakker, H. (1996) 'Construction and reconstruction of sacred space in Vārāṇasī', *Numen*, vol. 43, no. 1, pp. 32–55.
Bakker, H. and Isaacson, Z. (2004) 'A sketch of the religious history of Vārāṇasī up to the Islamic conquest and the New Beginning', in H. Bakker and Z. Isaacson (eds.) *The Skandapurāṇa, vol. IIA* (Adhyāyas 26–31.14), The Vārāṇasī Cycle, Egbert Forsten, Gröningen, pp. 19–82.
Bayly, C. A. (1992 [1983]) *Rulers, Townmen and Bazars: North Indian Society in the Age of the British Expansion, 1770–1870*, Cambridge University Press, Cambridge.
Bayly, C. A. (1985) 'The pre-history of "communalism"? Religious conflicts in India, 1700–1860', *Modern Asian Studies*, vol. 19, no. 2, pp. 177–203.
Bayly, S. (1999) *Caste, Society and Politics in India, from Eighteenth Century to the Modern Age*, in AA.VV. *The New Cambridge History of India, IV: 3*, Cambridge University Press, Cambridge.
Bendix, R. (2009) 'Heritage between economy and politics. An assessment from the perspective of cultural anthropology', in L. Smith and N. Akagawa (eds) *Intangible Heritage*, Routledge, London, pp. 253–269.
Cohn, B. S. (1987) *An Anthropologist Among the Historians and Other Essays*, Oxford University Press, New Delhi.

Couté, P. and Léger, J. M. (1989) *Bénarès. Un voyage d'architecture*, Editions Créaphis, Paris.

Dalmia, V. (1997) *The Nationalization of Hindu Traditions: Bharatendu Harischandra and Nineteenth-Century Banaras*, Oxford University Press, Oxford.

Derné, S., (1998) 'Feeling water: Notes on the sensory construction of time and space in Banaras', *Man in India*, vol. 78, no. 1&2, pp. 1–7.

Darrian, S. G. (1978) *The Ganges in Myth and History*, University Press of Hawaii, Honululu.

Dehejia, V. (2009) *The Body Adorned. Dissolving Boundaries Between Sacred and Profane in India's Art*, Columbia University Press, New York.

Desai, M. (2007) 'Resurrecting Banaras: Urban spaces, architecture and religious boundaries', PhD thesis, University of California, Berkeley.

Dodson, M. S. (2012) *Banaras: Urban Forms and Cultural Histories*, Routledge, New Delhi, New York, London.

Doron, A. (2013) *Life on the Gaṅgā. Boatmen and the Ritual Economy of Banaras*, Cambridge University Press, New Delhi.

Doron, A. et al. (2014) An Anthology of Writings on the Gaṅgā. Goddess and River in History, Culture and Society. Oxford University Press, New Delhi.

Eck, D. L. (1981) 'India's Tīrthas: "Crossings" in sacred geography', *History of Religions*, vol. 20, no. 4, pp. 323–344.

Eck, D. L. (1983) *Banāras: City of Light*, Penguin Books, London.

Freitag, S. B. (1989) *Culture and Power in Banaras. Community, Performance and Environment, 1800–1980*, University of California Press, Berkeley.

Freitag, S.B. (2008) 'Visualizing Cities by Modern Citizens: Banaras Compared to Jaipur and Lucknow', in M. Gaenszle and J. Gengnagel (eds) *Visualizing Space in Banaras: Images, Maps and the Practice of Representation*, Oxford University Press, New Delhi, pp. 233–254.

Gaenszle, M. and Gengnagel J. (2008), *Visualizing Space in Banaras. Images, Maps, and Practice of Representation*, Oxford University Press, New Delhi.

Gengnagel, J. (2011) *Visualized Texts: Sacred Spaces, Spatial Texts and the Religious Cartography of Banaras*, Harrassowitz Verlag, Wiesbaden.

Ginzburg, C. (2000) *Rapporti di forza. Storia, retorica, prova*, Feltrinelli, Milano.

Gutschow, N. (2006) *Benares: The Sacred Landscape of Vārānasī*, Edition Axel Menges, Stuttgart/London.

Jain-Neubauer, J. (1981) *The Stepwells of Gujarat*, Abhinav Publications, New Delhi.

Jalais, S. (2014) 'Walking the Ghāṭs: A measured approach to the Banāras Riverfront', in I. Keul (ed.) Banāras Revisited. Scholarly Pilgrimages to the City of Light, Harrassowitz Verlag, Wiesbaden, pp. 133–150.

Hazra, R. C. (1985) *Studies in the Puranic Records on Hindū Rites and Customs*, Motilal Banarsidass, Delhi

Kramrisch, S. (1946) *The Hindū Temple*, University of Calcutta, Calcutta.

Kumar, N. (1988) The Artisans of Banaras: Popular Culture and Identity, 1880–1986, Princeton University Press, Princeton.

Lazzaretti, V. (2013) 'Banaras jyotirlingas: constitution and transformations of a transposed divine group and its pilgrimage', in Kervan, International Journal of Afro-Asian Studies, edited by the University of Turin and Enna, N. 17, pp. 1–20.

Lazzaretti, V. (2015) 'Tradition versus urban public bureaucracy? Reshaping pilgrimage routes and religious heritage around contested places', in Y. Narayanan (ed.) *Religion and Urbanism: Reconceptualising 'Sustainable Cities' for South Asia*, Routledge, New Delhi, New York, London.

Lazzaretti, V. (in preparation) 'Water hides in front of you: voices from the invisible conflicts around a sacred well', in L. Cortesi and K. J. Joy (eds) *Split Waters: Examining Conflicts*

Related to Water and their Narration, Forum for Policy Dialogue for Water Conflict in India, Pune.

Madan, P. L. (2005) *The River Gaṅgā: A Cartographic Mystery*, Manohar, New Delhi.

Masani, R. P. (1918) *Folklore of Wells: Being a Study of Water-Worship in East and West*, Taraporevala sons & Co, Bombay.

Medhasananda S. (2002) *Varanasi at the crossroads, Vol. I–II*, The Ramakrishna Mission Institute of Culture, Kolkata.

Pandey G. (1990) *The Construction of Communalism in Colonial North India*, Oxford University Press, Oxford.

Prakash G. (2007) 'Secular nationalism, Hindūtva and the minority', in A. D. Needham and R. S. Rajan (eds) *The Crisis of Secularism in India*, Permament Black, Ranikhet, pp. 177–188.

Prinsep J. (2009 [1833]) *Benares Illustrated in a Series of Drawings*, Pilgrims Publishing, Varanasi.

Rötzer K., (1989) 'Wells, Pokhara, Ghāṭ and Hammam', in P. Couté and J. M. Léger (eds) *Bénarès, Un voyage d'architecture*, Editions Créaphis, Paris, pp. 89–97.

Schütte S. (2008) 'The Social Landscape of the washermen in Banaras', in M. Gaenszle and J. Gengnagel (eds) *Visualizing Space in Banaras: Images, Maps and the Practice of Representation*, Oxford University Press, New Delhi, pp. 279-302.

Sherring M.A. (1868) *Benares. The Sacred City of the Hindus: An Account of Benares in Ancient and Modern Times*, Trübner & Co, London.

Singh R. L. (1955) *Banaras: A Study in Urban Geography*, Nand Kishore & Bro, Banaras.

Singh, Rana P. B. (1994) 'Water symbolism and sacred landscape in Hindūism. A study of Benares (Varanasi)', Erdkunde, vol. 48, no. 3, pp. 210–227.

Singh, Rana P. B. (2002) *Towards the Pilgrimage Archetype. The Pancakrośi Yātrā of Banāras*, Indica Books, Varanasi.

Smith, T. L. (2007) 'Re-newing the ancient: the Kāśīkhaṇḍa and Śaiva Vārāṇasī', *Acta Orientalia Vilnensia*, vol. 8, no. 1, pp. 83–108.

Sukul, K. (1977) *Vārāṇasī Vaibhava*, Bihār Rāṣṭrabhāṣā Pariṣad, Paṭna.

Vyas, K. (1987) *Pañcakrośātmaka jyotirliṅga kāśīmāhātmya evaṃ kāśī kā prācīna itihāsa*, Khandelavala Press, Vārāṇasī.

Zara, C. (2011) 'Sacred journeys and profane travellers: Representation and spatial practice in Varanasi (India)', PhD thesis, University of London, London.

15 Resettling a River Goddess

Aspects of local culture, development
and national environmental movements
in conflicting discourses on Dhārī Devī
Temple and Srinagar Dam Project in
Uttarakhand, India.

Frances A. Niebuhr

Introduction

Dhārī Devī is a local version of the goddess Kali. Her statue and the temple are
situated on the river Alaknanda in the Himalayan state of Uttarakhand (UK).
Due to her popularity, mainly among the local population, but also among
pilgrims, the place receives a steady flow of worshippers, especially during
religious festivals. The river Alaknanda and with it the goddess's temple forms
part of the sacred waterscape of the Ganges. This fact – that the cultural identity
of the goddess is shaped by the water course[1] – recently resurfaced, when in
2008 the construction of a 330 MW hydropower project was taken up by the
GVK Company in its vicinity at Srinagar (UK). Since the deity's temple lay in
the submergence zone behind the envisaged dam (90 m height from deepest
foundation), plans were made to shift the statue onto a platform erected above
its original location. This prospect met with local resistance and evolved into a
sizeable culturally shaped environmental struggle comprising conflicting epistemologies about water.

The developments around Srinagar dam were fed by the paradigm rising
to dominance in the 1950s, when in the course of the secularization of the
political sphere, water became a part of the immanent paradigm shift and transformed into a mere economic resource, a source of energy, a symbol of energy
independence, as well as a vehicle of progress. In this light the case of Srinagar
dam and Dhārī Devī temple represents almost a metaphor for Nehru's then
euphoric exclamation of hydropower projects being the temples of modern
India (Aryal 1995; Swain 1997; Baghel 2014). The two antagonistic conceptions
of temples seem to be epitomized by the dam and the holy site of Dhārī Devī.

The case of Dhārī Devī, however, is surrounded not only by conflicting
epistemologies. The opposition to the relocation of the statue of the goddess
and her temple, and in its wake against the hydropower plant, is unfolding
against the backdrop of former anti-dam struggles in India. This chapter examines how elements of earlier dam struggles and obtained knowledge about

water are unfolding, reappearing and developing further, or deviating in the struggle revolving around Srinagar dam.

Over the past three decades, extensive research has focused on the conflicting perceptions about water that arise in the context of mega-dam projects in India. Most of these studies have focused on the social justice aspect of hydropower projects, with the issues of displacement and resettlement standing at the forefront of the discussion (e.g. Amte 1990; Kothari 1996; Baviskar 1995). At the micro-level, scholars have addressed the problems arising from certain aspects of dam projects, such as the impacts on specific segments of society – for example on indigenous people and land rights (Ghosh 2006; Baviskar 1995), or focusing on the effects of displacement on women (Thukral 1996; Rawat 2004; Drew 2014). Baghel (2014), on the other hand, highlighted the role of expert knowledge – while Bose (2004) dealt with the intellectual set of 'critics' in the dam discourse. Other authors shifted their attention to the dynamics of protest and protest practices. Gadgil and Guha (1994) and Swain (1997) especially depicted the trajectory of the Indian environmental movement and the culture of protest, starting from the colonial period via Chipko to Tehri, as well as the Narmada controversy. Furthermore, Sharma (2002, 2009, 2012) and Mawdsley (2006, also Williams & Mawdsley 2006) concentrated on the entanglement of politics within the Tehri movement by analyzing the role of Hindu right-wing forces seizing ecological topics.

Although Rawat (2004) and to a degree James (2004) have examined the legacy of the Chipko movement reappearing in other social movements in the state of Uttarakhand, gaps remain in understanding the interrelatedness of different dam protest movements in terms of protest practices and in the identification of agendas and epistemologies of the engaged actors. Additionally, the link still needs to be established from former movements to present-day dam conflicts unfolding among altered social, environmental and political circumstances. Thus, this chapter aims to contribute to the assessment of the dynamics of current dam conflicts by investigating the issue of Srinagar dam and Dhārī Devī temple in comparison to the two flagship examples of dam struggles in India, the Narmada and Tehri Movements.

These two objects of investigation – the Narmada and Tehri Movements – have been chosen because of their outstanding position in the history of dam building projects and related protest movements, but also because they seem to be the most significant in terms of their influence upon developments in Srinagar. Additionally, they constitute the most researched and analysed dam conflicts.[2] The first object of comparison, the organized resistance against the Narmada Valley Dam Projects, relates to a chain of dams on the river Narmada stretching through the three states of Madhya Pradesh, Gujarat and Maharashtra. Central to the conflict is the biggest undertaking: the Sardar Sarovar dam (163 m, 1450 MW). The Narmada Bachao Andolan (NBA), also known as the Save Narmada Movement, with its leaders Medha Patkar and Baba Amte, epitomizes the opposition to these projects. The second important conflict unfolded around the construction of a mega-dam on the river

Bhagirathi near the old town of Tehri in Uttarakhand. When the construction of the Tehri project (261 m, 2400 MW) started in 1978, the dam was designed to become the fifth largest project in the world. In the first year of its construction the Tehri Bandh Virodhi Sangharsh Samiti (Committee for the Struggle against the Tehri Dam, TBVSS) was founded by the veteran freedom fighter Virendra Datt Saklani, although the movement soon came to be mostly identified with Sunderlal Bahuguna.

Srinagar Dam and Dhārī Devī

Although the Srinagar Hydroelectric Power Project in the Garhwal district of Uttarakhand was approved in 1985, due to financial and administrative issues, the construction started as late as 2008 (Rawat 2015). By 2009 it turned into a contested site and the issue of Dhārī Devī became prevalent.[3]

The Srinagar dam's first reported problems were of an environmental nature, such as problems emerging from building activities or compensation issues, which extended through the entire process of the dam's construction. In 2009 the first violations of environmental norms were highlighted in a local newspaper. After instances of repeated irregularities and damages at the building site, some environmentalists[4] began to pay attention to the project and warned of its consequences. Concurrently Bharat Jhunjhunwala, one of the major activists, directed some local people to file a court case against the project at Nainital High Court in regard to the power project's upgrade from 200 to 330 MW without renewed environmental clearance.

In August 2009 the focus of local newspaper reports shifted towards the issue of the temple. In a meeting in Srinagar, the Hindu nationalist political party Vishva Hindu Parishad (VHP)[5] declared that religious sentiments would be hurt if the ancient character of Dhārī Devī temple was changed. The party threatened to start a people's agitation, which is the first occasion pointing towards the formation of a more organized protest. The engagement of the VHP resulted in the hunger strike of one of its young members, Anuj Joshi, at the Dhārī Devī temple. After this initial phase of resistance and with a growing threat to the temple, a movement built up to prevent the deity's relocation. Supporters founded the Dhārī Devī Bacāo Samiti (Save Dhārī Devī Committee), which performed 1,000 days of *dharnā* (sit-in) at changing venues: first at the temple, later in Srinagar, the next major city of the district. The Samiti consisted of religiously motivated individual local actors, often members of women groups, as well as of different political parties. Religious authorities from various branches, as well as several politicians and political groups across party lines expressed their solidarity.

One of the special characteristics of the protests revolving around Dhārī Devī and the dam was the fact that from July 2010 on, two protest movements started operating parallel to one another with opposing objectives. The second movement constituted a response to the former with the aim to support the hydropower project and the construction of a new temple for the statue of the deity. The pro-project movement consisted of local people and received its backing from the villages adjacent to the dam – several agitators were engaged with the project, be it as workers or contractors – and from several politicians of the state government who sided with them. Thus, as a peculiarity, the very people that would normally provide the basis of a protest movement against the construction of such a large-scale project were in this case the ones defending the project and the relocation of the statue. The two movements however were not merely parallel to one another, but rather defined, influenced and at times clashed with one another. While participants of the first committee were mostly in line with the tradition of environmental protests in India, i.e. conforming with the Gandhian principles of *satyagraha* marked by nonviolent actions, their opponents engaged in similar techniques, also staging *dharnās*, blocking the highway and performing hunger strikes. On the other hand, the opponents – the 'affected people' – didn't feel obliged to any tradition of non-violence and thus their expressions of opposition were at times marked by acts of aggression or violence.

While the mobilization of the Dhārī Devī Bacāo Samiti kept on growing, it also became more and more absorbed by broader campaigns of activists and already existing protest movements. In this context a *padyātrā* (foot march), led by a former home minister of the state government, who uses the spiritual name Swami Chinmayanand, came in a supportive move from Haridwar to Srinagar (*Amar Ujala* 27.6.2010). Renowned environmental activists like 'Waterman' Rajendra Singh from Rajasthan also took up the issue (*Amar Ujala* 28.3.2010). Likewise, G.D. Agarwal from Madhya Pradesh, an environmental engineer with a PhD from the University of California at Berkeley, who adopted the spiritual name Swami Gyan Swaroop Sanand, included Dhārī Devī into his agenda (*Amar Ujala* 26.5.2012). He had earlier been an activist against dam projects on the Ganges and with his extensive hunger strikes in 2009 managed to put two undertakings on the river Bhagirathi on hold. During the performance of another fast in February 2012, Dhārī Devī and the Srinagar dam constituted a part of his agitation (*Amar Ujala* 25.2.2012). Uma Bharti represented another major actor who involved herself in the agitation to save the Dhārī Devī temple. She is a Bharatiya Janata Party (BJP) politician from Madhya Pradesh, and a *sādhvī*[6] lately – in 2014[7] – appointed minister for water resources, river development and Ganga rejuvenation under the government of prime minister Narendra Modi. Along with these activists, the save Ganga movement, a movement in which saints and *sādhus* are dominant and with a stronghold in the ashrams of Haridwar and Rishikesh, included the preservation of the Dhārī Devī temple in its agenda. The religious people stood in complete opposition to dams on the river Ganges, seeing them as a threat to its sanctity. At its peak, the movement of saints held a protest rally at Jantar Mantar in Delhi in June 2012 (*Amar Ujala* 19.6.2012). They called it the Gaṅgā Mukti Mahāsangrām, the great battle for a free Ganges.

As a result of different court cases filed by the activists and the interventions of Uma Bharti and other religious actors, the project underwent several breaks in construction in 2010, 2011 and 2012. At the same time, several expert groups were set up in order to conduct surveys. Still, in spite of the official orders to stop construction, the power company continued building the project. In 2013 in the course of the disastrous Uttarakhand flood, the water behind the already erected dam rose until Dhārī Devī temple was on the verge of submergence. In a dramatic rescue operation the company and temple priests lifted the statue onto the established platform meant to house the modern version of a temple. Months after the flood and nationwide media coverage subsided, the hydropower station became operational in 2015.

Comparison to Tehri movement and Narmada Bachao Andolan

General Features

The struggle of the NBA exerted influence on the Srinagar issue in a more general way, whereas the Tehri dam conflict assumed greater importance due to its direct influence on the events in Srinagar. The formative influence of the NBA (Narula 2008)[8] is discernible by the fact that issues that emerged during its struggle are repeatedly resurfacing in other anti-dam protests, including Srinagar. The direct effect of the Tehri struggle results from an identical cultural setting, geographical closeness and geophysical similarity. The Srinagar conflict emerges as one of the successor battles of Tehri, with former activists and know-how directly merging into the later movement. Furthermore, the two projects form part of a shared imagined sacred space. The aspect of sacrality is specifically enhanced due to both projects' position on two different Ganges tributaries, but also by falling into the more encompassing divine territory of Uttarakhand termed as the *dev bhūmi,* the land of the gods.[9] However, while the construction site in Srinagar directly affects a local place of worship, the temple of the goddess Dhārī, the Tehri dam is positioned in a larger and more abstract sacred space.

The presence of resettlement issues, according to Dwivedi (2006) is one of the basic grievances that stimulates collective action, thus the formation of a movement. This phenomenon also stands in close connection to one of the distinct features of the Indian ecological movement, the inclusion of the social element: 'Third World environmentalism, in comparison [to Western environmentalism] has been occupied with environmental problems arising from social, political, and economic inequities ...' (James 2004, p. 363). In Srinagar only a small number of families were subjected to a resettlement process as compared to the dam projects in Tehri and Sardar Sarovar. Accordingly, different socio-cultural issues came to the forefront. Out of a submergence zone of approximately 475 ha[10] 324 ha (Rawat 2015) of private land, mostly farmland, came under water. Thus, in spite of social-cultural factors being present in different forms at Srinagar, it was in fact the goddess herself who became the main object undergoing resettlement – and she was promised a generous compensation in the form of a new and modern temple.[11] The majority of people did not feel a threat to their livelihoods, but instead had the vision of an ultimate improvement of their life circumstances. This naturally reduced the urgency for them to struggle against initially abstract threats of a dam, which might manifest in their full visibility at a later stage. What is more, the spirit to oppose might have been low from the beginning. The apathy of the local population that emerged in Tehri[12] may have extended to Srinagar. Apathy or a negative image of environmental movements in the state had however already started with Chipko,[13] as Sethi points out: 'Chipko, more so its glorification, came to be perceived as a conspiracy of city-based greens to foreground nature over people, environment over livelihoods' (Pathak *et al.* 2002). In spite of an obviously missing grassroots support, there had also been a fair degree of consent among the local population with the opposition to the deity/dam issue. However, these actors kept passive for several reasons. Causes included peer pressure and an observed missing identification with the native soil due to generations of out-migration of the male population (Pathak *et al.* 2002). Another feature detected by Drew (2014) in dam struggles in the higher reaches of Garhwal, is that people, especially women, did not entirely agree with the objective and course of action taken by leading (male) actors of protests, like Professor G. D. Agrawal.

Mechanisms of the movement

The discourse in the Sardar Sarovar struggle was framed under the dichotomy 'water versus displacement' (Narula 2008, p. 361), while the essence of the Tehri movement can be denominated as 'economy versus environment' (including its social stake). In Srinagar, through the integration of and eventually concentration on the religious symbol – and with the associated threat of losing traditional knowledge, culture and practices – the discourse shifted towards one of 'development versus tradition'.

The all-pervading Indian mantra of 'vikās' (development), which is dominant in all kinds of contemporary discourses and one of the most powerful arguments environmental movements in India have to cope with, underwent a water-based modification in Uttarakhand. Agrawal is expounding: 'Policymakers have a "grand vision" of turning Uttarakhand [...] into an *ūrjā pradēś* (energy state) [...] The ex-chief minister, Ramesh Pokhariyal Nishank of the Bharatiya Janata Party, tried to sell this dream to the people of the state, promising them employment and "development" [...]'. But also the current chief minister Vijay Bahuguna is in line with this programme, for he considers power projects 'pillars and symbols of development' and 'necessary to yield power' (Agrawal 2013, p. 14). Accordingly policymakers and the supporters of hydropower projects are rather framing the discourse in 'a grand future vision versus obstructing forces'. These promising future prospects managed to capture the minds of many people, especially of the younger generation.

During the implementation of Srinagar dam, some well-known patterns were repeated, which Jain describes as the standard model of dam building in India:

> First understate the height of the dam, the size of the auxiliaries and thus total cost to make it easier to obtain financial and other approvals; second, seal the approval with a foundation stone by a VIP. And then in calculated stages push up the scale of the project and its cost beyond recognition; and thus armed ward off all objections on the ground that the Prime Minister had already laid the foundation stone.
>
> (Jain 2001, p. 95 in Baghel 2014, p. 13)

Though sounding cynical this model can nevertheless be clearly observed in all three projects. It remains unclear who laid the foundation stone in Srinagar, but the later upgrade of the project without renewed environmental clearance did take place and became the basis for the first court case filed by Dr. Bharat Jhunjhunwala in 2009 (Seetharaman 2015). Another familiar pattern that emerged was that of 'creating facts' or a 'fait accompli'. This strategy could be found in Gujarat, where the Sardar Sarovar Narmada Nigam Ltd closed the sluice gates of the dam at a strategically convenient moment (Narula 2008). The submergence of Tehri was enforced even before the process of resettlement was in any way accomplished and not only this, a study of seismicity that was supposed to have been carried out beforehand was ignored in the course (Pathak *et al.* 2002).[14] In Srinagar two comparable situations occurred when the company kept the sluice gates of the dam shut in the course of two flood events. During the first such event in 2012 the goddess Dhārī was several times on the brink of submergence, while on the second occasion of the flood disaster in 2013 the deity, in order to be saved from the water, had to be shifted to the new platform. Observers alleged that the dam company intentionally created these situations, making use of the natural disaster in 2012 when the protest against the shifting of Dhārī Devī was at its height and in 2013 when an agreement about the auspiciously proper time for the relocation of the goddess couldn't be achieved. In fact, the local villagers kept on stalling the finalization of the deity's resettlement, since they began to suspect that the company might not fulfil its promises at a later stage.

A motif and dichotomy that explicitly reappears in the case of the Srinagar dam and Dhārī Devī protests is that of outsiders against locals. The dam supporters included this theme into their argumentation in order to polarize between 'them and us'. In Tehri and already during the height of the Chipko movement (James 2004, p. 368), the 'outsiders' were identified to be outside contractors threatening local order and livelihoods and exploiting the hill area for the benefit of the plains (Aryal 1995, p. 14; Drew 2014, p. 237). In the discourse surrounding Srinagar the pattern, on the contrary, was taken up under reversed conditions, namely that the perceived threat coming from outside was interfering non-local environmentalists trying to take away or obstruct the expected fruits of development from reaching the local community. The question is why even a south Indian executing company was credited with less mistrust than the alliance of protesting Swamis and environmental activists. After all, the mechanisms of earlier dam projects were the same – the electricity generated is mainly going to the state of Uttar Pradesh in the plains (88 per cent). The answer may be found in the fact that, although people knew from former struggles that promises made are regularly broken, the generous financial means distributed by the company in the beginning may have cemented a notion of a social agenda coming along with the cooperation. In fact, the company had beforehand promised (and later partly realized) several projects relating to improvements of infrastructure, such as new roads, street lightning, health services and beautification works along the newly formed lake.

The protest techniques applied during the Srinagar dam conflict were basically coherent with those observed in the preceding, formative struggles, which were the *dharnās, rāstā roko* (road blockage), hunger strikes of different shapes[15] and *padyātrās* (foot marches or pilgrimages). Court cases initiated by activists and following inquiry committees and studies concerning the goddess and/or the dam project slowed down or halted the construction of all three projects. Additionally, the self-drowning threat of *jal-samādhi* was uttered several times by activists from the religious faction that included Uma Bharti, whereas the women groups warned of a sacrificial mass suicide for Dhārī Devī. The prospect of a *jal-samādhi* bears similarity with the threats and half performed *jal samarpaṇ* (Narula 2008) of the NBA movement. However, *jal-samādhi* constitutes the more active form of self-immersion in water and is loaded with religious meaning, while *jal samarpaṇ*, the non-cooperation version[16] passively arises from resisting to leave one's villages or houses while the lake forming behind a dam is rising and flooding the place. Some aspects of direct action in accordance with *satyāgraha* were however lacking in Srinagar. Due to a lack of general resistance at the village level, elements like non-cooperation and civil disobedience (Nayak 2010) remained absent. An exception occurred in the blockage of project-related works by activists who were trying to save the goddess, and occasionally by inhabitants of the project-affected villages demanding compensation for damages and constraints during the construction process.

The distinct form of organized protest of the dam supporters was a novelty in Srinagar. The dam-supporting group also used some of the same protest techniques. Even if the NBA was faced with counter protests instrumented by the government, and similarly in Tehri a broad base of dam supporters emerged,[17] the unique phenomenon in Srinagar was that the dam supporters drew from Gandhian-influenced protest practices and argumentations. These actors however extended their protest to more aggressive methods. Notwithstanding that part of the counter protest had assumedly been staged by the power company, in Uttarakhand there also prevails a real–existing pro-dam attitude – accompanied by a hope for economic improvements.

The actors in the Srinagar dam conflict

The presence of the Vishva Hindu Parishad (World Hindu Council, VHP) in the Srinagar/Dhārī Devī controversy constitutes a commonality with the events in Tehri; in Srinagar, however, the party assumed only a minor role. Critical in Tehri was that the engagement of the VHP brought along its agenda of Hindutva,[18] or in this case the Ganges-specific version called 'Gangatva' (Rajalakshmi 2002). The following ideologically charged protest deterred many former supporters of the movement and eventually led to its weakening (Rajalakshmi 2002). In Srinagar, due to the centrality of a religious issue the VHP became the initiator of protests, yet later, this organization faded into the background. The dominant party in the protests was rather the BJP – which constituted from 2012 the opposition party in Uttarakhand.[19] However, the Dhārī Devī issue proved to be a politically difficult terrain for them. First of all, there was a divide on the matter between the Uttarakhand branch and the central leadership of the BJP. Apart from the fact that some of its local members engaged in the struggle to preserve Dhārī Devī temple, the central unit of the party also proved sympathetic, with the protestors defending the traditional version of the temple. As a consequence, the BJP in Uttarakhand slid into the precarious position of being identified with an anti-development[20] attitude and as such was at risk of losing future electoral votes. Due to this peculiarity the actors affiliated with the BJP, like Uma Bharti, but also the Dhārī Devī Bacāo Samiti, started to emphasize that their engagement was solely concentrated on saving the temple with the statue and had nothing to do with the project, stating respectively that they were not 'anti-dam'. This argument could gain little credibility because it was commonly known that the associated Swamis and Saints were on a mission against dam projects on the Ganges. Furthermore, it diluted the protestor's line of argumentation and eventually weakened the movement. Together with this kind of reasoning, the movement – at least this group of the agitators – also detached itself from the ecological implications of the dam project and stood in a contradicting position to the preceding struggles that had always emphasized an environmental agenda.

The protest against the Srinagar dam and against the shifting of the deity was influenced by dominant figures. Unlike with Tehri and the NBA movement, leadership was not concentrated in one person.[21] Several multifaceted key actors emerged during the protest; however, arguably the most influential person, Dr. Bharat Jhunjhunwala worked in the background, providing the movement with expert knowledge and advice and thus did not directly become a figure of identification. His area of activism concentrated on the masterminding of different court cases and establishment of media relations, as well as creating the intellectual and ecologically oriented background for the movement. In this regard he published several books on the matter of hydroelectric power. Although he did include spiritual aspects of the river flow and goddess Dhārī into his discourse, his main approach rather emphasized the scientific–environmental dimensions. The key actors in the protest were older persons coming from the urban middle class,[22] some of them with an expatriate Indian (Non Resident Indian, NRI) background, who later in life started to engage in either religious activities and/or to deal with social and environmental issues.[23] In Tehri and the NBA the defining actors had slightly different backgrounds – Medha Patkar originated from an urban middle-class family, but started her engagement at a young age, while Sunderlal Bahuguna is a local but upper-caste Brahmin.

For the Tehri and Sardar Sarovar movements the social base was among mobilised tribals and the poor peasantry (Gadgil and Guha 2015, p. 197), which can be specifically attributed to the NBA movement. In Srinagar the support of the grassroots was widely missing. At times it rather seemed to transform into a protest movement of religious VIPs. Although the presence of famous and media attention-drawing actors can be seen as an important element of a movement (Bose 2004), and was even welcomed and supported through networking activities by local activists, it appears that the balance in this case was lost. As a consequence it may have become more difficult for the movement to appeal to the interests of the average population, the *ām ādmī* (common man). In particular, figures of identification were not suited for the younger male population. This was said to have been part of the reason for the failure of the Tehri movement.

A further point which assumes importance in Srinagar falls into the category 'water as a women-dominated sector', as highlighted by Mukhopadhyay and Devi (this volume) in their study of cultural heritage discourses in Rajasthan. The gender perspective constitutes a decisive element of water discourses which relates to both traditional knowledge and development projects. With regard to Uttarakhand, the gender perspective has been employed by Drew (2014) in her work about the involvement of mountain women in anti-dam activities in the higher basins of the river Bhagirathi. She observed that women agreed only to a limited extent with the campaigns of one of the most famous activists, G. D. Agarwal. She further argues that in contrast to him, as one of the promoters of a 'pan-Indian "Hindu culture"' (2014, p. 240), the Mountain women's ideas about their sacred, social and economic waterscape exhibited a more local nature and their approach towards issues of development possessed a multidimensionality which didn't match the rather one-dimensional objective of the religious faction.

However, with regard to the ecological engagement of the saffron faction or the Hindu right, which drew much criticism with regards to the Tehri Dam,[24] I argue that in the case of Srinagar there is not such a clear picture of a 'saffron front' seizing the protests. Even though the VHP initiated the protest surrounding Dhārī Devī temple, and their youth organization, the Bajrang Dal, expressed support on the matter, too many other actors were involved in the agitation who cannot be classified in a simple Hindu right-wing category due to their multifaceted nature. For example, Pujya Swami Chidanand Saraswati from Parmath Niketan Ashram in Rishikesh is on the one hand a respected religious authority, but on the other hand also the founder of several eco-activist programmes. His Ashram is engaged in many international campaigns but also yoga festivals, which makes the location more of a mixed place of new-age Western esoteric and yoga movement that also conducts social activism on the basis of Hindu traditions.[25] Likewise, the religious leader Shankaracharya Swami Swaroopanand Saraswati, who was leading the protest of the *sādhus* in Delhi, launched several eco-oriented programmes and has rather close links with the more secular oriented Congress party.

Although the rhetoric of the involved religious people is mixed and does not necessarily conform to scientific ideals or standards,[26] these people deviated from the former purely religious and often incendiary rhetoric prevailing in Tehri. This indicates that such actors are shifting towards a more nuanced understanding of processes – that encompasses scientific knowledge as well as a global perspective on ecological issues. Nevertheless, even though the activists consisted of mixed groups with distinct approaches and only rarely advocate a radical Hindu worldview, by centring the protest on a Hindu symbol and using Hindu symbolism, minority groups, such as the Muslim part of society, are deterred. Moreover, this form of religious environmentalism is subject to criticism from environmental groups in India, which follow a more secular approach to environmental activism. In spite of shared targets, they prefer not to ally with the religiously conservative forces whose larger agenda they do not identify with (Del Bene 2014; Seth 2016). In fact, the ecological engagement of most of the saints and *sādhus* is focused on the Ganges. Thus, these actors inform an epistemology of water which remains in, and is restricted to, a space defined by the concept of the sacred.

Conclusion

This chapter has sketched the clashing imaginations concerning the trajectories of a present-day waterscape in Srinagar and beyond. In the course of the water-based conflict, actors coming from a wide religious, scientific, economic and political spectrum have employed a variety of arguments to express their respective knowledge systems. Concurrently these actors have supported either their personal or their affiliated group's conceptions of a reshaped waterscape by creating a meshwork of the involved epistemologies. Thus, actors from the religious sphere employed scientific arguments and those from a scientific background went into a spiritually backed argumentation to preserve the sacrality of the place and/or the bigger sacred space of the holy river Ganges. On a larger scale, lines of argumentation extended concerns to the ecological state of the river system with broader consideration for the ecological balance of the Himalayan region. Other actors, again placing an emphasis on an economic and developmental objective, integrated the temple and the goddess into their future vision of an emerging technically engineered waterscape. However, this vision was not homogeneously determined by economic interests, but included a comprehensive imagination of a reformed public and sacred space with the river at its centre.

Formerly established knowledges and practices connected to water conflicts re-emerge and become modified under the influence of contemporary dynamics and tendencies unfolding in a specific place and by the agendas of participating actors. The rhetoric of development and its specific local version, and in particular the role of a sacred space defined by water at the centre of an environmental struggle, have been highlighted. The instrumentalization and agency of an imagined space in the context of an environmental conflict proves to be powerful though contested (Seth 2016). Due to this distinctive feature, the protest movement in Srinagar, though influenced by the NBA, was even more swayed by the spatially and culturally closer Tehri movement.

The case of Dhārī Devī and Srinagar dam does display unique characteristics with an ecological protest movement laying its focus on a religious symbol, the emergence of a protest counter movement equally claiming the religious symbol for its own purposes and the rise of religious authorities as promoters of eco-activism. With this unique constellation, Srinagar is deviating from nation-wide practices to deal with dam projects. Thus, the case may be solely symptomatic for a certain local space, the attitude of people in Uttarakhand state and the political and social climate of a certain period. The challenge for future research is to investigate the possibilities of such detected features becoming new tendencies in a much-needed national water-based environmental dialogue.

Notes

1 Already the different origin stories of the goddess are defined by the watercourse. In one version the statue is said to have been washed up to its site on a lion-shaped rock during a massive flood in 1894.
2 Tehri Dam: e.g. James (2004), Rajalakshmi (2002), Sharma (2012, 2009, 2002), Rigzen (1997), Mawdsley (2006); Narmada Dam: Aryal (1995), Thukral (2000), Routledge (2003), Narula (2008), Peterson (2010), Pfaff-Czernecka (2007), Karan (1994); both projects: Gadgil and Guha (1994), Swain (1997).
3 The chronological sequence of protests has been established by articles of the local Hindi newspaper *Amar Ujala* (Garhwal edition) and through interviews with local informants and activists (also Lahiri 2011: 7).
4 In a prominent position Vimal Bhai, convenor of social organisation Matu Jan Sangathan.
5 English translation: World Hindu Council.
6 The female form of sādhu, a religious ascetic or holy person.
7 She assumed this position only after the shifting of Dhārī Devī temple and after the protests had receded.
8 The NBA not only shaped the Indian environmental movement but at the same time left its marks on global environment groups, as well as triggering policy changes in particular within the World Bank (Pfaff-Czernecka 2007).
9 A term used historically to refer to the mountainous area of Uttarakhand, nowadays also strongly promoted by the tourism industry.
10 See http://www.tmpsystems.net/s/LEGEND_IHydro_CaseStudyDatabase_Final-Draft-bxdk.xlsx (accessed 20 January 2016).
11 This practically constitutes the equivalent to the modern houses promised to the families that became resettled in the state capital Dehra Dun.

12 In short, analysts pointed out that the Tehri movement "failed to acquire a significant social presence both in the local and national domains" (Pathak *et al.* 2002). The reason was primarily found in the fact, that the movement was too much concentrated on its lead-figure Bahuguna and his conduct. His narrow concentration on the local character of protest broadly prevented effective networking activities with groups and experts coming from the outside. Eventually he sided with the VHP, the right wind Hindu party which infiltrated the protest movement and deterred political moderate forces" (Pathak *et al.* 2002).

13 Chipko emerged in 1973 as the 'tree-hugger movement' in Garhwal district of Uttarakhand. 'In their effort to save forest resources from exploitation by outside contractors, the protesters stopped commercial felling in nearby forests by placing themselves between tree and logger's saw' (Swain 1997: 821). '... As a result of its novel technique and Gandhian philosophy, the movement acquired quick fame' (Swain 1997: 821). A special feature of the Chipko movement was the participation of women as central actors (Gadgil and Guha 1994), a pattern that reproduced itself in further environmental st ruggles, especially in mountain areas.

14 Another example of this strategy in Choudhury (2010).

15 Although the hunger strikes in the two other projects are what, Gadgil and Guha (1994: 204) described as a method usually employed by a charismatic, outstanding figure, in Srinagar this technique is on several occasions turned into a collective action with the application of the so-called kramik anashan (relay hunger strike) during which different participants are alternating.

16 The translation for samarpaṇ would be surrender, devotion or dedication.

17 See also Choudhury (2010: 23).

18 The practice to promote a Hindu nationalist state.

19 VHP and BJP are both part of the Rashtriya Swayamsevak Sangh (RSS) family of organizations, with the former being religious and the latter being political in nature.

20 Kothari (1996) for example indicated that critics of dam projects are not only labelled by dam supporters and state organs as anti-project or anti-development, but also as antinational (1996: 1478).

21 The Tehri struggle is almost solely associated with the lead figure Sunderlal Bahuguna, while the NBA movement is linked to the outstanding personality Medha Patkar.

22 This constitutes another reappearing pattern which was coined during the time of the NBA; Baghel termed it as: 'the broadening of the opposition to the urban middle class that was not directly affected by the projects' (2014: 15).

23 E.g. Bharat Jhunjhunwala, G.D. Agarwal, Vimal Bhai, U. Bharti.

24 See Mawdsley (2006, 2010), Sharma (2002, 2009, 2012), Rajalakshmi (2002).

25 The Swami however is only indirectly associated with the Dh r Dev issue, as he participated in the protests for a free Ganges in Delhi

26 One Swami for example, was suggesting as an alternative to hydropower the establishment of more coal-fired power plants (*Amar Ujala* 26.5.2012).

References

Amar Ujala, Garhwal edition (n.d.), accessed online, http://epaper.amarujala.com.

Agrawal, R. (2013) 'Hydropower projects in Uttarakhand: Displacing people and destroying lives', *Economic & Political Weekly*, vol. 48, no. 29, July 20, pp. 14–16.

Amte, B. (1990) 'Narmada project: The case against and an alternative perspective', *Economic and Political Weekly*, vol. 25, no. 16, pp. 811–818.

Aryal, M. (1995) 'Dams: The vocabulary of protest', *Himal*, July/August, pp. 11–21.

Baghel, R. (2014) *River Control in India: Spatial, Governmental and Subjective Dimensions*, Springer, Dordrecht, Heidelberg, New York, London

Baviskar, A. (1995) *In the Belly of the River: Tribal Conflicts Over Development in the Narmada Valley*, Oxford University Press, New Delhi.

Bose, P. S. (2004) 'Critics and experts, activists and academics: Intellectuals in the fight for social and ecological justice in the Narmada Valley, India', *International Review of Social History*, vol 49, no. 12, pp. 133–157.

Choudhury, N. (2010) 'Sustainable dam development in India, between global norms and local practices', *Discussion Paper, Deutsches Institut für Entwicklungspolitik*, Bonn, Germany.

Del Bene, D. (2014) 'No resistance can win without an alternative vision, Ashish Kothari on India's activist agenda', Ejolt, www.ejolt.org/2014/11/no-resistance-can-win-without-an-alternative-vision-ashish-kothari-on-indias-activist-agenda/, accessed 24 October 2015.

Drew, G. (2014) 'Mountain women, dams, and the gendered dimensions of environmental protest in the Garhwal Himalaya', *Mountain Research and Development*, vol. 34, no. 3, pp. 235–242.

Dwivedi, R. (2006) *Conflict and Collective Action: The Sardar Sarovar Project in India*, Routledge, London.

Gadgil, M. and Guha, R. (1994) 'Ecological conflicts and the environmental movement in India', in P. Utting (ed.) Revisiting Sustainable Development, UNRISD, Geneva, Switzerland, pp. 187–216.

Ghosh, K. (2006) 'Between global flows and local dams: Indigenousness, locality, and the transnational sphere in Jharkhand, India', *Cultural Anthropology*, vol. 21, no. 4, pp. 501–534.

Jain, L. C. (2001) *Dam vs. Drinking Water: Exploring the Narmada Judgment*, Parisar, Pune, India.

James, G. (2004) 'The Environment and environmental movements in Hinduism', in R. Rinehart (ed.) *Contemporary Hinduism: Ritual, Culture and Practice*, ABC-CLIO, Santa Barbara, California, pp. 341–380.

Kothari, S. (1996) 'Whose Nation? The displaced as victims of development', *Economic and Political Weekly*, vol. 31, no. 24, June 15, pp. 1476–1485.

Lahiri, N. (2011) *Srinagar Hydroelectric Power Project, Report*, Ministry of Environment and Forests, Government of India, http://www.moef.nic.in/downloads/public-information/DOC200611.pdf, accessed 26 January 2016.

Mawdsley, E. (2006) 'Hindu nationalism, neo-traditionalism and environmental discourses in India', *Geoforum*, vol. 37, no. 3 pp. 380–390.

Mawdsley, E. (2010) 'The abuse of religion and ecology: The Vishva Hindu Parishad and Tehri Dam', in A. Guneratne (ed.) *Culture and the Environment in the Himalaya*, Routledge, London, New York, pp. 151–165.

Narula, S. (2008) 'The story of Narmada Bachao Andolan: Human rights in the global economy and the struggle against the World Bank', *New York University Public Law and Legal Theory Working Papers*. Paper 106, New York.

Nayak, A.K. (2010) 'Big dams and protests in India: A study of Hirakud Dam', *Economic & Political Weekly*, January 9, vol. 45, no. 2, pp. 69–73.

Pathak, S., Sethi, H. and Thakkar, H. (2002) 'Tehri: Is it curtains?' Down to Earth, www.downtoearth.org.in/blog/tehri-is-it-curtains-14080, accessed 14 June 2015.

Peterson, M. J. (2010) 'Narmada Dams Controversy', Project International Dimensions of Ethics Education in Science and Engineering, www.umass.edu/sts/ethics, accessed 25 October 2015.

Pfaff-Czernecka, J. (2007) 'Challenging Goliath: People, dams, and the paradoxes of transnational critical movements', in H. Ishii, D. N. Gellner, K. Nawa (eds) *Political and Social Transformations in North India and Nepal: Social Dynamics in Northern South Asia, Japanese Studies on South Asia, vol. 2, no. 76*, Manohar, Delhi, pp. 399–433.

Rajalakshmi, T. K. (2002) 'Trouble in Tehri: The closure of two tunnels of the Tehri dam in the Ganga basin heightens tensions in the Tehri valley', Frontline, vol. 19, no. 01, 5–18 January, www.frontline.in/navigation/?type=static&page=flonnet&rdurl=fl1901/19010330.htm, accessed 4 August 2015.

Rawat, B. S. (2015) 'Why the Srinagar Hydro Electric Project continues to remain a threat', SANDRP, https://sandrp.wordpress.com/2015/05/25/why-the-srinagar-hydro-electric-project-continues-to-remain-a-threat-2/, accessed 20 January 2016.

Rawat, R. (2004) 'Chipko's quiet legacy: Forest rights, women's empowerment, people's institutions, and new urban struggles in Uttarakhand, India', MA thesis, York University, Toronto, Canada.

Rigzen, T. (ed.) (1997) *Fire in the Heart, Firewood on the Back: Writings on and by Himalayan Crusader Sunderlal Bahuguna*, Parvatiya Navjeevan Mandal for Save Himalaya Movement, Silyara, Tehri, Uttarakhand, India.

Routledge, P. (2003) 'Voices of the dammed: Discursive resistance amidst erasure in the Narmada Valley, India', *Political Geography*, vol. 22, no. 3, pp. 243–270.

Seetharaman, G. (2015) 'India's water security crisis: Dams, pollution and climate change biggest threats facing Himalayan rivers', *The Economic Times*, www.articles.economictimes.indiatimes.com/2015-07-05/news/64112378_1_hydel-power-project-hydel-capacity-installed-power-capacity, accessed 7 August 2015.

Seth, B. L. (2016) '"Keep religion out of river movements": An interview with Indian activist Vimal Bhai', International Rivers, Berkeley, https://www.internationalrivers.org/blogs/328-28, accessed 2 February 2016.

Sharma, M. (2002) 'Saffronising Green', *Seminar*, vol. 516, pp. 26–30.

Sharma, M. (2009) 'Passages from nature to nationalism: Sunderlal Bahuguna and Tehri Dam opposition in Garhwal', *Economic and Political Weekly*, vol. 44, no. 8, pp. 35–42.

Sharma, M. (2012) *Green and Saffron: Hindu Nationalism and Indian Environmental Politics*, Permanent Black, Ranikhet.

Swain, A. (1997) 'Democratic consolidation? Environmental movements in India', Asian Survey, vol. 37, no. 9, pp. 818–832.

Thukral, E. G. (1996) 'Development, displacement and rehabilitation: Locating gender', Economic and Political Weekly, vol. 33, no. 24, pp. 1500–1503.

Williams, G. and Mawdsley, E. (2006) 'Postcolonial environmental justice: Government and governance in India', *Geoforum*, vol. 37, no. 5, pp. 660–670.

Index

adaptive management 5, 88
agriculture: commercial 134, 173, 198–9;
    failure 199; employment in 91; harvest
    198–9, 224; history of 60, 65, 213, 216;
    improvement 93, 98, 100, 110–11, 192,
    202–3; intensification 13, 192–4, 197–8;
    irrigated 70, 74–5, 87, 91, 110–11,
    180, 213, 222, 224; monsoon, relation
    to 36–7, 40–3, 47–9; organic 201–2;
    perception of 183–5, 199–201, 204–5;
    policy 183; water stress 1, 91, 194
Alaknanda river 246
Alwar district see Rajasthan
Anthropocene 1–5, 15–17
aridity see arid region
arid region 11–12, 71, 74–5, 87–105,
    136, 138–40, 213, 223; groundwater
    dependence 90; North Rajasthan 87;
    Thar desert 12, 109–10, 113, 130–1,
    133, 139; traditional water harvesting in
    116–19; water management in 103, 117
Arvari catchment 12, 87–90, 93–4, 96–104
Asia 1–3, 14, 36, 70–1, 73, 142–6, 157–8,
    197, 210–11, 213; Central Asia 71, 73;
    South Asia 6, 60, 71, 73, 92, 215, 232–3;
    Southeast Asia 36, 210–12, 213
Asian Development Bank 143, 155
Australia 36

Bali 3, 13, 192, 211
Baltistan 71, 74–5
Banaras see Varanasi
Bangladesh 53, 58–9, 64
Batukaru, Mount 198–9
Běitóu, Taipei 158
Bengal 11, 53, 57–65, 235
Bhumibol Adulyadej 210, 214–20, 222–4
biodiversity 40, 89
British: colonial rule 14, 53–7, 60–3, 130,
    135, 234–7, 239–41; Empire 60, 71

Brundtland Report 126
bureaucracy 71, 81–3, 89, 91, 105, 145,
    173, 211–24

Chao Phraya delta 213–14
citizen science 22–33
climate change 36–8, 41–2, 48; adaptation
    strategies 36, 48–9
climate phenomena (influencers) 36–9,
    41–8; El Niño Southern Oscillation 38,
    43–5, 47; Madden-Julian Oscillation 38,
    43; sea surface temperature 38, 44–5;
    Walker Circulation 45
colonialism: British see British colonial
    rule; impact on water knowledge 148–9,
    187, 234–7, 239–41; Japanese 161–2
common pool resource (CPR) 12, 88, 90,
    103–4, 118
common property 75, 186
community water harvesting 11–12,
    87–105, 112–17, 125–40
complex systems 29–30, 83, 176
consumerism 1, 121, 143–4, 146, 149, 154,
    173–6, 179, 182, 185, 187
contamination, radioactive 19–33
cosmology 2, 7–8, 113, 193–4, 197–205,
    210, 217
cultural values 7, 109, 113, 146, 148, 153;
    transformation of 120–21

dam 14, 95, 177, 225, 246–54, 257; anicut
    94, 96; hydropower 246–54, 257; protest
    movements 247–56
Delhi 12–13, 142–54, 174, 176–9, 187
democracy 72, 82–3, 194
desert see arid region; development: aid
    80, 93, 134, 214, 221; community 96–8;
    discourse 103, 143–4, 174, 151, 153–5,
    174–5, 197–8, 206, 210–18, 251–7;
    intervention see development projects;

practice 11, 70, 83, 119, 144; projects 7, 11–12, 14, 70–1, 79, 111–12, 200, 217, 222, 224, 246, 251–7; role of Thai king 216–21, 223–5; sustainable 2, 126, 201–5; urban 142, 151, 153–5, 230–1, 236–7, 241; water resource 4–5, 13, 63, 211–13, 216–19, 223

earth system science 1–2
ecology 58, 60–2, 71, 125, 127, 129–30, 139–40, 193–7; cultural ecology 56; deep ecology 63, 128; ecological Self 125, 128–9, 138–9; ecological turn 127; political ecology 6, 88, 175; traditional ecological knowledge (TEK) see knowledge
ecosystems 5, 8, 12, 27, 89, 97, 102–3, 127–9, 142, 194, 203; restoration 103
education 83, 126–7, 130, 135; higher education 70, 83, 127
environmental: awareness 72, 82, 201; conflict (protest movements) 14, 246–58; conservation 129, 200; determinism 55–6; disaster 11, 19–33, 180, 252; environmentalism 55, 77, 128, 250–2, 256; historians 54–6, 60, 64; politics 28, 246–58; problems 82, 126; science see studies; studies 52, 54–6, 63–4; sustainability 63, 126; technologies 126–7; thinking see environmentalism; wisdom 53
epistemic 52, 183; categories 52; community 3–10; divide 52; frames 3–10; violence 154–5
ethics see virtue ethics or water ethics
Eurocentrism 3, 7, 74

farming see agriculture
fisheries, governance of 10, 19–22 science 29
fishing 19–33, 59, 65, 94–5, 102, 114
floods 6, 11, 36, 53, 55, 57–61, 91, 99, 122, 214, 217–19, 223, 233, 250–3, 257
Fukushima nuclear disaster 10–11, 19–32
Fukushima see Fukushima nuclear disaster

Gadsisar lake 110–11, 113–14, 117, 120, 135
Gandhian principles 12, 88–9, 92–3, 95, 101, 103, 105, 125, 130, 139, 176, 249, 253, 258
Ganges (also Gaṅgā) river 58, 65, 230–1, 233–4, 240–1, 246, 249–50, 254, 256

Gangetic (also Ganges) delta 53, 58–9, 65
geography 52, 55–6
geography of knowledge 8
global environmental change 1–2, 36–49, 89
globalization 130, 185
global water system 1, 15, 186
Gorno-Badakhshan 71, 74
governance: community 103; complexity of 29; participatory 5, 88, 96, 103–4, 152; role of expert knowledge in 28–32; role of state 88; scale of 30–1, 87–8, 91, 103, 109; of uncertainty 19–33; urban 143–5
governmentality 6, 212, 218
Green Revolution 180–2,192, 197–9, 201–4
groundwater: dependence 90–2; depletion 92, 103; irrigation 87, 91, 180; management 87–9, 92, 113, 117, 136–7; overexploitation 92; recharge 12, 87, 89, 93–4, 98–9, 101, 103, 110–11, 117

hazards see natural disasters
heritage: area 162, 201; cultural 14, 110, 121–2, 136, 199, 204–7, 240–3; discourse 14, 231, 241, 255
Himalaya 58–9, 71, 246, 256
hybrid environment 29, 53–4, 59, 63
hybridity 52–7, 63–4
hydraulic society (Wittfogel) 13, 210–14, 219–25
hydropolitics 6, 87
hydropower 14, 173, 222, 225, 246, 248, 250–1, 258
hydrosocial see social hydrology

identity 28, 56, 129, 142, 149, 176–7, 181–2, 185–6
Indian Premier League (IPL) 179, 182–3, 184–5
India-Pakistan: border 108–9; relations 108
India; see also Delhi, Ladakh, Maharashtra, Rajasthan, Uttarakhand, Uttar Pradesh; government water policy 97, 109, 135, 246–8, 173–7, 182, 187, 251–2; history of 60–1, 64–5, 71, 91, 105, 108, 130, 187, 215, 232–3; water requirements of 87, 92, 178, 180, 187
Indira Gandhi Canal 12, 14, 110–11, 119–21, 135

Indonesia 11, 36, 39, 41–2, 44, 48–9, 192–206; East Flores 39–44, 46; Tabanan (Bali) 197–9
Integrated Water Resource Management (IWRM) 4, 5, 7, 72–3, 79, 82, 173,175
irrigation: customs 78, 118; farmer managed irrigation system (FMIS) 70–83; groundwater 87, 91, 180; institutions 71, 74; inundated 57; overflow irrigation 60, 62; ritual irrigation 192, 194–7, 199; service fees 11, 82; state control 187, 212–14; technology and infrastructure 73–4, 80–1, 111, 176, 193; tubewell 91, 180–81; turns 75, 78
Iwaki see Fukushima

Jaisalmer *see* Rajasthan
Japan 19–33
Jñānavāpī, sacred well of knowledge 230–1, 234–5, 237–42

Karakorum 71, 74, 79
Kargil district see Ladakh
Kāśī Viśvanāth 230, 237–8, 240–1
knowledge: complexity of 29; environmental 3, 6, 8, 9, 127; experts 3, 6, 8, 19–22, 26, 48, 54, 58, 63, 73, 83, 88, 126–7, 133–4, 136, 152, 184; fragmentation of 70, 72, 74, 83, 127; indigenous see local and traditional; knowledge systems 70, 82–3; local and traditional 9, 12, 37, 39, 70–1, 73–4, 77, 81–3, 103, 108–9, 112, 121, 133–4, 139, 251, 256; modern (also Western) 134, 197; scientific 70–2, 74, 79, 82, 256; situated 159, 164; technical 71, 83, 133, 174; traditional ecological knowledge (TEK) 12, 103, 121, 125, 129–30, 132–3, 135, 137, 140 transfer of 109
knowledge/power 6
Kuhnian perspective 72, 82
Kyrgyzstan 73

Ladakh 11, 71, 74–6, 81–3; Kargil 71, 75, 79–80, 83
lake see water, reservoir
landscape: cultural 7–8, 110, 116–18, 121, 194, 208; meaning of 52–3, 56–7, 60–1, 64; sacred 231, 234–40; urban 14, 144, 149, 157–8, 163, 230, 232, 241
land tenure 61
large dam see dam

liberalization 175, 179, 182
livelihood 27, 36, 40, 48, 55, 71, 76–7, 87, 90, 93, 97, 101–3, 112–14, 117, 153, 175, 177, 181, 183–4, 220, 251–2

Maharashtra 183–4
marine environment 19–33
market 125, 130, 134, 174–5, 177, 182–3, 186
Mekong 6
metaphysics 8, 127, 129, 140
middle class 145, 149–52, 174–9, 181–2, 184–5, 187
Millennium Development Goals 143
Mishra, Anupam 12, 125–7, 129–40
models: atmosphere-ocean circulation 37; climate models 37–9, 46–7; downscale 39, 41, 44, 47
monsoon 11, 36–40, 48–9, 53, 58–60, 64–5, 144; Australian-Indonesian 36–44, 46–9; duration 40, 42–3, 46–7; East Asian monsoon 38, 42; onset 40–2, 44, 46–8; South Asian monsoon 38, 42; variables see windspeed, rainfall intensity, rainfall volume, monsoon onset, monsoon duration
mountain valleys 70–1, 74, 81
Mumbai 183–4

Narmada 247
nationalism 175–76, 182, 185, 187
natural disasters 6,
Nehruvian water management 12, 88, 89, 105, 246
New Delhi see Delhi
nomenclature 135, 137
non-government agencies (NGOs) 70–4, 81–2, 183

ontology 127, 129, 140, 146, 199

Pacific Ocean 19–33, 157
Pakistan 71, 73–4, 79, 81
Pamir 71, 74, 79
paradigm 55, 72–4, 79, 82–3, 89, 125, 134, 174; alternative 125; dominant 125–6, 174; shift 55, 201, 246; water management 72–4, 79, 82, 89, 197
participation 71, 149, 154, 173
phenomenology 159
philosophy 127, 130, 132–33, 139
political ecology 6, 88, 175
privatization 102, 162, 173, 176, 186

property rights 175
public-private partnerships 173, 186

radioactivity, *see* contamination, radioactive
rainfall 38–45, 46–7, 59, 74, 80, 125, 130–
  31, 139–40; see also monsoon; annual
  91; deficit 74, 91; historic patterns 39;
  intensity 40, 43–4, 47; royal influence
  on 217–18; volume 40, 43–4, 46–7
Rajasthan: Alwar district 87–93, 97–105;
  Jaisalmer 108–17, 119–24, 131, 134–5;
  water scarcity 175; water traditions
  125–40
reductionism 2, 4
regime of truth (Foucault) 14, 218
rehabilitation, post-disaster 19–33, 216
rice 42, 49, 59–60, 65, 180–1, 193, 197,
  199, 200–4
risk 14, 27–32
Russian Empire 71

Sardar Sarovar Project 175, 247, 250–2,
  255
scale 37–48, 53, 55, 70–1, 82, 186; data
  37, 42–3, 47–8; epistemology of 37, 46,
  82; geographical 53, 55; global 37–8,
  41–8; household 37, 39–40, 42–8; of
  knowledge 10, 29–32; local 37, 39,
  43–8; regional 37–9, 41–8
scarcity, water 6, 11, 75, 91, 114–15, 117,
  125, 130, 132, 138–9, 152–3, 175,
  178–9, 217–18
science 40, 53–4, 63, 70, 72, 81–3,
  125–7, 129, 132–3, 137, 139,
  173–4, 176; -based policy 70, 72,
  83; environmental see environmental
  studies; mainstream 73; modern (also
  Western) 82, 125, 127, 129, 133, 137,
  139
science and technology studies 2, 4, 125–7
sense of place (SoP) 128, 139
sensory experience of water 12, 13, 158,
  160, 163–5, 195–6
Shiva, Vandana 125, 134
Singh, Rajendra 92, 249
social construction 54, 56, 111, 128
social hydrology 6, 142, 158, 160–2,
  165–8, 177, 185
social nature see socio-nature
sociohydrology see social hydrology
socio-nature 6, 54–6, 185
Soviet Union 71–3, 82
Srinagar dam see Uttarakhand

state legitimacy, role of water in 13, 88,
  210–19; see also Nehruvian water
  management
STEEP (social, technological, economic,
  environmental, political) model 89, 90
Stockholm Declaration 126
subak see water users association
sugarcane 174, 177, 179–85, 187
Sundarbans 53–4, 58, 61, 65
sustainability 5, 63, 87–8, 125–7, 130,
  132, 139–40, 201–4; epistemological
  foundations of 127; sustainable
  development see development

Taipei, Taiwan 13, 157, 160–4
Tajikistan 71, 73–4, 79, 81
tank see water reservoir
Tarun Bharat Sangh (TBS) 12, 88–9,
  92–101, 103–5
Tehri dam see Uttarakhand
territory 52, 55
Thailand 13, 210–17, 220–4
Thanarat, Sarit 214–16, 220–4
Thar desert see arid region
Tinsulanonda, Prem 220–1
tourism 20, 31, 110, 120–1, 161–2, 204–5;
  pilgrimage 234–40, 246, 253; UNESCO
  sites 110, 193, 201, 205
tradition 39, 56, 70, 74–5, 125, 129, 131–
  5, 137–40, 173–5, 182, 187, 233; see
  also water tradition, traditional ecological
  knowledge
trans-disciplinary research 36, 48, 54

Umwelt (von Uexküll) 8, 195
urbanisation 1, 12, 143
urban planning 109, 143–6
urban-rural divide 13
Uttarakhand 14; Srinagar dam 246, 248–56;
  Tehri dam 247–8, 250–4
Uttar Pradesh 59, 180–2

Varanasi 14, 230
virtue ethics 127, 130, 140
Vishva Hindu Parishad (VHP) 248, 254

water: access to 115, 132, 173–8, 182;
  allocation 72, 75, 78–9, 142–8, 150,
  153, 193; architecture 232–3, 235;
  canal 60, 75, 180, 197, 213, 230; see also
  farmer managed irrigation system, Indira
  Gandhi Canal; catchment see water
  catchment; class differences in 12–13,

92, 109, 143–52, 153–6, 174–9, 181–2, 184–7; conflict 13–14, 94, 102, 187, 192, 247, 250, 253–7; conflict resolution 88, 104; distribution 71, 73, 78, 118–19, 143–8, 150–5, 173, 186, 217–18, 225; domestic 12, 87, 111, 113, 143, 152, 173–4, 186; drinking 109, 111, 114, 131, 151–2, 174–5, 177–8, 183, 185–6; ethics 7, 125–7, 129–30, 140, 149; harvesting *see* water harvesting; human right 174–5; infrastructure *see* water infrastructure; management *see* water management; neoliberalisation 12, 88, 92, 102–3, 162, 168, 174, 205; policy *see* water policy; pond 52, 60, 93, 130–1, 134, 138; pricing 178; rain water 93, 98, 131–2, 134, 140; regimes 7, 88, 103, 174; regulatory framework 109; reservoir (*also* lake, tank) 52–3, 60, 116–17, 130–1, 134–5, 140, 176, 187; resources management *see* water management; rights 73–4, 77–9, 174–5, 177; scarcity 75, 125, 130, 132, 138–9, 175, 178–9; security 125, 174; shortage *see* scarcity; social relationships 109, 158, 165–8; supply 75–8, 175–6, 178–9; thermal baths 159, 160–5; tradition 62, 74–5, 125, 129–32, 134–7, 139; tubewell 180–1; types of 131, 136; urban context 109, 142–86, 148–55, 162, 237–41, 233–37; well 117, 130–1, 135, 180–1, 230, 232–5, 237–41
water catchment 12, 88–90, 94–9, 101–4, 109–10, 117
water harvesting 11–12, 87–105, 113, 129–30, 132, 139; structures 12, 93–7, 99–101, 116–19; traditional knowledge 103, 113, 132
water infrastructure 73–4, 80–1, 105, 143; see also Indira Gandhi Canal; communal 11, 103, 130–2, 236–7; engineering 14, 73, 109, 111, 146–7, 217, 223; modernisation of 108–9, 236, 253; planning of 109, 146
water management 120; see also Integrated Water Resource Management; community based 70, 84, 88, 91, 103, 200; cultural aspects 109, 200; local 103–5, 109; Paradigms 72, 82, 175; role of women 5, 97, 112, 118–19, 151, 248, 258
water policy 48, 70, 72–4, 83, 101, 125–6, 144–6, 173–4, 177, 179–80, 183–5
waterscape 7, 13–14, 53, 58–60, 63–4, 147, 192–9, 203–6, 231–7, 240–2, 246, 256
watershed 7, 90, 107, 193–5, 198, 200
water users association 73, 80–1, 173, 193–4
Western knowledge 12, 133–4, 139, 159; see also Eurocentrism
wet theory 52–5, 57, 63
windspeed 39, 40–1, 45–7
World Bank 6, 143, 186
worship 192, 232–3, 236, 240; ritual 2, 3, 7, 114, 116, 161, 192, 194, 196, 199, 232–5, 237–40; temple 117, 194, 230, 232, 235, 237, 246, 248

Milton Keynes UK
Ingram Content Group UK Ltd.
UKHW040443071024
449327UK00020B/963